Lecture Notes in Computer Science 9052

Commenced Publication in 1973
Founding and Former Series Editors:
Gerhard Goos, Juris Hartmanis, and Jan van Leeuwen

More information about this series at http://www.springer.com/series/7409

An Liu · Yoshiharu Ishikawa
Tieyun Qian · Sarana Nutanong
Muhammad Aamir Cheema (Eds.)

Database Systems
for Advanced Applications

DASFAA 2015 International Workshops, SeCoP, BDMS, and Posters
Hanoi, Vietnam, April 20–23, 2015
Revised Selected Papers

 Springer

Editors
An Liu
Soochow University
Suzhou
China

Yoshiharu Ishikawa
Nagoya University
Nagoya
Japan

Tieyun Qian
Wuhan University
Wuhan
China

Sarana Nutanong
University of Hong Kong
Hong Kong
China

Muhammad Aamir Cheema
Monash University
Clayton, VIC
Australia

ISSN 0302-9743   ISSN 1611-3349   (electronic)
Lecture Notes in Computer Science
ISBN 978-3-319-22323-0   ISBN 978-3-319-22324-7   (eBook)
DOI 10.1007/978-3-319-22324-7

Library of Congress Control Number: 2015944507

LNCS Sublibrary: SL3 – Information Systems and Applications, incl. Internet/Web, and HCI

Springer Cham Heidelberg New York Dordrecht London

Printed on acid-free paper

Springer International Publishing AG Switzerland is part of Springer Science+Business Media
(www.springer.com)

Message from the Workshop Chairs

Welcome to proceedings of the workshops held in conjunction with the 20th International Conference on Database Systems for Advanced Applications (DASFAA 2015), which took place in Hanoi, Vietnam.

The objective of the workshops associated with DASFAA 2015 is to give participants the opportunity to present and discuss emerging hot topics related to database systems and applications. To this end, we selected the following two workshops:

1. The Second International Workshop on Semantic Computing and Personalization (SeCoP 2015)
2. The Second International Workshop on Big Data Management and Service (BDMS 2015)

These proceedings contain selected workshop papers. We would like to express our sincere thanks for the hard work of the individual workshop organizers for organizing their workshops, handling the paper submissions, reviewing, and selecting workshop papers to achieve a set of excellent programs.

An Liu
Yoshiharu Ishikawa

Message from the Poster Chairs

Welcome to the proceedings of the Poster Session of the 20th International Conference on Database Systems for Advanced Applications (DASFAA), held during 20–23 April, 2015, in Hanoi, Vietnam.

The call for posters resulted in a total of 28 submissions from China, India, Korea, Israel, Taiwan, Japan, and Germany. Every submission was assigned to three Program Committee members for review. The Program Committee decided to accept 12 posters. In addition, there were five papers transferred from the research track. Hence, we had 17 papers in total for the poster session. These papers cover an interesting and diverse set of topics in database systems, scalable data management, and data mining.

We would like to thank the Program Committee members who devoted their time contributing to the strong poster program. We would also like to extend a big thank you to the main conference organizers, especially Matthias Renz, Trinh Vu Tuyet, Nguyen Hong Phuong, and Muhammad Aamir Cheema, for coordinating with us.

June 2015

Sarana Nutanong
Tieyun Qian

DASFAA 2015 Workshop Organizers

Workshop Co-chairs

An Liu   Soochow University, China
Yoshiharu Ishikawa   Nagoya University, Japan

Publication Chair

Muhammad Aamir Cheema   Monash University, Australia

Second International Workshop on Semantic Computing and Personalization (SeCoP 2015)

Workshop Co-organizers

Fu Lee Wang   Caritas Institute of Higher Education,
                             Hong Kong SAR, China
Yi Cai   South China University of Technology, China
Haoran Xie   Caritas Institute of Higher Education,
                             Hong Kong SAR, China

Program Committee

Ho Wai Shing   Caritas Institute of Higher Education,
                             Hong Kong SAR, China
Jianfeng Si   Institute for Infocomm Research, Singapore
Wei Chen   Agricultural Information Institute of CAAS, China
Xudong Mao   Alibaba, China
Rao Yanghui   Sun Yat-Sen University, China
Raymond Y.K. Lau   City University of Hong Kong, Hong Kong SAR,
                             China
Rong Pan   Sun Yat-Sen University, China
Yunjun Gao   Zhejiang University, China
Shaojie Qiao   Southwest Jiaotong University, China
Jianke Zhu   Zhejiang University, China
Neil Y. Yen   University of Aizu, Japan
Derong Shen   Northeastern University, China

Jing Yang	Research Center on Fictitious Economy & Data Science CAS, China
Yuqing Sun	Shangdong University, China
Raymond Wong	Hong Kong University of Science and Technology, Hong Kong, China
Jie Tang	Tsinghua University, China
Jian Chen	South China University of Technology, China
Wong Tak Lam	Hong Kong Institute of Education, Hong Kong, SAR China
Xiaodong Zhu	University of Shanghai for Science and Technology, China
Zhiwen Yu	South China University of Technology, China
Wenjuan Cui	China Academy of Sciences, China
Shaoxu Song	Tsinghua University, Beijing, China
Tao Wang	South China University of Technology, China

Second International Workshop on Big Data Management and Service (BDMS 2015)

Workshop Co-organizers

Xiaoling Wang	East China Normal University, China
Kai Zheng	The University of Queensland, Australia
Guanfeng Liu	Soochow University, China

Program Committee

Muhammad Aamir Cheema	Monash University, Australia
Cheqing Jin	East China Normal University, China
Qizhi Liu	Nanjing University, China
Bin Mu	Tongji University, China
Xuequn Shang	Northwestern Polytechnical University, China
Weiwei Sun	Fudan University, China
Yan Wang	Macquarie University, Australia
Lizhen Xu	Southeast University, China
Xiaochun Yang	Northeastern University, China
Kun Yue	Yunnan University, China
Dell Zhang	University of London, UK
Xiao Zhang	Renmin University of China, China

DASFAA 2015 Posters Organizers

Poster Co-chairs

Sarana Nutanong   City University of Hong Kong, Hong Kong SAR, China

Tieyun Qian   Wuhan University, China

Program Committee

Muhammad Aurangzeb Ahmad	University of Minnesota, USA
Zoheb Borbora	University of Minnesota, USA
Mohammed Eunus Ali	Bangladesh University of Engineering and Technology, Bangladesh
Kuo-Wei Hsu	National Cheng Chi University, Taiwan
Cheqing Jin	East China Normal University, China
Peiquan Jin	University of Science and Technology, China
Shin'ichi Konomi	The University of Tokyo, Japan
Guohui Li	Huazhong University of Science and Technology, China
Qi Liu	University of Science and Technology, China
Yubao Liu	Sun Yat-Sen University, China
Nishith Pathak	Ninja Metrics Inc., USA
Daling Wang	Northeastern University, China
Huayu Wu	Institute for Infocomm Research (I2R), Singapore
Wei Wu	Institute for Infocomm Research (I2R), Singapore
Xiaoyin Wu	Wuhan University, China
Hairuo Xie	University of Melbourne, Australia
Guandong Xu	University of Technology Sydney, Australia
Yuanyuan Zhu	Wuhan University, China

Contents

**The Second International Workshop on Big Data Management
and Service (BDMS)**

Posters

The Second International Workshop on Semantic Computing and Personalization (SeCoP)

A Novel Method for Clustering Web Search Results with Wikipedia Disambiguation Pages

Zhi Huang[1], Zhendong Niu[1,2,3]([✉]), Donglei Liu[1],
Wenjuan Niu[1], and Wei Wang[1]

[1] School of Computer Science, Beijing Institute of Technology, Beijing 100081, China
[2] Information School, University of Pittsburgh, Pennsylvania 15260, USA
zniu@bit.edu.cn
[3] Beijing Engineering Research Center of Massive Language Information Processing
and Cloud Computing Application, Beijing Institute of Technology,
Beijing 100081, China

Abstract. Organizing search results of an ambiguous query into topics can facilitate information search on the Web. In this paper, we propose a novel method to cluster search results of ambiguous query into topics about the query constructed from Wikipedia disambiguation pages (WDP). To improve the clustering result, we propose a concept filtering method to filter semantically unrelated concepts in each topic. Also, we propose the top K full relations (TKFR) algorithm to assign search results to relevant topics based on the similarities between concepts in the results and topics. Comparing with the clustering methods whose topic labels are extracted from search results, the topics of WDP which are edited by human are much more helpful for navigation. The experiment results show that our method can work for ambiguous queries with different query lengths and highly improves the clustering result of method using WDP.

Keywords: Ambiguous query · Web search result clustering · Wikipedia disambiguation pages · Concepts filtering

1 Introduction

Search engines are playing a more and more important role in browsing the Web. By sending a query to the search engine, we can get a list of high related results, as long as the query can exactly describe what we want. However, for most of the queries are single words or simple phrases, the query sometimes is ambiguous and has multiple meanings. It has been estimated that about 4.0 % of queries and 16.4 % of the most frequent queries on Web are ambiguous [1,2].

However, search engine typically returns a ranked list of results, about 10 search results a page. It will be highly inefficient to find the results with the less used meanings of an ambiguous query which has a lot of topics. For example, the query "Phoenix" has several topics and subtopics (Table 1), but the first page

© Springer International Publishing Switzerland 2015
A. Liu et al. (Eds.): DASFAA 2015 Workshops, LNCS 9052, pp. 3–16, 2015.
DOI: 10.1007/978-3-319-22324-7_1

results can just cover small part of them (Table 2). There are another two situations: (i) User A knows nothing about the query word "Phoenix". She searches the word on the Internet to find out what the word means. In this situation, a topic directory of "Phoenix" can help A to quickly take a overview of all the meanings of the query; (ii) User B wants to get the results of a special topic of "Phoenix", e.g., the topic "Military", but she can't find an appropriate word to describe the topic or just forget how to spell the topic word. In this situation, a topic directory of "Phoenix" can help B to quickly find out the special topic and corresponding results.

Table 1. Topics of "Phoenix" according to Wikipedia disambiguation pages

Topic	Subtopic of topic 2	Subtopic of topic 3
1. Mythology	2.1. Canada	3.1. Comics
2. Places	2.2. Ireland	3.2. Theatres
3. Media and entertainment	2.3. Kiribati	3.3. Film
4. Astronomy	2.4. Mauritius	3.4. Broadcasters
5. Computing	2.5. South Africa	3.5. Television
6. Education	2.6. United States	3.6. Fictional entities
7. Military		3.7. Literature
8. Plants and animals		3.8. Music
9. Ships		3.9. Musicians
		3.10. Albums
		3.11. Periodicals
		3.12. Sports
		3.13. Video games

From the above examples, we can see that organizing search results of an ambiguous query into a topic directory can facilitate information search on the Web. The typically solution is clustering the results, and then navigating them by the labels of the topic in each cluster. The main issues are: (i) how to cluster the search results, (ii) how to label the topic of each cluster.

A lot of work has been studied for web search results clustering, such as SRC methods, WSI-based methods and WSD-based methods, these methods will be described in next section. Most of existed methods, such as SRC methods and WSI-based methods, extract words or phrases with high scores in search results as topic labels. This may lead to low predictive labels, which means that the labels are not good indicators of the results they contain [3]. Using topics of a Web taxonomy, such as Open Directory Project (ODP)[1], WordNet[2] and WDP[3],

---

[1] http://www.dmoz.org/.
[2] http://wordnet.princeton.edu/.
[3] http://en.wikipedia.org/wiki/Category:Disambiguation_pages.

WSD-based methods can get predictive (meaningful and uniform) topic labels for ambiguous query. However, these methods always have bad clustering result, e.g. low F_1 measure [2], because both the snippets of search results and the descriptions of topics in the taxonomy just have few words and assigning results to relevant topics is difficult.

Table 2. The first page of search results of "Phoenix" by Google (google.ca)

[Ad] We Make Learning Easier for Working Adults Like You	6
Information on Greater Phoenix, including visitor information	2.6
Phoenix is the capital, and largest city, of the state of Arizona	2.6
In Greek mythology, a phoenix or phenix is a long-lived bird	1
Phoenix is an alternative rock band from Versailles, France	3.9
[In the news] Joaquin Phoenix, 40, said Monday night in	3.9
In Phoenix's world, nothing comes easy. Their meticulousness	3.9
TO THE SPORTS-MAD, Phoenix probably looks a lot like	3.12
Reminder for TomDispatch Readers: Despite the introduction	3.7
CityHall20140911	3.10
OUR NEW ALBUM 'BANKRUPT' OUT NOW!	3.9
[Images for Phoenix]	1
Official site containing news, scores, audio	3.12
[Right side] Phoenix (City in Arizona): Weather: 13C	2.6
[Right side] Phoenix (Rock band) Members: Thomas Mars	3.9

In this paper, we propose a novel method called CWD (Clustering search results with Wikipedia Disambiguation pages) to cluster Web search results of ambiguous query with WDP. The method first constructs topics from WDP and then assigns search results to relevant topics. We try to improve the clustering result of CWD by two components: (i) a concept filtering method proposed to filter noisy concepts based on a vector model [4], (ii) the Top K Full Relations (TKFR) algorithm proposed to assign search results to relevant topics.

The rest of this paper is organized as follows: Sect. 2 presents a brief summary of related work. Section 3 formally defines the problem. Section 4 describes our method, namely clustering Web search results with Wikipedia disambiguation pages. Section 5 contains the experiments and results. We finish with some conclusions in Sect. 6.

2 Related Work

A popular solution to solve the ambiguous query issues are mainly based on document clustering algorithm, i.e. the search result clustering (SRC) methods.

Mandhani et al. [5] propose a matrix density based algorithm to hierarchically cluster documents. In their method, clusters can be the submatrix made up of hight values in the vector model for documents. Kummamuru et al. [3] present a new hierarchical monothetic clustering algorithm to build a topic hierarchy Web search results. At each level of the hierarchy, the method identifies topics by maximizing the coverage while maintaining distinctiveness of the topics. Zamir and Etzioni [6] use STC (Suffix Tree Clustering) algorithm to group the Web documents into clusters labeled by phrases extracted from snippets of the search results. Bernardini et al. [7] present a new search results clustering algorithm based on extraction and merging of key phrases, and implement it into the KeySRC Web clustering engine. More related works can be found in [8–10].

Some other studies try to get topics first, and then assign each result to the most related topic. These methods can be divided into two categories, i.e. the Word Sense Induction (WSI) methods and Word Sense Disambiguation (WSD) methods. WSI-based methods first acquire various word sense of ambiguous query and then use the word sense to cluster the Web search results. Schütze and Pedersen [11] consider building a vector representation model by lexical co-occurrence for WSI. Marco et al. [2] present a novel WSI-based approach to cluster Web search results and integrate several graph based methods to their clustering framework. WSD-based methods use existing topics (word senses) to clustering the Web search results. The topics are edited by human and can form a taxonomy. Gabrilovich et al. [12] present a novel method using Wikipedia to build a high-dimensional space of concepts for texts. The method can perform word sense disambiguation by considering ambiguous words in the context of their neighbors. Marco et al. [2] present a simple WSD-based method as a baseline of their work. They use the meanings of the query in WDP as the clusters and assign each result to the meaning with maximum word overlap with the snippet of the result.

Clustering search results with just snippets may lead low clustering result caused by the sparse feature, and building feature from the original documents lead much more time cost. Bao et al. [13] show that social annotations can be used to find latent semantic association between queries and measure the quality of a web page. Xie et al. [14] present a community-aware method to construct resource profiles via social filtering. Furthermore researches have been studied for user profile enrichment [15] and finding user communities [16]. All these information can enrich the feature of web search results from social networks (such as Facebook, Google Plus and Weibo etc.). Another general solution to solve the sparse feature issue is using semantic enhancing. That means learning word vectors on a large corpus first [4,17,18], and then using word vectors for word similarity computing.

3   Problem Formulation

Given a query q, search engine will return a ranked list $R = [r_1, r_2, \ldots, r_n]$ and the WDP information of q is a topic tree. Figure 1 describes the topic tree of ambiguous query "Phoenix". The top topics are "Mythology", "Places" etc., and the topic "Places" has more refined topics, such as "Canada", "Ireland" etc. Each leaf topic is an example of its parent topic, e.g., the topic "Phoenix, British Columbia" is an example of topic "Canada". In our clustering method, we treat the topic tree, excluding leaf topics, as the topic directory for search results and take all the leaf topics to form the final clustering $C = [C_1, C_2, \ldots, C_m]$. Each cluster C_i will have a description, the corresponding information of leaf topic.

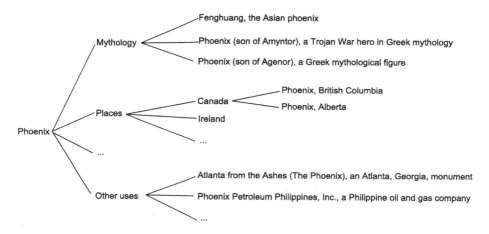

Fig. 1. The topic tree of "Phoenix" from WDP

To computing the sematic relatedness between two concepts (words or phrases), we introduce a word vector language model. In the model, word vectors are learned from a huge corpus through a neural network which can map concepts to a continuous vector space and keep striking semantic properties [4,19]. The vectors used in this paper are learned from Google news dataset and contains 300-dimensional vectors for 3 million words and phrases.[4] The similarity between two concepts is then measured by calculating the cosine value of their vectors:

$$sim\,(w_{i'}, w_{j'}) = \cos\,(v_{i'}, v_{j'}) \tag{1}$$

where $v_{i'}$ is the normalized vector of concept $w_{i'}$ and $v_{j'}$ is the normalized vector of concept $w_{j'}$ (For phrase which doesn't have a corresponding vector, we average the vectors of words in the phrase as the vector of the phrase.).

4   Method

In this paper, We use WDP as source of topics for ambiguous query and propose a novel search result clustering method, i.e. CWD. The CWD method mainly has

---

[4] http://code.google.com/p/word2vec/.

two steps: (i) construction of clustering structure: extracting important concepts from the description for each cluster, (ii) assignment of search results: assigning the search results to relevant clusters.

Figure 2 describes the workflow of our search result clustering method.

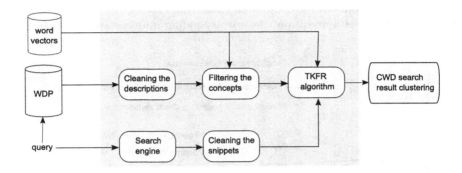

Fig. 2. The workflow of CWD search result clustering method.

4.1   Construction of Clustering Structure

To construct clustering structure with high related concepts, we need process each cluster C_i with two steps: (i) cleaning the description to get a collection of concepts, (ii) filtering noisy concepts by their similarities to the topic of C_i and the query q.

Cleaning the Description. The cleaning step contains three sub steps:

(a) For each description, we extract noun phrases and single words that are either nouns or adjectives that are most predictive of the content of the description.
(b) Then, we remove repeated words, stopwords and single words less than 3 characters from the terms got from a). Stopwords are the words collected from descriptions which are general in all descriptions, e.g., the query word.
(c) We index each term got from b) in the dictionary of vectors and remove the terms without corresponding vectors.

Table 3 shows the results of each steps for a cluster description of "Phoenix". Finally, for each cluster C_i, we get a list of concepts $a_i = [w_1, w_2, \ldots, w_{|a_i|}]$, i.e. the terms remained.

Filtering the Concepts. The "Cleaning the description" step doesn't use any semantic information, thus a_i will contain lots of noisy concepts which are unrelated to the corresponding topic or the query. In this paper, we propose a concept filtering method to filter the concepts by their similarities to the topic and query. For each a_i, the method contains four sub steps:

Table 3. An example of "Cleaning the description"

Step	Result of each step
Initial cluster description	Phoenix (1798), performed one voyage to India for the British East India Company between 1798 and 1800, and then as a merchant vessel was damaged in 1824 and turned into a prison hulk in Sydney
a)	Phoenix, voyage, India, East, India, Company, merchant, vessel, prison, hulk, Sydney
b)	Voyage, India, East, Company, merchant, vessel, prison, hulk, Sydney
c)	Voyage, India, East, Company, merchant, vessel, prison, Sydney

(a) First, we build a graph with each concept in a_i as a vertex, then add an undirected edge to each pair of vertices $w_{i'}$ and $w_{j'}$ if the similarity between them satisfied:

$$sim\,(w_{i'}, w_{j'}) \geq \delta_1, \tag{2}$$

where δ_1 is experimentally tuned threshold. Then, the main meaning of the topic can be captured by the concept of vertex which has the maximum degree in the graph, represented by cen_i (if more than one vertices have the maximum degree, we will compare the sum of edge weights of each vertex, the edge weight is the corresponding similarity between two vertices.). Table 4 shows the similarity matrix of concepts in the example used in Sect. 4.1. Figure 3 is the corresponding graph built and the final cen_i is "vessel" when we set $\delta_1 = 0.15$.

(b) Secondly, we compute the importance of each concept $w_{i'}$ in a_i by combining its similarities to cen_i and the query q:

$$p_{i'} = \lambda \times sim\,(w_{i'}, cen_i) + (1 - \lambda) \times sim\,(w_{i'}, q), \tag{3}$$

Table 4. The similarity matrix of concepts in the example

Similarity	Voyage	India	East	Company	Merchant	Vessel	Prison	Sydney
voyage	1.000	0.078	0.030	0.077	0.121	0.498	0.074	0.111
India	0.078	1.000	0.124	0.073	0.071	0.019	-0.019	0.231
East	0.030	0.124	1.000	0.037	0.068	-0.009	-0.034	0.013
Company	0.077	0.073	0.037	1.000	0.188	0.093	0.030	-0.012
merchant	0.121	0.071	0.068	0.188	1.000	0.220	0.043	0.040
vessel	0.498	0.019	-0.009	0.093	0.220	1.000	0.099	0.080
prison	0.074	-0.019	-0.034	0.030	0.043	0.099	1.000	0.046
Sydney	0.111	0.231	0.013	-0.012	0.040	0.080	0.046	1.000

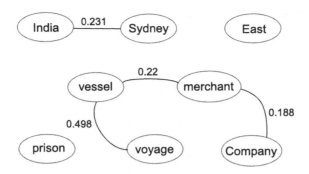

Fig. 3. The graph built for the concepts in the example.

where λ is experimentally tuned parameter. Then, we sort the concepts in a_i in descending order by their importances, represent the sorted importances as:

$$P_i = [p_1, p_2, \ldots, p_{|a_i|}], \tag{4}$$

let l_1 to be the position of the last element in P_i which has a value not less than δ_2, where δ_2 is experimentally tuned threshold. In a_i, the concepts after the l_{1th} concept can be filtered, for these concepts have less relatedness with the topic or the query.

(c) Then, we calculate the gradients of P_i as:

$$G_i = [g_1, g_2, \ldots, g_{l_1 - 1}], \tag{5}$$

where $g_{i'} = p_{i'} - p_{i'+1}$. Let l_2 to be the position of the element with the second largest value in G_i. Then in a_i, the concepts after the l_{2th} concept can be filtered, for a big gradient indicates that the concepts after are much less important than the concepts before. We don't use the largest value in G_i as a split point because the largest value in G_i is always the first element, for the most important concept is always cen_i and the corresponding importance value is much larger than other concepts as long as λ does not take a very small value.

(d) Finally, we take the first l concepts in a_i as the final concepts $a'_i = [w_1, w_2, \ldots, w_{|l|}]$, where l is calculated as follows:

$$l = \max(l_2, N), \tag{6}$$

and N is experimentally tuned parameter, means the minim number of concepts should return.

For the example description, if we set $\lambda = 0.5$, the importances of the concepts in a_i are showed in Table 5.

The sorted concepts a_i = ['vessel', 'voyage', 'Sydney', 'merchant', 'prison', 'India', 'Company', 'East'], if we set $\delta_2 = 0.05$, then $P_i = [0.493, 0.244, 0.181,$

Table 5. The importances of concepts in the example

Concept	Voyage	India	East	Company	Merchant	Vessel	Prison	Sydney
Importance	0.244	0.075	0.019	0.054	0.115	0.493	0.083	0.181

0.115, 0.083, 0.075, 0.054], $l_1 = 7$, $G_i = [0.249, 0.063, 0.066, 0.032, 0.008, 0.021]$ and $l_2 = 3$.

And if we set $N = 4$, then $l = 4$ and the final concepts $a_i' =$ ['vessel', 'voyage', 'Sydney', 'merchant'].

4.2 Assignment of Search Results

In this section, we describe the TKFR algorithm used to assign search results to relevant clusters. In our method, we just use the snippets of search results. The key idea of TKFR algorithm is computing the similarities between search results and clusters and assigning each result to the most similar cluster. To do that, for each result r_j, we process the snippet s_j with the "Cleaning the description" step described in Sect. 4.1) and transform it into a bag of concepts b_j. Then, the similarity between a result r_j and a cluster C_i can be calculated by comparing the similarities between the concepts in b_j and a_i'. The TKFR algorithm mainly contains three steps:

(a) For each cluster C_i and search result r_j, we calculate the similarities between concepts in a_i' and b_j:

$$U_{i,j} = \{sim(w_{i'}, w_{j'}) : w_{i'} \in a_i', w_{j'} \in b_j\}. \tag{7}$$

(b) Then, we take the mean value of the top K similarities in $U_{i,j}$ as the similarity between a_i' and b_j, i.e.:

$$sim\left(a_i', b_j\right) = \frac{\sum Top_K\left(U_{i,j}\right)}{K}, \tag{8}$$

where K is experimentally tuned parameter.

(c) The final cluster of r_j is the cluster which has the most similar concepts to b_j, i.e.:

$$cli\left(r_j\right) = \begin{cases} \arg\max_{i=1,2,\ldots,m} sim\left(a_i', b_j\right) & \text{if } \max_{i=1,2,\ldots,m} sim\left(a_i', b_j\right) \geq \delta_3 \\ 0 & \text{else} \end{cases}, \tag{9}$$

where δ_3 is experimentally tuned threshold. $cli\left(r_j\right) = 0$ means that r_j doesn't belong to any clusters in C. We simply add C_0 to C as a cluster of "Other uses" and remove empty clusters after all the assignment of results.

5   Experiments and Results

We construct two experiments in this paper. The first experiment evaluates the clustering results of CWD on two datasets with different average query lengths, and the second one compares CWD with other state-of-the-art search result clustering methods. The compared methods are SRC methods (STC and KeySRC), WSI-based methods (SquaT++E and HyperLex), and a WSD-based method proposed by Marco et al. [2] as a baseline in their work. The last method also uses WDP to cluster search results, we call it WikiC.

5.1   Datasets

We use two datasets of ambiguous queries, i.e. AMBIENT and MORESQUE. These two datasets have the same structure. Each query in the datasets contains three types of information: the topics of the query in WDP, the top 100 Web search results from Yahoo! and the golden standard clustering (i.e. search results with the most appropriate topics tagged by human).

- AMBIENT (AMBIguous ENTries)[5]: a dataset contains 44 ambiguous queries. The average length of queries in this dataset is 1.27 and the average cluster number of the gold standard clustering is 17.9.
- MORESQUE (MORE Sense-Tagged QUEry Results)[6]: a dataset contains 114 ambiguous queries. The average length of queries in this dataset is 2.86 and the average cluster number of gold standard clustering is 6.6.

5.2   Measurements

Given a clustering result C and the gold standard G. To compare with SRC methods and WSI-based methods which don't use topics in WDP as clusters, we define the topic of a cluster C_i as the majority topic t in C_i, which means C_i contains the maximum number of results tagged with t in the gold standard G, just the same as paper [2] did. Then, the precision P can be calculated as follows [2]:

$$P = \frac{\sum_{C_i \in C} |C_i^t|}{\sum_{C_i \in C} |C_i|} \qquad (10)$$

where t is the majority topic in C_i, and C_i^t is the set of results in C_i which are tagged with topic t in the gold standard G. The recall R, instead, calculated as:

$$R = \frac{\sum_{t \in T} |\bigcup_{C_i \in C^t} C_i^t|}{\sum_{t \in T} n_t} \qquad (11)$$

where T is the set of topics in the gold standard G for given query q, C^t is the subset of clusters of C whose majority topic is t, and n_t is the number of results

---

[5] http://credo.fub.it/ambient/.

[6] http://lcl.uniroma1.it/moresque.

tagged with topic t in the gold standard G. The F_1 measure is calculated as the harmonic mean of P and R:

$$F_1 = \frac{2PR}{P+R} \qquad (12)$$

In the first experiment, we evaluate P, R and F_1 of CWD on two databases with different average query lengths, and in the second experiment, we evaluate F_1, $\#cl.$ and $|\#cl.-\#gcl.|$ for different methods where $\#cl.$ is the cluster number of C and $|\#cl.-\#gcl.|$ is the difference between the cluster number of C and G. Then, for each measurement, we average the results of each query in the test set as the overall result.

5.3   Results

To tune the parameters of CWD, we construct a dataset of ambiguous queries[7] which are not in AMBIENT and MORESQUE. The tunset have the same three types of information with AMBIENT and MORESQUE. A slight difference is that, in tunset, we label search results with the top topics of topic tree which are more useful for navigation. Table 6 shows the optimal parameters of CWD to maximize F_1 measure on tunset.

Table 6. Optimal parameters of CWD on tunset

Papameter	δ_1	δ_2	δ_3	λ	K	N
Value	0.12	0.038	0.207	0.418	7	4

Result for Datasets with Different Query Lengths. Table 7 shows the results of CWD on AMBIENT and MORESQUE. From Table 7, we can see that CWD gets quite similar results on AMBIENT and MORESQUE, although the average query length and average cluster number of gold standard clustering of the two datasets are quite different (1.27, 17.9 for AMBIENT and 2.86, 6.6 for MORESQUE). That means CWD is quite robust to different lengths of queries.

Table 7. Results of CWD on AMBIENT and MORESQUE

Dataset	$P/\%$	$R/\%$	$F_1/\%$
AMBIENT	41.09	79.43	54.16
MORESQUE	**44.53**	**81.28**	**57.53**

Table 7 also shows that all the measurements on MORESQUE are higher than those on AMBIENT which means longer query length and less sub topics can get a better result with CWD. This is consistent with our intuition, i.e., a method should do the clustering better when the query is less ambiguous.

---

[7] https://github.com/HUANG-Zhi/SimpleAmbiguousQueryDataset.

Result for Different Methods. In this experiment, we merge AMBIENT and MORESQUE, then test the method on the merged dataset, just the same as [2] did. We use the same evaluator provided by [2] and reference some results of the compared methods in the article.[8] The average length of the queries in merged dataset is 2.86 and the average cluster number of gold standard clustering is 9.75.

Table 8 shows the results of different methods. We can see that the F_1 of CWD (56.60 %) is much higher than WikiC (14.33 %) which also uses WDP to cluster Web search results. Although both KeySRC (63.11 %) and SquaT++E (59.39 %) methods outperform CWD in terms of F_1, the cluster number #cl. of KeySRC (18.5) is rather large which may lead to many small clusters (about 5 results per cluster) and the cluster number #cl. of SquaT++E (2.7) is rather small which may lead to large clusters (about 37 results per cluster). And from all the $|\#cl. - \#gcl.|$ measures in Table 8, we can find that CWD get the minimum difference between the cluster number of clustering result C and gold standard G (1.37).

Table 8. Results of different methods on AMBIENT+MORESQUE

Algorithm	$F_1/\%$	#cl.	$|\#cl. - \#gcl.|$
$SquaT_{++E}$	59.39	2.7	7.05
$HyperLex$	**65.41**	13.0	3.25
STC	14.96	2.0	7.75
$KeySRC$	63.11	18.5	8.75
$WikiC$	14.33	5.7	4.05
CWD	56.60	8.38	**1.37**

6   Conclusions

In this paper, we propose a novel method using WDP to cluster Web search results of ambiguous query. Given an ambiguous query, the method first constructs clustering structure from WDP and then assigns search results to relevant clusters with the proposed TKFR algorithm. In addition, we propose a concept filtering method to filter noisy concepts by their similarities to the topic and query.

We construct two experiments on public datasets of ambiguous queries. The result of the first experiment shows that our method can work for ambiguous queries with different query lengths. In the second experiment, we compare our method with other state-of-the-art methods, the result shows that our method

---

[8] The authors organized a task for evaluation of clustering methods on AMBIENT and MORESQUE, http://www.cs.york.ac.uk/semeval-2013/task11/.

has a better average cluster size than other methods and highly improves the clustering result of the method using WDP.

In addition, comparing with the clustering methods whose topic labels are extracted from search results, the topic labels of our method which are constructed from WDP are uniform for the same topics of different queries and much more meaningful for reading and navigation. This will be helpful for browsing the search results of an ambiguous query with lots of sub topics (e.g., the example query "Phoenix").

Acknowledgement. This work is supported by the National Natural Science Foundation of China (No. 61370137), the International Corporation Project of Beijing Institute of Technology (No. 3070012221404) and the 111 Project of Beijing Institute of Technology.

References

1. Sanderson, M.: Ambiguous queries: test collections need more sense. In: Proceedings of the 31st Annual International ACM SIGIR Conference on Research and Development in Information Retrieval, pp. 499–506. ACM (2008)
2. Di Marco, A., Navigli, R.: Clustering and diversifying web search results with graph-based word sense induction. Comput. Linguist. **39**(3), 709–754 (2013)
3. Kummamuru, K., Lotlikar, R., Roy, S., Singal, K., Krishnapuram, R.: A hierarchical monothetic document clustering algorithm for summarization and browsing search results. In: Proceedings of the 13th International Conference on World Wide Web, pp. 658–665. ACM (2004)
4. Mikolov, T., Chen, K., Corrado, G., Dean, J.: Efficient estimation of word representations in vector space (2013). arXiv preprint. arXiv:1301.3781
5. Mandhani, B., Joshi, S., Kummamuru, K.: A matrix density based algorithm to hierarchically co-cluster documents and words. In: Proceedings of the 12th International Conference on World Wide Web, pp. 511–518. ACM (2003)
6. Zamir, O., Etzioni, O.: Web document clustering: a feasibility demonstration. In: Proceedings of the 21st Annual International ACM SIGIR Conference on Research and Development in Information Retrieval, pp. 46–54. ACM (1998)
7. Bernardini, A., Carpineto, C.: Full-subtopic retrieval with keyphrase-based search results clustering. In: IEEE/WIC/ACM International Joint Conferences on Web Intelligence and Intelligent Agent Technologies, wi-iat 2009 (2009)
8. Cutting, D.R., Karger, D.R., Pedersen, J.O., Tukey, J.W.: Scatter/gather: a cluster-based approach to browsing large document collections. In: Proceedings of the 15th Annual International ACM SIGIR Conference on Research and Development in Information Retrieval, pp. 318–329. ACM (1992)
9. Krishna, K., Krishnapuram, R.: A clustering algorithm for asymmetrically related data with applications to text mining. In: Proceedings of the Tenth International Conference on Information and Knowledge Management, pp. 571–573. ACM (2001)
10. Lawrie, D., Croft, W.B., Rosenberg, A.: Finding topic words for hierarchical summarization. In: Proceedings of the 24th Annual International ACM SIGIR Conference on Research and Development in Information Retrieval, pp. 349–357. ACM (2001)
11. Schütze, H., Pedersen, J.O.: Information retrieval based on word senses (1995)

12. Gabrilovich, E., Markovitch, S.: Computing semantic relatedness using wikipedia-based explicit semantic analysis. IJCAI **7**, 1606–1611 (2007)
13. Bao, S., Xue, G., Wu, X., Yu, Y., Fei, B., Su, Z.: Optimizing web search using social annotations. In: Proceedings of the 16th International Conference on World Wide Web, pp. 501–510. ACM (2007)
14. Xie, H.R., Li, Q., Cai, Y.: Community-aware resource profiling for personalized search in folksonomy. J. Comput. Sci. Technol. **27**(3), 599–610 (2012)
15. Xie, H., Li, Q., Mao, X., Li, X., Cai, Y., Rao, Y.: Community-aware user profile enrichment in folksonomy. Neural Netw. **58**, 111–121 (2014)
16. Xie, H., Li, Q., Mao, X., Li, X., Cai, Y., Zheng, Q.: Mining latent user community for tag-based and content-based search in social media. Comput. J. **57**(9), 1415–1430 (2014)
17. Schütze, H.: Word space. In: Advances in Neural Information Processing Systems 5. Citeseer (1993)
18. Bengio, Y., Schwenk, H., Senécal, J.S., Morin, F., Gauvain, J.L.: Neural probabilistic language models. In: Holmes, D.E., Jain, L.C. (eds.) Innovations in Machine Learning. Studies in Fuzziness and Soft Computing, vol. 194, pp. 137–186. Springer, Heidelberg (2006)
19. Mikolov, T., Yih, W.T., Zweig, G.: Linguistic regularities in continuous space word representations. In: HLT-NAACL, pp. 746–751 (2013)

Integrating Opinion Leader and User Preference for Recommendation

Dong Wu[1], Kai Yang[2], Tao Wang[2], Weiang Luo[2],
Huaqing Min[2(✉)], and Yi Cai[2]

[1] School of Information Science and Technology, Lingnan Normal College,
Chikan District, Zhanjiang 524048, Guangdong, China
[2] School of Software Engineering, South China University of Technology,
Panyu District, Guangzhou 510006, Guangdong, China
`hqmin@scut.edu.cn`

Abstract. Collaborative filtering (CF) is one of the most well-known and commonly used technology for recommender systems. However, it suffers from inherent issues such as data sparsity. Many works have been done by used additional information such as user attributes, tags and social relationships to address these problems. We proposed an algorithm named *OLrs* (Opinion Leaders for Recommender System) based on the trust relationships. Specifically, the opinion leaders who have a strong influence for the active user and an accurate evaluation of the recommend item will be identified. The prediction for a given item is generated by ratings of these opinion leaders and the active user. Experimental results based on Epinions data set demonstrated that the prediction accuracy of our method outperforms other approach.

Keywords: Recommender systems · Data sparsity · Opinion leader · Matrix factorization

1 Introduction

As the rapid development of information technology and extensive application of information services, people's daily behavior such as news reading, buddy communication, shopping and payments can be made online. It gained unprecedented convenience, but also caught in a serious problem how to find valuable information or find goods to meet the requirements. This is known as information overload problem. In order to effectively solve the problem, the recommender system was designed and has begun to be applied to various fields. Especially, it has played a good role in the large-scale e-commerce shopping guide website [1].

Collaborative filtering (CF) is one of the most well-known and commonly used technology for recommender systems. Many successful cases show that good results can be got using CF. There is a basic assumption that items appreciated by people similar to someone will also be appreciated by that person. As pointed in [1] that if this assumption is not satisfied, there will be some defects and data

© Springer International Publishing Switzerland 2015
A. Liu et al. (Eds.): DASFAA 2015 Workshops, LNCS 9052, pp. 17–28, 2015.
DOI: 10.1007/978-3-319-22324-7_2

sparsity is the most serious one. The data sparsity problem is the problem of having too few ratings, and hence, it is difficult to identifying similar users for the target user or similar items for the target item [2]. It was pointed out in [3] that the data sparsity problem is certainly exist when CF depends only on the user-item rating. Because only a small number of users will give item rating when using the actual recommender systems.

Currently, there are many methods have been proposed and data sparsity problem is resolved with the additional information such as context information [4], tags [5], social relationships [6]. The performance of these methods is better than CF.

Actually, there are opinion leaders in social networks especially trust networks. Some behavior or opinions of these opinion leaders often influence the behavior or comments of others in the network [7]. A choice made by ordinary people in trust networks is in fact a combination of views of opinion leaders and themselves. We can use this influence as the additional information to solve data sparsity problem in the recommender system. Thus, we proposed an algorithm named *OLrs* (Opinion Leaders for Recommender System) in this paper.

To determine whether an item such as a book is recommended to *Alan*, we have to predict the rating of this book *Alan* will be given. If the rating is highly, we can make recommend. The rating of the book given by *Alan* consists of two parts: one is based on *Alan's* own experience and another is based on the views of the opinion leaders who has a strong influence for *Alan* and accurate evaluation of the book. In our work, the former was calculated using tag-based recommender algorithm and the latter was calculated by doing the following things. First, the trust network of *Alan* was established based on trust relationships between users in the recommender system. Second, opinion leaders who can make an accurate evaluation on the book were found out from the most influential opinion leaders. Finally, the ratings of these opinion leaders are aggregated. We only used the views of the opinion leaders who have a strong influence for *Alan* and an accurate evaluation of the book. That is the difference between our method and the former algorithms in which the views of all opinion leaders who have influence for *Alan* are used.

The rest of this paper is organized as follows. Section 2 gives a brief overview of related works. Section 3 presents the *OLrs* method and Sect. 4 to verify the performance of present work, and compare with the state-of-the-art methods. We conclude the study in Sect. 5.

2   Related Work

Many works have been done in the literature to resolve the data sparsity problem, which is cause by lack of user-item ratings. These proposed methods can be classified into two categories: one is using CF method after filling the missing ratings with the average of existing ratings or the median rating, and another is merge various additional information, such as user attributes [8], multimedia content [9], and location information [10] into CF to improve recommendation performance. Many works show that the latter is effective than the former. Our proposed method in this paper is follow in the second way as well.

Recently, latent factor model has become popular. As the Netflix Prize competition has demonstrated, Matrix factorization model (MF) [11] are superior to classic CF for producing product recommendations. Salakhutdinov and Mnih [12] propose a new model name Probabilistic Matrix Factorization (PMF) based on Bayesian deduction. Singh and Gordon [13] propose collective matrix factorization (CMF), which simultaneously factorizes the user-item rating matrix and user attributes matrix to address the data sparsity problem better. However, these latent factor models face real difficulties to explain predictions.

Except for the attributes of user or item, there is other information created by user, such as Tags can be used to strengthen the recommendation. Tags are short labels used to mark the item [5, 14]. They are created according to the users' needs. The frequency of using some tags can reflect the user's interest in hobbies. This feature is used by researchers to improve the quality of recommendation. Tags are used to connect users and items in [15–17]. User-item rating matrix and tag-based user-user similarity matrix are used to factorize in the algorithm proposed in [18].

So many social relationships such as friendship, membership and social trust are applied to the CF with the emergence of a large number of social networking sites. And social trust is proved to be better than other social relationships to improve the performance of CF. Trust can be classified into two categories: implicit trust (e.g., [19, 20]) and explicit trust (e.g., [21, 22]). The former is inferred from user behaviors such as ratings whereas the latter is directly specified by users. Researchers considered that explicit trust is more reliable to implicit trust. TidlTrust [23] is a modified breadth-first search algorithm in a trust network. In this algorithm, it was considered highly trusted users should appear in the search of the shortest path. And the user's rating on item could be calculated by the ratings and degree of be trusted of these people. MoleTrust [24] is similar to TidlTrust. But depth-first search is used and the users whose depth smaller than a threshold are trusted in this algorithm. TrustWalker [25] is an algorithm in which random walk method is used to search trusted people in trust network. In all these algorithms there is one thing in common and it is searching trusted people in trust network. These trusted people are actually opinion leaders and their behavior or comments become the reference of other users in the network. But when users rate one item, they will make a conclusion not only according to the comment of opinion leaders, but also according to their own experience. And this is not taken account in the above work. The main idea of our work is getting the users final rating according to the ratings of opinion leaders and their own experience to improve data sparsity problem. These opinion leaders must have a strong influence for the user and an accurate evaluation of item.

3   The *OLrs* method

In this section, we propose an effective approach named *OLrs*, in which integrating the opinion leaders and user preference to predict the rating of active user for an item. Two steps are taken to make recommendations. First, the opinion leaders of the active user are identified. Second, recommendations are predicted by the aggregation of ratings of opinion leaders and the active user's rating calculated by tag-based method. Detailed descriptions as well as the insights of the *OLrs* are given in the subsequent sections.

3.1   Preliminaries

We first introduce some notations to model the recommendation problem. There are a set of users U, a set of items I, and a set of ratings R. We use the symbols u, v for the users and i, j for the items. Then $r_{u,i}$ represents a rating given by user u on item i. At the same time, user u will tagging item i use a set T_u of tags he likes, and item i have a set T_i of tags represent its properties as well. Otherwise, in a trust-aware recommender system, $t_{u,v} = 1$ represents user u trusted user v. And, the active user u may have some trusted neighbors, and these neighbors have some neighbors as well. Then, a trust network will construct around user u, and we can identify a set of opinion leaders in that network. Hence, the recommendation problem can be described as: given a set of user ratings $(u, i, r_{u,i})$, a set of user trust $(u, v, t_{u,v})$, and some set of tags of T_u and T_j, predict a best prediction $(u, j, r_{u,j})$ for user u on item j.

3.2   Identifying Opinion Leaders

There often are some authorities, who are the opinion leaders [26]. Comments on certain items they give can affect other people's attitude. The rating of user u on item i is affected by the ratings of opinion leaders on item i in *OLrs*. Thus, how to find the opinion leader k who has a strong influence for user u and an accurate evaluation of item i is a key work. In our presented method, the nodes of trust network are the users in recommender system. If user u trust user v, then there is an edge between these two nodes and this edge is directed from u to v. Directed trust network is constructed by the users and the trust relations between these users in recommender system.

Constructing the trust network of all users in recommender system is very exhausting. Fortunately, in order to find the opinion leaders of user u, we only need to construct the trust network of user u because the opinion leaders must be exit in this trust network.

We start from user u and use the breadth-first search algorithm to search d layers. Then the trust network of user u will be constructed by the nodes and the edges between nodes of d layers. The set of users in trust network is marked TN_u. Note that the greater d is, the more trusted neighbors will be inferred. However, the more cost will be taken and more noise is likely to be incorporated. According to the theory of six-degree separation [27], that is, any two users in the social network can be connected (if possible) within small (less than six) steps. In this work, we restrict $d \leq 4$ to prevent meaningless searching and save computational cost for large-scale data sets. We will discuss the influence of d in Sect. 4.

After the trust network of user u is constructed, the first step is finding out the opinion leaders who have a strong influence for user u. There are two types of influence opinion leaders in our method, one is the direct influence opinion leaders and another is the indirect influence opinion leaders. The formal definition of these influence opinion leaders is given in the following.

Definition 3.1. The direct influence opinion leader group of user u denoted by TNL_u is a set of users, as following:

$$TNL_u = \{v | P_v < P_{in}, t_{uv} = 1 \ and \ v \in TN_u\} \tag{1}$$

where TN_u is a set of users who trust by the active user u, it is sorted (in descending order) by the number of each user's trustees. t_{uv} represent the relationship between the active user u and his truster v, if t_{uv} equal to 1 means v is direct trust by u. P_v is the position of truster v in TN_u, and P_{in} is a threshold be used to determine how many truster will selected.

Definition 3.2. The indirect influence opinion leader group of user u denoted by TNG_u is a set of users, as following:

$$TNG_u = \{v | P_v < P_{in}, t_{uv} = 0 \ and \ v \in TN_u\} \tag{2}$$

where TN_u, t_{uv}, P_v and P_{in} are as well as definition 3.1, however t_{uv} equal to 0 means v is indirect trust by u.

Because the opinion leaders have their own expertise, they can't have proper evaluation for each class of goods. For example *Bobe* is a computer science professor and he must be the opinion leader in field of computer. *Bobe* can give a fair evaluation of the books in field of computer. But if the book is in field of military, his judge on this book maybe suspect. Then, we need to identify someone in TNL_u or TNG_u for their evaluation ability on the item i which will be recommended to user u. The ratings of these opinion leaders on item i can be the important reference for user u.

Next, we use Matrix Factorization [11] and Euclidian distance to compute the evaluation ability of an opinion leader about the item i. Matrix Factorization decomposes the ratings matrix into two lower dimension matrices $P \in R^{|U| \times d}$ and $Q \in R^{|I| \times d}$ which contain corresponding vectors with length L for every user and item. The resulting dot product $Q_i^T P_k$ captures the interaction between opinion leader k and item i, the leader's overall interest in the item's characteristics.

$$\hat{r}_{ki} = Q_i^T P_k \tag{3}$$

To determine the latent feature vectors, the system minimizes the regularized squared error on the set of observed ratings:

$$\min_{P^*,Q^*} \sum_{(k,i) \in R_c} (r_{k,i} - Q_i^T P_k)^2 + \lambda(\|P_k\|^2 + \|Q_i\|^2) \tag{4}$$

where R_c is the set of the (k, i) pairs for which r_{ki} is observed. Thus, Matrix Factorization characterizes every opinion leader and item by assigning them a latent feature vector P_k. We use the user feature vector to represent each opinion leader, and use the item feature vector Q_i to represent each item.

Additionally, we use the Euclidian distance to compute the evaluation ability of an opinion leader k about item i.

$$d_{ki} = \sqrt{(Q_{i1} - P_{k1})^2 + (Q_{i2} - P_{k2})^2 + \cdots + (Q_{iL} - P_{kL})^2} \tag{5}$$

According to definitions 3.1 and 3.2, here we have two types of opinion leaders in our method, one is the direct opinion leaders and another is the indirect opinion leaders. The formal definition of these opinion leaders is given in the following.

Definition 3.3. The direct opinion leader group of user u denoted by OLL_u is a set of users, as following:

$$OLL_u = \{k | P_k < P_o, k \in TNL_u\} \tag{6}$$

where TNL_u is defined in definition 3.1 and sorted (in descending order) by d_{ki}. P_k is the position of opinion leader k in TNL_u, and P_o is a threshold be used to determine how many opinion leader will selected.

Definition 3.4. The indirect opinion leader group of user u denoted by OLG_u is a set of users, as following:

$$OLG_u = \{k | P_k < P_o, k \in TNG_u\} \tag{7}$$

where TNG_u is defined in definition 3.2 and sorted (in descending order) by d_{ki}. P_k and P_o are as well as definition 3.3.

After the opinion leaders who have a strong influence for the user and an accurate evaluation of item have been identified, we present details of predicting ratings in the following section.

3.3   Predicting

In social life when user u ratings item i, he often refer to the views of his influential people i.e. opinion leaders. But certainly not taken action totally under the opinion leader's suggestions, he will take his own judgment in the end. For this reason, the rating of user u consists of two parts in $OLrs$. One is r_{oli} the ratings given by the opinion leaders, and another is r'_{ui} the rating of user u himself. The rating of user u is given by tag-based algorithm. Hence, $\hat{r}_{ui}$ is computed as a linear combination of the two parts:

$$\hat{r}_{ui} = \alpha \times r_{oli} + (1 - \alpha) \times r'_{ui} \tag{8}$$

where parameters α indicate the extent to which the combination relies on opinion leaders evaluation.

The rating of user u on item i influenced by opinion leaders is related to T_{uk} the degree of be trusted of opinion leaders. In our work, there are two types of opinion leaders. According to definition 3.3 and 3.4, the trust T_{uk} user u gives to direct opinion leaders is explicit trust directly given by user u, otherwise the trust user u give to indirect opinion leaders is indirect trust and it is calculated using the following formula.

$$T_{uk} = \frac{1}{d_{u,k}} \times T'_{uk} \tag{9}$$

where T'_{uk} denotes the inferred trust value by the MoldTrust [24] algorithm, d_{uk} is the shortest distance between user u and opinion leader k, which is no more than search depth as defined in Sect. 3.2. With the increase of d_{uk}, the trust user u give to opinion leader k will decline.

Having the trusted value of user u and each of his opinion leaders, we can calculate r_{oli} as follows:

$$r_{oli} = \frac{\sum\limits_{k \in O} T_{uk} \times r_{ki}}{\sum\limits_{k \in O} T_{uk}} \tag{10}$$

where r_{ki} is the rating on item i given by the opinion leader k, O is the opinion leader set OLL_u if we use the direct opinion leaders, or OLG_u when we use the indirect opinion leaders.

Inspired by [28], we will use tag-based method to compute the rating of user u on item i. Tag is a short label used to mark the item which rating by user u. If a tag is frequently used and the items marked by this tag have a high rating, it deducted that user u like using this tag to mark his favorite items. In other words, we can infer the user's rating on some item based on tags. The rating of user u on item i is calculated according to the following formula.

$$r'_{ui} = \frac{1}{|F(u,i)|} \sum_{f \in F(u,i)} \omega_u^f \tag{11}$$

where $F(u,i)$ is a set of tag which used by user u and mark the item i, $|F(u,i)|$ is the number of tags in $F(u,i)$. ω_u^f is the weight of tag f, and it is calculated as following.

$$\omega_u^f = \frac{1}{|u(f)|} \sum_{j \in u(f)} r_{u,j} \tag{12}$$

where $u(f)$ is the set of item which user u used tag f to mark, $|u(f)|$ is the number of time user u using tag f.

4   Evaluation

In order to verify the effectiveness of the present $OLrs$ method, we conduct experiments on a real-world data set. We want to address some questions: (1) how the performance of $OLrs$ in comparison with other counterparts; (2) how do the search depth d affect the recommendation results; (3) how the type of opinion leaders, direct or indirect affect the recommendation quality.

4.1 Data Acquisition

We use the Epinions data set to evaluate the present method, as this data set has been widely used in previous work such as [29, 30]. We obtain 47,064 ratings, assigned by 965 users on 6730 items. Each user has rated at least 2 items, and the ratings follow the 1 (bad) to 5 (excellent) numerical scale. The sparsity level of the data set is $1 - \frac{47,064}{965 \times 6,730}$, which is 0.9928. And there are 47,064 reviews, which can be used to extract user preference.

4.2 Evaluation Metrics

To measure the predicting accuracy, we use the effective metric mean absolute error (MAE), which is defined as the average absolute difference between predicting ratings and actual ratings. It is computed by the following formula.

$$MAE = \frac{\sum_u \sum_i |r_{ui} - \hat{r}_{ui}|}{N} \qquad (13)$$

where N is the number of testing ratings, r_{ui} is an actual user-specified rating on an item, and $\hat{r}_{ui}$ is the prediction for a user on an item given by the recommender system. A lower MAE value means that the prediction is close to the ground truth. Hence, for MAE values of a recommendation algorithm, the smaller the better.

4.3 Experimental Settings

In the experiments, we compare the performance of *OLrs* with the well-known user-based CF and the TGER [28]. CF computes user similarity with Pearson Correlation Coefficient and selected the Top N neighbors whose similarity is larger. For TGER method, we adopt the settings as the author report in our experiments. Otherwise, we divide the Epinions data set into the training set and the testing set, the former consist 80 % records and the latter consist 20 % records. We obtain the recommendation predictions based on the training set and use testing set to evaluate the accuracy of recommend algorithms.

4.4 Results and Analysis

Impact of the Parameter α. As mentioned in Sect. 3, users always took the views of their opinion leaders into consideration before judge the item. How can opinion leaders affected the users attitude, it is depends on users personal features in real world. But in *OLrs*, this influence is control by the parameter α. Then, we intend to determine the best α for *OLrs* method. In the first experiment, we vary the parameter α from 0 to 1 with step 0.1. The results are illustrated in Fig. 1.

According to Fig. 1, we found that α has large impact on the recommendation accuracy. With the parameter increases from 0 to 0.3, the value of MAE decreases

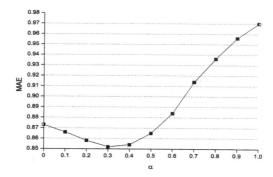

Fig. 1. The MAE of *OLrs* with different parameter α

from 0.873 to 0.852. It is because the views of opinion leaders have been taken into consideration, and help user done a good judge. However, as the parameter increase dramatically, MAE become more worsens. The worst value of MAE is 0.97 when α is equal to 1. This process can be explained by user's judge is not totally under the opinion leaders suggestion. We select the parameter such that the value of MAE is lower. Therefore, in the other experiments the parameter α is set to 0.3.

Effect of the Type of Opinion Leaders and the Search Depth D. According to definitions 3.3 and 3.4, the direct opinion leaders were pike up from the users who are direct trust by the active user, while the indirect opinion leaders were identified from the users who are indirect trust by the active user. As we know that, the direct opinion leaders may be affect the active user hugely in real world, for the reason that these direct opinion leaders are the most trust person. However, it is not sure that the most trust people can make an accurate evaluation of recommended item. There are may be some indirect opinion leaders can have a good judge about the recommended item. We analyze which type of opinion leaders would have better impact on the active user in this section.

In addition, the identified of indirect opinion leaders is also related with search depth d. It is considered that trust between users is transitive [23]. By propagating trust, more trusted people will be found and they can be used to improve recommended performance. The search depth d will decide the degree of propagating trust in *OLrs*. Then, the selected value of d is very important. If this value is too small, no trusted people will be found. If this value is too large, too many opinion leaders of user u in trust network will be found and more noise is likely to be incorporated, and their impact on user u will be reduced. In this section we analyze the impact of search depth d meanwhile.

As illustrated in Table 1, the value of MAE of *OLrs* based on direct opinion leaders is all 0.852. It is show that the search depth d affects nothing on direct opinion leaders, because this type of opinion leaders is directly trusted by the active user. While the search depth d affects the MAE of the algorithm based

Table 1. The MAE of OLrs with different type of leaders

	Type of leaders	search depth d			
		1	2	3	4
MAE	direct opinion leaders	0.852	0.852	0.852	0.852
	indirect opinion leaders	0.852	0.852	0.851	0.854

on indirect opinion leaders. We can found that the best value of MAE is 0.851 if the search depth d is 3. Thus, *OLrs* have a better performance if the indirect opinion leaders be used.

Comparison with State-of-the-Art Methods. Table 2 shows the MAE of *OLrs* and other compared methods. According to these values, we can found that *OLrs* is the best method for its MAE is 0.851, and CF is the worst method. The MAE of CF indicates that the data sparsity problem has been deteriorated the recommend accuracy, and we need some additional information beside user-item ratings to make improvement. TGER is the method which uses Tag to enhance recommends performance. The MAE of TGER is 0.873 which better than CF. Furthermore, *OLrs* achieves best performance than other methods by taking into account opinion leaders and user preference.

Table 2. The MAE of *OLrs* and other methods

method	OLrs	TGRE	CF
MAE	0.851	0.873	1.42

5   Conclusions and Future Works

To address the data sparsity problem, in this paper we have proposed *OLrs*, a recommend algorithm that combines opinion leaders and user preference in a unified frame. Experimental results based on the real-world data set demonstrate that our method outperforms other counterparts in terms of accuracy. In conclusion, we proposed an effective method to improve the performance of recommender systems. However, the present work depends on explicit trust which maybe not so easy to extract from social networks due to the privacy. Hence, one possible future work is to use the implicit trust in *OLrs*.

Acknowledgments. The authors are grateful to the anonymous reviewers and the helpful suggestion given by the partners. The research was supported by the National Natural Science Foundation of China (no. 61300137),the Foundation for Distinguished Young Teachers in Higher Education of Guangdong(no.Yq2014117), the Technology Project of Zhanjiang (no. 2013B01148), the Natural Science Foundation of Lingnan Normal College (no.QL1307, no.QL1410).

References

1. Adomavicius, G., Tuzhilin, A.: Toward the next generation of recommender systems: a survey of the state-of-the-art and possible extensions. IEEE Trans. Knowl. Data Eng. **17**(6), 734–749 (2005)
2. Huang, Z., Chen, H., Zeng, D.: Applying associative retrieval techniques to alleviate the sparsity problem in collaborative filtering. ACM Trans. Inf. Syst. (TOIS) **22**(1), 116–142 (2004)
3. Shi, Y., Larson, M., Hanjalic, A.: Collaborative filtering beyond the user-item matrix: a survey of the state of the art and future challenges. ACM Comput. Surv. (CSUR) **47**(1), 3 (2014)
4. Moshfeghi, Y., Piwowarski, B., Jose, J.M.: Handling data sparsity in collaborative filtering using emotion and semantic based features. In: Proceedings of the 34th International ACM SIGIR Conference on Research and Development in Information Retrieval, pp. 625–634. ACM (2011)
5. Robu, V., Halpin, H., Shepherd, H.: Emergence of consensus and shared vocabularies in collaborative tagging systems. ACM Trans. Web (TWEB) **3**(4), 14 (2009)
6. Leskovec, J., Huttenlocher, D., Kleinberg, J.: Signed networks in social media. In: Proceedings of the SIGCHI Conference on Human Factors in Computing Systems, pp. 1361–1370. ACM (2010)
7. Summers, J.O.: The identity of women's clothing fashion opinion leaders. J. Mark. Res. **7**, 178–185 (1970)
8. Gantner, Z., Drumond, L., Freudenthaler, C., Rendle, S., Schmidt-Thieme, L.: Learning attribute-to-feature mappings for cold-start recommendations. In: IEEE 10th International Conference on Data Mining (ICDM), pp. 176–185. IEEE (2010)
9. Davidson, J., Liebald, B., Liu, J., Nandy, P., Van Vleet, T., Gargi, U., Gupta, S., He, Y., Lambert, M., Livingston, B., et al.: The youtube video recommendation system. In: Proceedings of the fourth ACM Conference on Recommender Systems, pp. 293–296. ACM (2010)
10. Böhmer, M., Hecht, B., Schöning, J., Krüger, A., Bauer, G.: Falling asleep with angry birds, facebook and kindle: a large scale study on mobile application usage. In: Proceedings of the 13th International Conference on Human Computer Interaction with Mobile Devices and Services, pp. 47–56. ACM (2011)
11. Koren, Y., Bell, R., Volinsky, C.: Matrix factorization techniques for recommender systems. Computer **42**(8), 30–37 (2009)
12. Mnih, A., Salakhutdinov, R.: Probabilistic matrix factorization. In: Advances in Neural Information Processing Systems, pp. 1257–1264 (2007)
13. Singh, A.P., Gordon, G.J.: Relational learning via collective matrix factorization. In: Proceedings of the 14th ACM SIGKDD International Conference on Knowledge Discovery and Data Mining, pp. 650–658. ACM (2008)
14. Xie, H., Li, Q., Mao, X., Li, X., Cai, Y., Zheng, Q.: Mining latent user community for tag-based and content-based search in social media. Comput. J. **57**(9), 1415–1430 (2014)
15. Sen, S., Vig, J., Riedl, J.: Tagommenders: connecting users to items through tags. In: Proceedings of the 18th International Conference on World Wide Web, pp. 671–680. ACM (2009)
16. Xie, H.R., Li, Q., Cai, Y.: Community-aware resource profiling for personalized search in folksonomy. J. Comput. Sci. Technol. **27**(3), 599–610 (2012)
17. Cai, Y., Li, Q., Xie, H., Min, H.: Exploring personalized searches using tag-based user profiles and resource profiles in folksonomy. Neural Netw. **58**, 98–110 (2014)

18. Zhen, Y., Li, W.J., Yeung, D.Y.: Tagicofi: tag informed collaborative filtering. In: Proceedings of the third ACM Conference on Recommender Systems, pp. 69–76. ACM (2009)
19. O'Donovan, J., Smyth, B.: Trust in recommender systems. In: Proceedings of the 10th International Conference on Intelligent user Interfaces, pp. 167–174. ACM (2005)
20. Seth, A., Zhang, J., Cohen, R.: Bayesian credibility modeling for personalized recommendation in participatory media. In: De Bra, P., Kobsa, A., Chin, D. (eds.) UMAP 2010. LNCS, vol. 6075, pp. 279–290. Springer, Heidelberg (2010)
21. Ray, S., Mahanti, A.: Improving prediction accuracy in trust-aware recommender systems. In: 2010 43rd Hawaii International Conference on System Sciences (HICSS), pp. 1–9. IEEE (2010)
22. Chowdhury, M., Thomo, A., Wadge, W.W.: Trust-based infinitesimals for enhanced collaborative filtering. In: COMAD (2009)
23. Golbeck, J.A.: Computing and applying trust in web-based social networks (2005)
24. Massa, P., Avesani, P.: Trust-aware recommender systems. In: Proceedings of the 2007 ACM Conference on Recommender Systems, pp. 17–24. ACM (2007)
25. Jamali, M., Ester, M.: Trustwalker: a random walk model for combining trust-based and item-based recommendation. In: Proceedings of the 15th ACM SIGKDD International Conference on Knowledge Discovery and Data Mining, pp. 397–406. ACM (2009)
26. Watts, D.J., Dodds, P.S.: Influentials, networks, and public opinion formation. J. Consum. Res. **34**(4), 441–458 (2007)
27. Watts, D.J.: Six Degrees: The Science of a Connected Age. WW Norton and Company, New York (2004)
28. Cai Yi, Liu Yu, Z.G.C.J.M.H.: Tag group effect-based recommendation algorithm for collaborative tagging systems. Journal of South China University of Technology (Natural Science Edition) **41**(9), 65–70 (2013)
29. Guo, G., Zhang, J., Thalmann, D., Basu, A., Yorke-Smith, N.: From ratings to trust: an empirical study of implicit trust in recommender systems. In: Proceedings of the 29th Annual ACM Symposium on Applied Computing, pp. 248–253. ACM (2014)
30. Yang, B., Lei, Y., Liu, D., Liu, J.: Social collaborative filtering by trust. In: Proceedings of the Twenty-Third International Joint Conference on Artificial Intelligence, pp. 2747–2753. AAAI Press (2013)

Learning Trend Analysis and Prediction Based on Knowledge Tracing and Regression Analysis

Yali Cai[1], Zhendong Niu[1,2,3(✉)], Yingwang Wang[1], and Ke Niu[1]

[1] School of Computer Science and Technology, Beijing Institute of Technology,
Beijing 100081, China
zniu@bit.edu.cn
[2] Information School, University of Pittsburgh, Pittsburgh, PA 15260, USA
[3] Beijing Engineering Research Center of Massive Language Information Processing and Cloud
Computing Application, Beijing Institute of Technology, Beijing 100081, China

Abstract. Estimating students' knowledge is a fundamental and important task for student modeling in intelligent tutoring systems. Since the concept of knowledge tracing was proposed, there have been many studies focusing on estimating students' mastery of specific knowledge components, yet few studies paid attention to the analysis and prediction on a student's overall learning trend in the learning process. Therefore, we propose a method to analyze a student's learning trend in the learning process and predict students' performance in future learning. Firstly, we estimate the probability that the student has mastered the knowledge components with the model of Bayesian Knowledge Tracing, and then model students' learning curves in the overall learning process and predict students' future performance with Regression Analysis. Experimental results show that this method can be used to fit students' learning trends well and can provide prediction with reference value for students' performances in the future learning.

Keywords: Student modeling · Learning trend analysis · Learning performance prediction · Knowledge tracing

1 Introduction

Since the development of computer-based education, Intelligent Tutoring System has attracted popular attention. Currently, online education has become popular in the field of Internet and Intelligent Tutoring System has become one of the current focuses. Estimating students' knowledge is a fundamental and important task for student modeling in intelligent tutoring systems. Since the concept of knowledge tracing was proposed, there have been many studies focusing on estimating students' mastery of specific knowledge components, yet few studies paid attention to the analysis and prediction on a student's overall learning trend in the learning process, based on the student's knowledge predicted or predicted students' future performance and achievements taking advantage of the student's learning trend.

It is very significant to analyze trends and predict performances for students' learning. Firstly, we can know students' overall learning process and trends, and

© Springer International Publishing Switzerland 2015
A. Liu et al. (Eds.): DASFAA 2015 Workshops, LNCS 9052, pp. 29–41, 2015.
DOI: 10.1007/978-3-319-22324-7_3

discover learning progress or transition from the learning curve. Besides, with their learning trend, we can predict students' future learning status, identify problems, alert students and draw teachers' attention in prior, so that students would get the help and guidance in advance. Furthermore, we can assess students more reasonably and comprehensively, knowing both students' current learning levels and future development potential. In addition, it would be useful for teachers to understand the current status of the entire course of teaching, with reference to the prediction of all students.

For current studies, either most of them focus on estimating students' mastery of specific knowledge components, or a few of them focus on predicting students' post-test scores, but none of them analyze students' overall learning trend based on students' mastery of knowledge or predict based on students' learning trend. Considering those study vacancies, we propose a method based on Bayesian Knowledge Tracing and Regression Analysis to analyze a student' overall learning trend based on the student's mastery of knowledge predicted and predict the student's future performance and achievements with the student's learning trend analyzed.

Considering that learning is a long ongoing process, although we do not have any learning data of students to support us when predicting students' future learning performance, however, there is certain tendency and regularity in students' learning process, so we use this tendency as the basis to predict students' future performances in this paper. There are many potential factors that will cause a stable trend, such as students' relatively stable learning attitude, including their earnestness and self-requirements; relatively stable cognitive model, including ways of understanding, thinking characteristics; relatively stable personality characteristics, stable changes in learning interest as well as tending to learn in a community or not [1–3]. These factors will all change stably and they all have significant impact on learning. Therefore they will be bound to be reflected in the performances of the students' learning process as a trend or regularity. Analyze the regularity in these performances, then we can obtain students' overall learning trends and can predict students' future performances with it.

2   Related Work

2.1   Knowledge Tracing Model

In 1995, Corbett and Anderson proposed Bayesian Knowledge Tracing Model [4] to predict the probability that a student has mastered a knowledge component, by building a Bayesian network taking the knowledge level of a student as hidden variables. This model has been widely used to predict students' mastery of knowledge, and after this model, many studies made some improvements on Bayesian Knowledge Tracing model and many new knowledge tracing models came into being.

Many researchers have improved the Bayesian Knowledge Tracing Model, largely taking the form of relaxing the assumption that parameters vary by skill, but are constant for all other factors. In Beck's Help Model [5], help use is not treated as direct evidence of not knowing the skill. Instead, it is used to choose between parameters. This model did not lead to better prediction of student performance, but is useful for understanding effects of help. In 2010, Pardos and Heffernan modeled individualization of L0 in a

Bayesian networks implementation of knowledge tracking [6], which performed much better on some experiments but much worse on others. In 2008, Baker, Corbett and Aleven tried to build a more accurate student model through contextual estimation of Slip and Guess probabilities in Bayesian Knowledge Tracing [7]. However, the effect on future prediction of the new model is very inconsistent [8, 9] but it's predictive of long-term outcomes. In 2011, Baker, Goldstein and Heffernan built a new model by detecting learning moment-by-moment [10]. Although it didn't lead to overall better prediction of student performance, but it is found to have a surprising number of other uses. There are other types of extension to BKT including modifications to address multiple skills or include item difficulty [11, 12].

While many of the studies focus on improving the BKT model, many new knowledge tracing model have been built, mainly including Performance Factor Analysis [13, 14] and Item Response Theory [15]. The Performance Factor Analysis Model (PFA Model) predict students' performances in exercise by modeling logistic regression for item difficulty and performance of solving problems on time series, which is an alternative to BKT addressing some of the limitations of BKT, but doesn't have all of the nice features of BKT. However, it is also that PFA beats BKT sometimes but on the contrary in other experiments. The Item Response Theory is based on the idea that the probability of a correct response to an item is a mathematical function of student and item parameters. The typical use of IRT is to assess a student's current knowledge of topic X. However, it is not used quite so often in online learning because of its limitations.

From the analysis of those models above, it can be seen that these advanced BKT models or new knowledge tracking models do not appear to lead to overall improvement on predicting within-tutor performance, even though they are better than BKT sometimes.

2.2 Post-Test Score Prediction

While most of the studies are focusing on estimating students' mastery of specific knowledge components, a few studies pay attention to students' overall achievements.

Some studies consider whether ensembling can produce better prediction than individual models, when ensembling is performed at the post-test level [16], and some focus on whether finer-grained skills models is useful to the prediction of students' post-test score [17]. However, these studies are more focused on students' mastery of specific knowledge components.

Except for predicting based on students' mastery of knowledge components, there are also studies predicting students' post-test score based on students' comprehensive information and materials of their learning process [18].

Although there are studies predicting based on students' learning trend in their learning process, they are by means of periodic benchmark tests for students [19], or predict after when students have learned all of the knowledge components yet. All of these studies didn't analyze a student's overall learning trend based on students' mastery of knowledge components and predict based on the learning trends or they all didn't predict in the middle of students' learning without having learned all of the knowledge components.

3   Problem Analysis

Through the analysis of the theoretical basis for the prediction, we can see that students' learning trend can be reflected in students' performances and behaviors during their learning process, so if we want to analyze students' learning trends, first of all, we need to track students' performances through the learning process, in order to analyze students' mastery of knowledge components that they have learned, i.e. their learning status in the past, and then, to analyze students' learning trends based on their learning in the past and predict students future performances and achievements according to the trends.

Therefore, the problem studied in this paper can be divided into two sub-problems:

Estimating Students' Mastery of Knowledge. To solve this sub-problem, we need to collect students' behavioral data of learning in the learning process, build appropriate estimating model, estimate students' mastery of knowledge components they have learned, which can be used to analyze students' learning trend and predict students' performance.

Estimating students' mastery of knowledge components based on students' leaning behavior in the tutoring system is just the main work of Knowledge Tracing Model mentioned previously. From the analysis above about the related works, we can see that although there have been many advanced BKT models and new knowledge tracing models after the BKT model was proposed, they didn't lead to overall improvement on predicting, even though they are better than BKT in some experiments. Therefore, we use the standard BKT model to estimate students' mastery of knowledge in our scheme.

Trend Analysis and Performance Prediction. To solve this sub-problem, we need to model the implicit content in the data using predictive modeling techniques, based on the data of students' knowledge estimated previously using BKT model, in order to analyze students' learning trend and predict students' future learning performances and achievements.

There are a lot of predictive modeling techniques, such as Neutral Networks, Support Vector Machine, Association Rules and Regression Analysis, and each model has its own characteristics and advantages. Among them, the Regression Analysis is a kind of numerical prediction technique, which can be used to fit students' learning curves as well as predict students' future numerical achievements, so it can meet demand in this study well and therefore we use Regression Analysis in this part to do certain analysis and prediction.

4   Algorithm Design

Based on what we analyzed in chapter 3, we will elaborate the algorithm design in this chapter, including the definition of course structure which is the foundation of student modeling and including the part of knowledge tracing, regression modeling as well as learning prediction of the model we proposed.

4.1   Course Structure Definition

In this study, the student modeling is performed in a course in the tutoring system, and therefore, to analyze the students' learning data, first of all, we need to define the course structure required in this study. We define the required course structure as follows:

- *Course C*: the course students are going to learn.
- *Knowledge Component Model*: the knowledge components that students need to learn in Course C, such as KC_1, KC_2, ..., KC_n.
- *Problem Sets* of Couse C. There are a series of Problems corresponding to each KC(Knowledge Component) in each problem set.
- The hierarchy in problems of Course C:

- *Problem Hierarchy*, indicates which part or which Problem Set the Problem belongs to.
- *Problem*, refers to a problem or question for students to solve or answer in Course C, and the *Problem* may correspond to several KCs of Course C.
- *Step*: there are a series of steps needed when solving a Problem, and every *Step* corresponds to a KC of Course C.

4.2   Knowledge Tracing

Bayesian Knowledge Tracing Model. As analyzed above, we use Bayesian Knowledge Tracing Model [4] to estimate students' knowledge in our scheme. This model is a method that iteratively predicts the probability that students has mastered a knowledge component, according to a series of performances when a student is answering questions relating to the knowledge component.

The standard Bayesian Knowledge Tracing Model is a Bayesian network taking the knowledge level of students as hidden variables, as depicted in Fig. 1, which contains two kinds of nodes. There are four parameters in BKT model: $P(L_0)$, $P(T)$, $P(S)$, and $P(G)$. The parameters of BKT are learned from existing data, originally using curve-fitting [4], but more recently using expectation maximization (BKT-EM) [20] or brute force/grid search (BKT-BF) [6, 9]. We use BKT-BF in this study.

Input and Output of BKT part. According to the needs of the model and the data. Structure defined above, the input data of the BKT model in this study is the performance data when students solving a series of Steps relating to each knowledge component. And an example of input data is shown in Table 1.

Then, according to the input data and BKT model, estimate the probability that each student has learned each knowledge component and the probability that each student will answer a new relevant question correctly. And an example of output data is shown in Table 2.

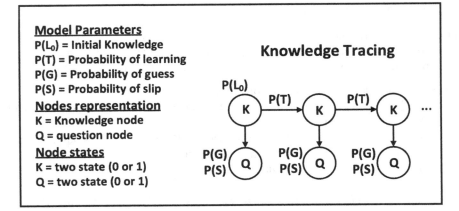

Fig. 1. Bayesian Knowledge Tracing

Table 1. An example of input data of BKT Model

Num	Student Id	Step Name	Correctness	KC	Opportunity
01	Student 001	Step 01	0	KC A	1
02	Student 001	Step 05	1	KC A	2
03	Student 001	Step 03	0	KC A	3
04	Student 001	Step 04	1	KC B	1

Table 2. An example of output data of BKT Model

Num	Student Id	Knowledge Component	Probability Learned	Probability Correct(n+1)
01	Student 001	KC A	0.78	0.89
02	Student 001	KC B	0.36	0.56
03	Student 001	KC C	0.89	0.90
04	Student 002	KC A	0.75	0.85

4.3 Regression Modeling

The task of this part is to establish the appropriate regression model taking the output of the BKT model elaborated above as the basic data, which can be regarded as the learning curve of students, representing the learning trend of students.

In the scheme, the regression model we plan to build here is how the trend of a student's knowledge state goes with the student's learning process. As what elaborated above, there is a KC Model in the course learned and for each KC in the model, every student will get a probability of mastering. And the learning of the KC sequence is chronological. Therefore, the regression model needing to be built here is the way the trend of the probability of a student's knowledge state goes with the KC sequence.

However, the probability that a student will master a KC, is not only affected by factors related to the student's learning state, but also affected by the difficulty of the KC which will largely affect the prediction results. Therefore, we consider the difficulty of KCs as another important factor in the regression model and it should be a good solution to take the mean of all students' mastery probabilities of the KC as its difficulty coefficient, which is the just the way we use.

Then, build the regression model, according to the standard steps.

1. In this model, the dependent variable y is the probability that the student has mastered a KC or will answer the question relating to the KC correctly. There are two independent variables. One independent variable x is the number of the KC in the chronological sequence of KCs that the student learns. Another independent variable h is the difficulty coefficient of the KC, which is the mean of all students' mastery probabilities of the KC.
2. The data required in the modeling is the output of the BKT model mentioned above.
3. As for choosing the assumed format of regression model, we will view the scatter diagram of students' mastery probabilities of each KC, to choose the appropriate regression equation and then make some adjustments to the assumption with reference to the result of hypothesis test and model prediction. Because of the feature of taking the mean of all students' mastery probabilities as the difficulty degree of the KC, it will represent that the difficulty of the KC changes faster, when the value of the difficulty coefficient gets closer to its interval boundary. By experience, we know that the cube polynomial equation can fit this feature well. So, we advise to choose Formula 1 as a prior choice of the regression equation.

$$y = ax + b_3h^3 + b_2h^2 + b_1h + c \qquad (1)$$

Where the variable y, x and h represents respective meaning as described in step 1 and the coefficients a, b_3, b_2, b_1 and c are regression coefficients which needs to be calculated with actual data in the process of regression.

4. Using the data collected, based on the least square method, estimate the value of each regression parameter.
5. Carry out statistical inspections for the regression model built, to inspect its validity and effect of data fitting. The inspections include t-test for regression coefficients, F-test for regression equation, test of fitting effect.
6. Modify the regression model according to the results of statistical tests and model prediction.
7. Finally, obtain the most appropriate regression model.

4.4 Trend Analysis and Learning Prediction

Learning Trend. The regression equation obtained above is about the relationship between the probability of the student's knowledge state and the KC sequence, ie the learning curve of trend in the student's learning process.

Through the learning curve, we can understand the overall learning trend in the student's learning process, which can help understand the student's learning state and predict the future state.

Model Predicting. For the KC Model the student learns, we can estimate the mastery probability of the knowledge components that the student has learned with reference to the BKT part of our scheme and predict the mastery probability that the student is going to learn in the future with reference to the regression analysis part.

We denote the probability that the student answers the questions correctly relating to KC_1, KC_2, ..., KC_n, by P_1, P_2, ..., P_n, and denote the value assigned of each KC in the post-test examination by V_1, V_2, ..., V_n. Then, the mathematical expectation of the score that the student will obtain in the post-test can be predicted as follows:

$$Score = P_1 V_1 + P_2 V_2 + \cdots + P_n V_n \tag{2}$$

5 Experiment Design and Results Analysis

5.1 Experiment Data Selection

In our experiment, we used the '*Lab study 2012 (cleanedLogs)*' dataset accessed via *DataShop* (Koedinger et al., 2010) [21].

The dataset records the students' learning data when they are doing Fractions Lab Experiment. And its content and structure meets the requirements and definitions in this study. The dataset totally includes 74 students, 14959 steps and there are 19 KCs in the KC Model adopted.

5.2 Experiment Design

The purpose of this experiment is to test and evaluate the prediction effect of the whole model proposed in this paper, i.e. the effect of prediction of students' post-test scores. We design the experiment, based on the model proposed and the data used.

Data Processing. According to the requirement of the model proposed in this paper, the original dataset should be processed as follows:

1. Sort the rows of the data successively by student ID, knowledge component and time, and select the row of the last Step relating to each knowledge component, which will be combined into the Post-test Paper denoted by Test A as the TestData of this study.
2. Sort the KCs that each student have learned by time, forming the knowledge component list of each student.

3. According to a certain percentage, choose the previous part from the KC list of each student as *Knowledge List A* and regard the related rows as the learning data of the knowledge that the student has learned, and by contrast, the rows not selected as the learning data of the knowledge components (denoted by *Knowledge List B*) that the student is to learn afterwards,. The rows relating to the KCs in *Knowledge List A* will be used as the training data denoted by TrainData.

Experiment Procedure. According to the requirements of the scheme and the experiment in this paper, we design the experiment procedures as follows:

1. According to the input data format of the model, organize the TrainData into the format required, as shown in Table 1.
2. With reference to BKT model, calculate the probabilities that each student has mastered each KC with TrainData.
3. On the basis of regression analysis, build the regression models for each student as elaborated above, and then predict the probabilities for each student that he or she would master each KC in *Knowledge List B*.
4. Based on the section of model predicting, calculate each student's score of *Test A*.
5. According to the actual performances of students in TestData, calculate each student's actual score of *Test A*.
6. Use appropriate evaluation methods to evaluate the effect of the model, i.e. evaluate the deviation between students' actual scores and the scores predicted by the model we proposed.

Experiment Evaluation. In order to facilitate the observation of the effect of the model we proposed, we use two types of measurement methods to evaluate the result, i.e. the type of total deviation and prediction accuracy.

In addition, in order to make the results more informative, before using the measurement methods to evaluate, we standardize the scores first, i.e. the full marks of the paper is standardized to be 1 point and the scores that the students achieved are mapped to the interval [0,1].

For the measurement of total deviation, we use the method of root mean square error (RMSE), the formula is as follows:

$$RMSE = \sqrt{\frac{1}{n}\sum_{t=1}^{n}\left(y_t - \hat{y}_t\right)^2} \tag{3}$$

Where y_t represents the actual score of each student, $\hat{y}$ represents the score predicted and n represents the total number of students.

For the measurement of prediction accuracy of the model proposed, we use the Theil Inequality Coefficient μ, the value of which is in the interval [0,1] and it represents a better effect of our model if μ is closer to 0. The formula is as follows:

$$\mu=\frac{\sqrt{\frac{1}{n}\sum_{t=1}^{n}\left(y_t-\hat{y}_t\right)^2}}{\sqrt{\frac{1}{n}\sum_{t=1}^{n}y_t^2}+\sqrt{\frac{1}{n}\sum_{t=1}^{n}\hat{y}_t^2}} \tag{4}$$

Where y_t represents the actual score of each student, $\hat{y}$ represents the score predicted and n represents the total number of students.

5.3 Results and Analysis

First, using the dataset accessed via *DataShop* mentioned above, with the BKT part of our model, calculate the probabilities that students has mastered each KC and draw the scatter diagrams, as shown in Fig. 2.

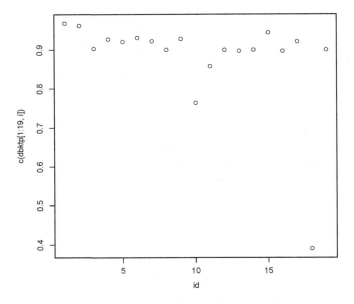

Fig. 2. The probability that a student has mastered each KC

And then, in the part of regression and prediction, with reference to students' scatter diagrams, we tried a variety of different models and calculate its values of statistical tests and evaluation indicators, as shown in Table 3.

In Table 3, the model O is the reference model, estimating students' mastery status of 19 KCs totally using BKT, with no prediction of students' future learning performances. In BKT, the result about students' knowledge is also a estimation from students' performances, which means there is also some deviation in BKT. Besides, the test paper used also has some limitations because of limited questions. Therefore, the model O itself also has some deviation, even though it is relatively small.

Table 3. Comparing the effects of different models

No.	Model	KC number selected	mean p-value of F-test	mean R-squa-red	RMSE	μ
O	BKT (no prediction, only as reference)	19/19	NA	NA	0.0977	0.0571
A	$y = ax^2 + bx + c$	11/19	0.1288	0.4846	0.2138	0.1319
B	$y = ax + c$	11/19	0.0939	0.3997	0.1029	0.0615
C	$y = ax + bh + c$	11/19	0.0970	0.6020	0.0998	0.0585
D	$y = ax + b_2h^2 + b_1h + c$	11/19	0.0762	0.7637	0.1027	0.0599
E	$y = ax + b_3h^3 + b_2h^2 + b_1h + c$	11/19	0.0701	0.8637	0.1002	0.0585
F	$y = ax + b_3h^3 + b_2h^2 + b_1h + c$	14/19	0.0658	0.7647	0.0965	0.0564
G	$y = (a_1x + a_0)(b_3h^3 + b_2h^2 + b_1h + b_0)$	11/19	NA, parameter estimation doesn't converge			

Then, we do the comparative analysis of each model in Table 3.

Comparing model A, B with model C, D, E, we can see that before the difficulty coefficient h is added into model, no matter for model A or model B, the effect of regression fitting and model prediction are both poor. And the effect is significantly improved after involving h, which proves the importance of difficulty coefficient h just as we analyzed above.

After involving the difficulty coefficient h into model, the prediction effect of model C, D, E has been very close to model O, the reference model mainly using BKT, which means the model proposed in this paper has got pretty good prediction results, showing that the theoretical analysis and specific scheme for learning trend analysis represented in this paper is very reasonable.

Among model C, D and E, for all the evaluation indicators, model E all gets better effects, and especially the R-squared value 0.8637 means that the fitting effect of model E is pretty good, which verifies what we analyzed above when choosing the assumed format of regression model.

Comparing model E and F, we can see that when the number of KC selected increases, although fitting results declined, however, the effect of model predicting obtains a promotion.

Therefore, model E in Table 3 will be the chosen regression model in this experiment.

6   Conclusion

Considering the study vacancy at the topic of analyzing students' overall learning trend in the field of student modeling, we propose a model of learning trend analyzing and predicting using BKT and regression, based on the viewpoint that students' learning is a process with stable tendency and regularity because of the impact of many latent stable factors. In the model proposed, BKT is used to estimate students' mastery status of knowledge and regression is used to fit students' overall learning trend and predict students' future performance using the trend.

In the experiment using the dataset accessed via *PSLC Datashop*, the model we proposed is able to fit students' learning trends very well and gets pretty good prediction effect, which verifies the rationality of our theoretical analysis and scheme, proves that there is stable tendency in students' learning process and proves we can model students' learning trends by analyzing learning data and get valuable result when predicting using learning trends, showing that it is feasible and very significant to analyzing students' overall learning trends.

Acknowledgement. This work is supported by the National Natural Science Foundation of China (Project Nos. 61370137), the International Corporation Project of Beijing Institute of Technology (No. 3070012221404) and the 111 Project of Beijing Institute of Technology.

References

1. Xie, H.R., Li, Q., Cai, Y.: Community-aware resource profiling for personalized search in folksonomy. J. Comput. Sci. Technol. **27**(3), 599–610 (2012)
2. Xie, H., Li, Q., Mao, X., Li, X., Cai, Y., Rao, Y.: Community-aware user profile enrichment in folksonomy. Neural Netw. **58**, 111–121 (2014)
3. Xie, H., Li, Q., Mao, X., Li, X., Cai, Y., Zheng, Q.: Mining latent user community for tag-based and content-based search in social media. Comput. J. **57**(9), 1415–1430 (2014)
4. Corbett, A.T., Anderson, J.R.: Knowledge tracing: modeling the acquisition of procedural knowledge. User Model. User-Adap. Interact. **4**, 253–278 (1994)
5. Beck, J.E., Chang, K.-M., Mostow, J., Corbett, A.T.: Does help help? introducing the bayesian evaluation and assessment methodology. In: Woolf, B.P., Aïmeur, E., Nkambou, R., Lajoie, S. (eds.) ITS 2008. LNCS, vol. 5091, pp. 383–394. Springer, Heidelberg (2008)
6. Pardos, Z.A., Heffernan, N.T.: Modeling individualization in a bayesian networks implementation of knowledge tracing. In: De Bra, P., Kobsa, A., Chin, D. (eds.) UMAP 2010. LNCS, vol. 6075, pp. 255–266. Springer, Heidelberg (2010)
7. Baker, R.S., Corbett, A.T., Aleven, V.: More accurate student modeling through contextual estimation of slip and guess probabilities in bayesian knowledge tracing. In: Woolf, B.P., Aïmeur, E., Nkambou, R., Lajoie, S. (eds.) ITS 2008. LNCS, vol. 5091, pp. 406–415. Springer, Heidelberg (2008)
8. Baker, R.S., Corbett, A.T., Aleven, V.: Improving contextual models of guessing and slipping with a truncated training set. Human-Computer Interaction Institute 17 (2008)
9. Baker, R.S., Corbett, A.T., Gowda, S.M., Wagner, A.Z., MacLaren, B.A., Kauffman, L.R., Mitchell, A.P., Giguere, S.: Contextual slip and prediction of student performance after use of an intelligent tutor. In: De Bra, P., Kobsa, A., Chin, D. (eds.) UMAP 2010. LNCS, vol. 6075, pp. 52–63. Springer, Heidelberg (2010)
10. Baker, R.S., Goldstein, A.B., Heffernan, N.T.: Detecting learning moment-by-moment. Int. J. Artif. Intell. Educ. **21**, 5–25 (2011)
11. Pardos, Z.A., Heffernan, N.T.: KT-IDEM: Introducing item difficulty to the knowledge tracing model. In: Konstan, J.A., Conejo, R., Marzo, J.L., Oliver, N. (eds.) UMAP 2011. LNCS, vol. 6787, pp. 243–254. Springer, Heidelberg (2011)
12. Gowda, S.M., Rowe, J.P., de Baker, R.S.J., Chi, M., Koedinger, K.R.: Improving models of slipping, guessing, and moment-by-moment learning with estimates of skill difficulty. EDM **2011**, 199–208 (2011)

13. Pavlik Jr., P.I., Cen, H., Koedinger, K.R.: Learning Factors Transfer Analysis: Using Learning Curve Analysis to Automatically Generate Domain Models (2009)
14. Pavlik Jr., P.I., Cen, H., Koedinger, K.R.: Performance Factors Analysis–A New Alternative to Knowledge Tracing (2009)
15. Senturk, D.: Explanatory item response models: A generalized linear and nonlinear approach. Technometrics **48**, 568–569 (2006)
16. Pardos, Z.A., Gowda, S.M., de Baker, R.S.J., Heffernan, N.T.: Ensembling predictions of student post-test scores for an intelligent tutoring system. In: EDM, pp. 189–198 (2011)
17. Pardos, Z.A., Heffernan, N.T., Anderson, B., Heffernan, C.L., Schools, W.P.: Using fine-grained skill models to fit student performance with Bayesian networks. Handbook of educational data mining 417 (2010)
18. Heffernan, N., Koedinger, K., Junker, B., Ritter, S.: Using Web-based cognitive assessment systems for predicting student performance on state exams. Research proposal to the Institute of Educational Statistics, US Department of Education. Department of Computer Science at Worcester Polytechnic Institute, Massachusetts (2001)
19. Anozie, N., Junker, B.W.: Predicting end-of-year accountability assessment scores from monthly student records in an online tutoring system. Educational Data Mining: Papers from the AAAI Workshop. Menlo Park, CA: AAAI Press (2006)
20. Chang, K.-M., Beck, J.E., Mostow, J., Corbett, A.T.: A bayes net toolkit for student modeling in intelligent tutoring systems. In: Ikeda, M., Ashley, K.D., Chan, T.-W. (eds.) ITS 2006. LNCS, vol. 4053, pp. 104–113. Springer, Heidelberg (2006)
21. Koedinger, K.R., Baker, R., Cunningham, K., Skogsholm, A., Leber, B., Stamper, J.: A data repository for the EDM community: The PSLC DataShop. Handbook of educational data mining 43 (2010)

Intensive Maximum Entropy Model
for Sentiment Classification of Short Text

Yanghui Rao[1], Jun Li[1(✉)], Xiyun Xiang[1], and Haoran Xie[2]

[1] School of Mobile Information Engineering,
Sun Yat-sen University, Zhuhai, China
raoyangh@mail.sysu.edu.cn,
{lijun223,xiangxy3}@mail2.sysu.edu.cn
[2] Caritas Institute of Higher Education, Hong Kong, China
hrxie2@gmail.com

Abstract. The rapid development of social media services has facilitated the communication of opinions through microblogs/tweets, instant-messages, online news, and so forth. This article concentrates on the mining of emotions evoked by short text materials. Compared to the classical sentiment analysis from long text, sentiment analysis of short text is sometimes more meaningful in social media. We propose an intensive maximum entropy model for sentiment classification, which generates the probability of sentiments conditioned to short text by employing intensive feature functions. Experimental evaluations using real-world data validate the effectiveness of the proposed model on sentiment classification of short text.

Keywords: Sentiment classification · Short text analysis · Intensive maximum entropy model

1 Introduction

Nowadays, the thriving of short text-based social media services has provided us with torrents of news on diverse topics and entities. Leveraging by the convenience of communication among online users, microblogs/tweets and instant-messages are constantly filled with opinions towards a large amount of topics such as politics, sports and other prevailing topics. Thus, it is important for us to identify and classify sentiments from short texts automatically.

Different from the normal documents, the number of words is few and most words only occur once in each short text. Thus, for the tasks of sentiment classification and annotation, it is usually impossible to classify or annotate emotions consistently using the limited information contained in short text [1]. Most existing approaches enriched the context of short text by retrieving relevant

Jun Li—The research work described in this article has been substantially supported by "the Fundamental Research Funds for the Central Universities" (Project Number: 46000-31610009).

© Springer International Publishing Switzerland 2015
A. Liu et al. (Eds.): DASFAA 2015 Workshops, LNCS 9052, pp. 42–51, 2015.
DOI: 10.1007/978-3-319-22324-7_4

long documents from the web, which may result in waste of both memory and computation time. In light of these considerations, we first propose to model sentiments and the limited words using intensive feature functions, Then, sentiments of unlabeled short text is classified according to the principle of maximum entropy [2]. Experimental results show that the proposed model can effectively integrate prior conditions from the sparse text.

The remainder of this paper is organized as follows. In Sect. 2, we firstly summarize the related work in sentiment classification of short text. Then, we propose our model and conduct experimental analysis in Sects. 3 and 4, respectively. Finally, we present conclusions and future works in Sect. 5.

2   Related Work

In this section, we review some previous works on sentiment classification and short text analysis, which shed light on sentiment classification of short text for our work.

Sentiment Classification. Sentiment classification mainly concentrate on extracting emotions from reviews, messages and news documents, which convey the opinion of writers or readers. The existing methods of sentiment classification can be divided into three categories primarily: lexicon-based, supervised and unsupervised learning strategies. The lexicon-based method [3–5] classified sentiments by constructing word- or topic-level emotional dictionaries. The supervised learning strategy used existing classification algorithms to split the emotional orientation of words or phrases into positive and negative, which included naïve Bayes, maximum entropy and support vector machines [6]. An unsupervised learning technique was also utilized to classify the emotional orientation of users' reviews (e.g., reviews of movies, travel destinations, automobiles and banks), which computed the overall polarity of the review by counting the occurrence of positive and negative terms [7]. However, those lexicon-based, supervised or unsupervised learning strategies were mainly designed for long text.

Short Text Analysis. The main feature of short text is the sparsity of words. Due to the fact that most words only occur once in each short text, it is difficult to accurately conduct classification, clustering, retrieval and other tasks.

In a preliminary work, Sahami and Heilman [8] defined a web-based kernel function to measure the semantic similarity of short text snippets. The general process was as follows: First, each short text was inputted to the web search engine as a query. Second, a large amount of retrieved long documents was combined to enrich the context of each short text. Third, the vector space model [9] was used to estimate the similarity values. Banerjee et al. [10] also proposed a method of enhancing the representation of short text by including the additional features from Wikipedia. However, due to the high-dimensional vectors, those methods may be memory or time-consuming. To address this issue, a collapsed Gibbs sampling algorithm for the Dirichlet multinomial mixture model was proposed [11]. Unfortunately, the model was designed for short text clustering rather than classification.

Our Work. Research into sentiment classification of short text began with the "affective text" in SemEval-2007 tasks [1], which aimed to annotate news headlines according to the predefined emotions. In the SWAT system [1], a word-emotion mapping dictionary was first constructed, in which, each word was scored according to multiple emotion labels. Then, the dictionary was used to classify the emotions of unlabeled news headlines. Recently, the emotion-term(ET) algorithm and the emotion-topic model(ETM) [12] were proposed to improve the performance of existing systems. ET is a variant of the naïve Bayes classifier and ETM is model associating emotions with topics jointly. Nevertheless, experimental results have shown that the performance of sentiment classification of short text is limited for SWAT, ET and EMT [3]. The reason may be that short documents lack enough context from which statistical conclusions can be drawn early [13]. We here develop an intensive maximum entropy model to classify sentiments of short text, which has a concentrated representation for modeling emotions and the limited words.

Table 1. Notations of frequently-used variables.

Symbol	Description
V	Number of unique word tokens
N	Number of short text
M	Number of emotion labels
Φ	Set of all word-emotion pairs
Θ	Set of distinct word-emotion pairs

3   Maximum Entropy Model via Intensive Feature Functions

In this section, we propose the intensive maximum entropy model(IMEM), a maximum entropy model via intensive feature functions for short text sentiment classification. The problem is first defined, including the relevant terms and notations, and then the IMEM is presented in detail. Finally, we describe the estimation of parameters.

3.1   Problem Definition

For convenience of defining the issue of short text sentiment classification, and describing our intensive maximum entropy model, we here defined the following terms and notations:

A document collection consists of N short text $\{t_1, t_2, \ldots, t_N\}$ with word tokens and an emotion label. We represent the list of word tokens and emotion labels by $\{w_1, w_2, \ldots, w_V\}$ and $e = \{e_1, e_2, e_3, \ldots, e_M\}$, respectively, where V is the numble of unique terms, and M is the amount of emotion labels. The common instances of emotion labels are "joy", "anger", "fear", "surprise", "touching", "empathy", "boredom", "sadness", "warmness", etc. Table 1 summarizes the notations of these frequently used variables.

3.2 Intensive Maximum Entropy Model

In this section, we first briefly introduce the principle of entropy and maximum entropy model [2], and then present out intensive maximum entropy model for sentiment classification of short text.

Entropy is the average amount of information contained in each message. The general idea is that the less likely an event is, the more information it provides when it occurs, i.e., the larger entropy it has. The probability distribution of the events, coupled with the information amount of each event, forms a random variable whose expected value is the average amount of information, generated by this distribution.

The principle of maximums entropy indicates that when predicting the probability distribution of a random event, the distribution should satisfy all our prior conditions and knowledges (e.g. the training set that expressed testable information), and make none subjective assumptions about the unknown case. Under this condition, the probability distribution has the largest value ors entropy, and the error of prediction could be minimized.

Table 2. Samples of the training set.

Short text	Word tokens	Emotion Label
t_1	$\{w_3, w_4\}$	e_1
t_2	$\{w_1, w_2\}$	e_2
t_3	$\{w_1, w_3\}$	e_3
t_4	$\{w_2, w_3, w_4\}$	e_1

In our task of short text sentiment classification, the prior conditions are the co-occurrences of word tokens and emotion labels. Table 2 shows the samples of a training set, where $N = 4, V = 4, M = 3$. To make a concentrated representation of modeling emotion labels and the limited word tokens in short text, we propose and intensive feature function as follows:

$$f(w, e) = \begin{cases} 1 & w \in \{w_1, w_2, \ldots, w_V\}, e \in \{e_1, e_2, \ldots, e_M\} \text{ and } (w, e) \in \Theta \\ 0 & \text{otherwise.} \end{cases} \quad (1)$$

where Θ is the set of distinct word-emotion pairs that occurred in the training set, i.e., $\Theta = \{(w_1, e_2), (w_1, e_3), (w_2, e_1), (w_2, e_2), (w_3, e_1), (w_3, e_3), (w_4, e_1)\}$. We then define the empirical probability distribution of the training set $\bar{p}(w, e)$, as follows:

$$\bar{p}(w, e) = \frac{1}{|\Phi|} \times count(w, e) \quad (2)$$

where Φ is the set of all word-emotion pairs in the training set, $|\Phi| = 9$, and $count(w, e)$ is the number of times that (w, e) occur in Φ.

The expected value of our feature functions $f(w, e)$ with respect to the empirical distribution can be estimated by:

$$\bar{E}(f) = \sum_{w,e} \bar{p}(w, e) f(w, e) \tag{3}$$

The expected value of with respect to the probability of emotion label e conditioned to word token w, i.e., $p(w|e)$ is derived as follows:

$$E(f) = \sum_{w,e} \bar{p}(w) p(e|w) f(w, e) \tag{4}$$

where $\bar{p}(w)$ is the empirical distribution of w in the training set. Thus, the first constraint condition of the IMEM is as follows:

$$E(f) = \bar{E}(f) \tag{5}$$

According to Eqs. (3) and (4), we get

$$\sum_{w,e} \bar{p}(w) p(e|w) f(w, e) = \sum_{w,e} \bar{p}(w, e) f(w, e) \tag{6}$$

A mathematical measure of the uniformity of the conditional distribution $p(e|w)$ is provided by the conditional entropy:

$$H(p) = -\sum_{w,e} \bar{p}(w) p(e|w) \log p(e|w) \tag{7}$$

Then, the IMEM is formulated as the following optimization problem:

$$\text{maximize } H(P) = \text{argmax} \sum_{w,e} \bar{p}(w) p(e|w) \log \frac{1}{p(e|w)} \tag{8}$$

subject to

$$E(f) - \bar{E}(f) = 0 \tag{9}$$

$$\sum_{e} p(e|w) - 1 = 0 \text{ for all } w \tag{10}$$

To estimate the value of $p(e|w)$ that maximizes $H(p)$, we resolve the above primal optimization problem to a unconstrained dual optimization problem by introducing the Lagrange parameters λ, as follows:

$$p_\lambda(e|w) = \frac{1}{Z_\lambda(w)} exp(\sum_{i=1}^{|\Phi|} \lambda_i f_i(w, e)) \tag{11}$$

$$Z_\lambda(w) = \sum_{e} exp(\sum_{i=1}^{|\Phi|} \lambda_i f_i(w, e)) \tag{12}$$

3.3   Parameter Estimation

To estimate the parameters of IMEM, i.e., λ, we use an iterative method as shown in Algorithm 1.

Algorithm 1. Iterative algorithm for IMEM

Input:
  Feature functions $f_1, f_2, \ldots, f_n$;
  Empirical distribution $\bar{p}(w, e)$s
Output:
  Optimal values of each parameter λ_i;
  Set $\lambda_i^{(0)}$ to some initial values, e.g.: $\lambda_i^{(0)} = 0$.
  repeat
    $\lambda_i^{(r+1)} = \lambda_i^{(r)} + \frac{1}{C} \log \frac{\bar{E}(f_i)}{E^{(r)}(f_i)}$
  until convergence
  where t is the iteration index and the constant C is defined as follows:

$$C = \max_{w,e} \sum_{i=1}^{n} f_i(w, e)$$

After estimating the optimal values of each parameter λ_i, predicting the emotion label of unlabeled short text is straightforward.

Table 3 presents an example of the testing set. Given and unlabeled short text t with three words tokens $\{w_1, w_2, w_3\}$, and two predefined emotion labels $\{e_1, e_2\}$, we get six intensive feature functions in total.

Table 3. Samples of the testing set.

Predifined emotion label	w_1	w_2	w_3
e_1	f_1	f_5	f_6
e_2	f_2	f_3	f_4

According to Eqs. (9) and (10), we have:

$$p_\lambda(e_1|t) = \frac{1}{Z_\lambda(w)} exp(\lambda_1 f_1 + \lambda_5 f_5 + \lambda_6 f_6)$$

$$p_\lambda(e_2|t) = \frac{1}{Z_\lambda(w)} exp(\lambda_2 f_2 + \lambda_3 f_3 + \lambda_4 f_4)$$

where

$$Z_\lambda(w) = exp(\lambda_1 f_1 + \lambda_5 f_5 + \lambda_6 f_6) + exp(\lambda_2 f_2 + \lambda_3 f_3 + \lambda_4 f_4)$$

4 Experiments

In this section, we evaluate the performance of the IMEM for sentiment classification of short text. We designed the experiments to achieve the following two goals: (i) to analyze the influence of number of iterations on the accuracy of sentiment classification for the IMEM, and (ii) to conduct comparative analysis with various baselines.

4.1 Data Set

To test the effectiveness of the proposed model, we collected 4570 news headlines from the society channel of Sina (news.sina.com.cn/society/). The news headlines, and user ratings across eight emotions (i.e.,touching, empathy, boredom, anger, amusement, sadness, surprise, and warmness) were gathered. The publishing dates of the news headlines range from January to April of 2012. To ensure that the stability of user ratings, the data set was crawled from half a year after the publishing date. Table 4 summarizes the statistics for each emotion label of the data set. The number of titles for each emotion label represents the amount of the headlines that had the highest ratings for that emotion. For example, there are 749 news headlines that had the highest user ratings for "Touching", with a total number of ratings of 41,796 for that emotion. Figure 1 presents an example of emotion labels and user ratings, in which multiple emotion labels were voted by 3064 users for a particular news text.

In the preprocessing step, a Chinese lexical analysis system (ICTCLAS) is used to perform the Chinese word segmentation. ICTCLAS is an integrated Chinese lexical analysis system based on multi-layer HMM. We random select 80 percent of news headlines as the training set and the rest as the testing set. The existing SWAT [1], emotion-term (ET) and emotion-topic model (ETM) [12] were implemented for comparison. All hyper parameters were set at default. We also included a dummy algorithms, MaxC as the baselines. MaxC always picks the emotion label with the largest total user ratings in the training set. To make an appropriate comparison with other models, the accuracy was employed as the indicator of performance [12]. The accuracy is essentially the micro-averaged $F1$ measure, which equally weights precision and recall.

4.2 Influence of the Number of Iterations

To evaluate the influence of iterative times, we varied the number of iterations from 1 to 400 (the amount of different iterative times tested was 49 in total).

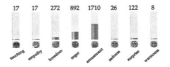

Fig. 1. Example of emotion labels and user ratings

Table 4. Statistics of the data set.

Emotion label	Number of titles	Number of ratings
Touching	749	41,796
Empathy	225	23,230
Boredom	273	21,995
Anger	2048	138,167
Amusement	715	43,712
Sadness	355	37,162
Surprise	167	11,386
Warmness	38	7,986

Figure 2 shows the performance of IMEM when using different numbers of iterations. The results indicate that the IMEM converges to its asymptote in less than 150 iterations. After its convergence, the variance of accuracy is noticeably less than that of initial values, Based on the above observation, we choose 300 iterations as the default setting unless otherwise specified.

4.3   Comparison with Baselines

In this section, we measure and compare the performance of different models on sentiment classification of short text comprehensively. The accuracy of IMEM and four baselines is present in Table 5. Compared to the baselines of ET, SWAT, ETM, and MaxC, the accuracy of IMEM improved 31.70 %, 15.64 %, 11.58 %, and 19.74 %, respectively.

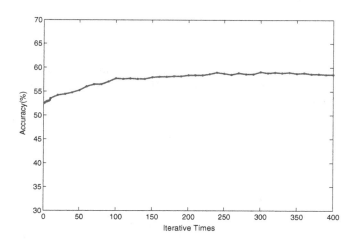

Fig. 2. Performance with different iterative times

Table 5. Statistics of different models.

Models	Accuracy(%)	Improvement(%)
IMEM	59.08	—
ET	44.86	31.70
SWAT	51.09	15.64
ETM	52.95	11.58
MaxC	49.34	19.74

5 Conclusion

Sentiment classification is helpful for understanding the preferences and perspectives of online users, and therefore can facilitate the provision of more relevant and personalized services, including hybrid search in social media [14], construction of user profiles [15], financials analysis [16], emotion-based document retrieval [17], and social emotion detection [18]. In this paper, we have proposed an intensive maximum entropy model for sentiment classification of short text. We evaluate our model on real-world data and compare it to four existing models. The result show that our approach outperforms those baselines.

In our subsequent study, we plan to extend out method to multi-label conditions, and employ topic modeling to generate concept-level feature functions.

Acknowledgements. The authors are thankful to the anonymous reviewers for their constructive comments and suggestions on an earlier version of this paper. The research described in this paper has been supported by "the Fundamental Research Funds for the Central Universities" (46000-31121401), and a grant from the Research Grants Council of the Hong Kong Special Administrative Region, China (UGC/FDS11/E06/14).

References

1. Katz, P., Singleton, M. and Wicentowski, R.: Swat-mp: the semeval-2007 systems for task 5 and task 14, In: The 4th International Workshop on Semantic Evaluations. ACL, pp. 308–313 (2007)
2. Ratnaparkhi, A.: Maximum entropy models for natural language ambiguity resolution. Encyclopedia of Machine Learning, pp. 647–651(2010)
3. Rao, Y.H., Lei, J.S., Wenyin, L., Li, Q., Chen, M.L.: Building emotional dictionary for sentiment analysis of online news. World Wide Web Internet Web Inf. Syst. **17**, 723–742 (2014)
4. Rao, Y.H., Li, Q., Mao, X.D., Wenyin, L.: Sentiment topic models for social emotion mining. Inf. Sci. **266**, 90–100 (2014)
5. Rao, Y.H., Lei, J.S., Wenyin, L., Wu, Q.Y., Quan, X.J.: Affective topic model for social emotion detection. Neural Netw. **58**, 29–37 (2014)
6. Pang, B., Lee, L., Vaithyanathan, S.: Thumbs up? sentiment classification using machine learning techniques. In: Empirical Methods in Natural Language Processing. ACL, pp. 79–86 (2002)

7. Turney, P.D.: Thumbs up or thumbs down? semantic orientation applied to unsupervised classification of reviews. In: Proceedings of the 40th Annual Meeting of the Association for Computational Linguistics (ACL), Philadelphia, pp. 417–424 (2002)
8. Sahami, M., Heilman, T.D.: A web-based kernel function for measuring the similarity of short text snippets. In: Proceedings of the 15th International Conference on World Wide Web, WWW 2006, Edinburgh, Scotland, UK, 23–26 May 2006
9. Salton, G., Wong, A., Yang, C.S.: A vector space model for automatic indexing. Commun. J. ACM **18**(11), 613–620 (1975)
10. Banerjee, S., Ramanathan, K., Gupta, A.: Clustering short texts using wikipedia, In: Proceedings of the 30th Annual International ACM SIGIR Conference on Research and Development in Information Retrieval, pp. 787–788 (2007)
11. Yin, J.H., Wang, J.Y.: A Dirichlet multinomial mixture model-based approach for short text clustering. In: The 20th ACM SIGKDD International Conference on Knowledge Discovery and Data Mining, pp. 233–242 (2014)
12. Bao, S.H., Xu, S.L., Zhang, L., Yan, R., Su, Z., Han, D.Y., Yu, Y.: Mining social emotions from affective text. IEEE Trans. Knowl. Data Eng. **24**(9), 1658–1670 (2012)
13. Song, Y., Wang, H., Wang, Z., Li, H., Chen, W.: Short text conceptualization using a probabilistic knowledgebase. In: Proceedings of the 22nd International Joint Conference on Artificial Intelligence, pp. 2330–2336 (2011)
14. Xie, H.R., Li, Q., Mao, X.D., Li, X.D., Cai, Y., Zheng, Q.R.: Mining latent user community for tag-based and content-based search in social media. Comput. J. **57**(9), 1415–1430 (2014)
15. Xie, H.R., Li, Q., Mao, X.D., Li, X.D., Cai, Y., Rao, Y.H.: Community-aware user profile enrichment in folksonomy. Neural Netw. **58**, 11–121 (2014)
16. Li, X.D., Xie, H.R., Chen, L., Wang, J.P., Deng, X.T.: New impact on stock price return via sentiment analysis. Knowl.-Based Syst. **69**, 14–23 (2014)
17. Wang, Q.S., Wu, O., Hu, W.M., Yang, J.F., Li, W.Q.: Ranking social emotions by learning listwise preference. In: Proceedings of the 1st Asian Conference on Pattern Recognition, ACPR, pp. 164–168 (2011)
18. Lei, J.S., Rao, Y.H., Li, Q., Quan, X.J., Wenyin, L.: Towards building a social emotion detection system fro online news. Future Gener. Comput. Syst. **37**, 438–448 (2014)

Maintaining Ranking Lists in Dynamic Virtual Environments

Mingyan Teng[1]($\boxtimes$), Denghao Ma[2], and Xiaoyong Du[2]

[1] Beijing Institute of Information and Control, Beijing 100049, China
tengmingyan@gmail.com
[2] Information School, Renmin University of China, Beijing 100872, China

Abstract. Preference queries serve for retrieving a small set of tuples with top aggregated scores over multiple features, from a large set of tuples. We consider the problem of maintaining the ranking lists of items for preference queries in dynamic virtual environments, which is very useful for avatars in virtual environments to continuously monitor interesting items surrounding them. Traditional solutions on preference queries utilize the pre-computed materialized ranking lists to efficiently find top items by only retrieving a prefix of the ranking lists. However, for preference queries in virtual environments, items (tuples) to be ranked change frequently due to the movements and updates of avatars. Creating and maintaining materialized ranking lists in such dynamic scenarios will be extremely expensive. In this paper, we address the problem by proposing a solution as a marriage of continuous range query and continuous top-k query. A preference query is continuously processed by dynamically adding and removing the perceived items of an avatar. Extensive experimental studies show that the proposed techniques are very efficient in handling the continuous updates of ranking lists.

1 Introduction

Online community systems with virtual environments (VEs) are increasing in popularity, with examples such as Second Life [1] and Warcraft (WoW) [2]. Players are attracted by such systems because many community activities (e.g., gathering, meeting, shopping, gaming) can be conducted in virtual environments (VEs) with very vivid three-dimensional environments. A huge number of people may log into such systems via a client and play games simultaneously in an online manner. Each player controls the behaviors of an avatar in the VE, and through the avatar, the player experiences the virtual world by walking around, talking and interacting with other players.

We consider a data intensive VE where a huge number of items (goods) are designed to entertain players. Each item exhibits some design properties that can benefit some players in certain cases. Each player holds a number of items

---

Supported by the Fundamental Research Funds for the Central Universities, the Research Funds of Renmin University of China No. 14XNLQ06.

A. Liu et al. (Eds.): DASFAA 2015 Workshops, LNCS 9052, pp. 52–65, 2015.
DOI: 10.1007/978-3-319-22324-7_5

which can be used in certain situations. There are many buildings (e.g., shops) containing many items for players to explore. In the regions where many avatars concentrate, there exist a large number of items held by various owners (avatars or shops). In such a VE, we assume that players are often interested in items that are very useful to them. They can obtain interesting items by exchanging them with avatars around them, buying or even robbing the items from nearby shops or avatars. As there is often a large number of items perceived by an avatar, a player will benefit from a short list of the most interesting items around his avatar. A search process is therefore necessary to retrieve only a small number (supposed to be k) of items that a player may be interested in. This presents the problem of searching localized top-k items for players.

Retrieving top-k items whose scores are measured from a monotonic aggregate function over multiple features is a typical top-k information ranking problem which has been widely studied [3,8,10]. Fagin et al. [8] have given a detailed description of two representative algorithms: No Random Access (NRA) and Threshold Algorithm (TA). The basic idea of these algorithms is to reduce the number of sequential and random I/Os by materializing a ranking list for each feature as a means of pre-computation. The retrieval of top-k items is conducted by sequentially scanning the items of ranking lists in parallel, until a stop condition is satisfied. However, these top-k algorithms assume the ranking lists are static, or will hardly change once created. In the problem of monitoring top-k interesting items in VEs, the list of items perceived by an avatar will update frequently, which are extremely costly to be maintained. Moreover, traditional top-k algorithms fail to continuously monitor top-k interesting items. They have to be periodically executed even though the top-k items of a query may not change. As a result, a lot of preference queries will be frequently presented. Maintaining a large number of ranking lists based on preference queries will be a big challenge.

Because of the movements of avatars, the problem of continuously maintaining ranking lists (CRL) is a kind of continuous queries conducted over moving objects. There have been a large stream of studies on continuous queries over moving objects, such as continuous nearest neighbor (CNN) queries [17] and continuous range (CR) queries [11], in which the query logic is simply focused on the positions of objects. Studies on continuously processing top-k queries over streaming data have been recently tried in [6,15]. However, these methods assume that all the queries are conducted over the same evolving data set. However, in the CRL problem, each query is conducted over a separated evolving item list. Maintaining the complex facilities for the perceived item list of each player will be extremely expensive. In this paper, we address the CRL problem by proposing a solution as a marriage of continuous range query [9,11,14] and continuous top-k query [20].

The rest of the paper is organized as follows: Sect. 2 gives the preliminaries of CRL problem. Section 3 presents our solution to the CRL problem. The experimental studies are described in Sect. 4. Related work and the conclusion are given in Sects. 5 and 6 respectively.

2   Continuously Ranking List Maintenance

Searching interesting items over the whole virtual world may not be practical because the owner of the interesting items may be far away from the requestor in the virtual world. Moreover, searching all items in the entire virtual world means that all items are open to every player. A player will thus get the same results when presenting the same query, and will lose interest in "exploring" the virtual world. Therefore, it is reasonable to constrain the items that can be perceived by an avatar so that a player may only discover interesting items among those items that can be perceived by his avatar.

Let an item i have an interest vector $\mathbf{i} = (\mathbf{i}^1, \ldots, \mathbf{i}^d)^T \in \Re^d$, where $\mathbf{i}^j \geq 0$. The preference of an avatar q is modelled as a d-dimensional vector $\mathbf{q} = (\mathbf{q}^1, \ldots, \mathbf{q}^d)^T \in \Re^d$, where $\mathbf{q}^j \geq 0$ and $\sum_{j=1}^{d} \mathbf{q}^j = 1$. The feature j is a constituent feature of q if $\mathbf{q}^j > 0$. The personalized interest score of an item i to an avatar q can then be computed by a preference function $s(q, i) = \mathbf{q} \cdot \mathbf{i} = \sum_{j=1}^{d} \mathbf{q}^j \cdot \mathbf{i}^j$ [10]. By simply tuning the weighting vector, players can easily adjust their interest incline. An avatar q has a perceiving region constraining the items that it perceives. We denote the set of all items perceived by q as I_q. The problem of preference query can then be described as: for a query $\mathbf{q}$ presented by an avatar q, we want to find a list $\mathbb{I}_q \subseteq I_q$ of k items, such that for any item $i \in \mathbb{I}_q$ and any item $i' \in I_q - \mathbb{I}_q$, $s(q, i) \geq s(q, i')$. Ties are broken arbitrarily. If we want to continuously maintain $i \in \mathbb{I}_q$ when I_q is continuously updated, the problem becomes the continuous ranking list (CRL) maintenance problem.

When an avatar walks around in the virtual world, it can perceive some surrounding avatars and buildings which are owners of certain items. To simplify the perceiving relationship between avatars and items, we assume that an avatar can perceive an item as long as it can perceive the owner of the item. Figure 1 shows an example where an avatar q_1 perceives a surrounding avatar q_{908}. It therefore perceives all items of q_{908}. In this way, all items perceived by an avatar q, I_q, can be derived by combining all items of owners that the avatar perceives. The localized top-k query of an avatar q can then be processed by ranking the interest scores of all items in I_q.

To our best knowledge, there is no study on the CRL problem over moving objects applications. According to the existing studies on top-k queries, we first give two baseline algorithms for addressing the CRL problem.

2.1   Preference Query by Global Ranking Lists

To apply the traditional top-k algorithms [8], one possible solution is to maintain d global ranking lists of all items. The changes of the perceiving relationships (shorted as PRs hereafter) between avatars and items will not affect the global ranking lists. When processing a localized preference query for a player, a typical NRA or TA algorithm is applied on the d global ranking lists. However, such a solution has some disadvantages. First, because a localized preference query only conducted over I_q, most items scanned from the global ranking lists by a top-k

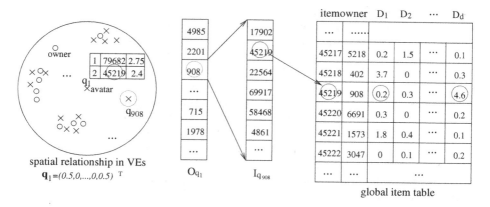

Fig. 1. Illustration of localized preference queries

algorithm will not belong to I_q, which results in much redundant computational cost. Second, global ranking lists are based on the assumption that the ranking lists are almost static. However, the updates of items may happen frequently in VE applications, which brings challenges for efficient maintenance of the global ranking lists. Third, a preference query has to be periodically executed to handle the updates of PRs caused by the movements of avatars.

2.2   Preference Query by Brute Force

Another way to avoid the expensive maintenance of ranking lists in CRL is to dynamically compute and rank the interested scores of all items in I_q. Given a query **q**, the top-k interesting items are straightforwardly discovered by checking the interest scores of all items in the dynamic item set I_q. In this way, no ranking lists are required any more. However, a brute force way of searching top-k items has to be frequently executed to allow the updates of PRs and item lists. Preference queries are also not incrementally computed in such a brute force approach.

3   A Solution of CRL

3.1   Maintaining the Perceiving Relationships

The PRs between avatars and owners are the basis of preference queries because they determine the scope of items involved in preference queries. If an avatar q perceives an owner o, there will be a PR denoted as $r = q \triangleright o$. In our problem, a PR is simply related to the positions of the avatar q and the owner o. The maintenance of PRs can be addressed by continuous range queries, i.e., each avatar continuously monitors its perceived owners within a certain distance to it. Grid has been widely used to support efficient continuous queries of moving objects [9,11,14]. We therefore apply it on maintaining the PRs between avatars and owners. However, when the number of perceived owners of an avatar is large (e.g., in hundreds) and the speed of avatars is large enough, continuous range queries of a large number (e.g., in thousands) of moving avatars will be expensive. However, in VE applications, the PRs do not need to be precisely defined based on the exact positions of avatars and owners. They can be simplified based on the grids that avatars and owners fall in. Assume that the whole range of the game space is $(0 : x_m, 0 : y_m)$. It is partitioned into grids of equal-sized square cells of width ω. We use $c[i, j]$ to denote the cell covering the game space $(i\omega : (i + 1)\omega, j\omega : (j + 1)\omega)$. A point (x, y) in the game space will therefore fall in the cell $c[\lfloor \frac{x}{\omega} \rfloor, \lfloor \frac{y}{\omega} \rfloor]$. Given an owner o (or an avatar q), the cell that the owner (or the avatar) falls in is denoted as c_o. A cell always maintains a list of owners falling within it. The list of a cell is updated only when some avatars (or owners) move into or out of the cell. The perceiving region of an avatar q is defined as a number of cells surrounding the cell c_q. The set of all the perceived cells of q when q is in the cell c is denoted as $C_{q,c}$. The avatar q can perceive all owners falling in the perceived cells C_{q,c_q}.

The movement of an avatar q may result in the updates of PRs related to q. To exactly maintain the PRs, an update process is triggered once an avatar q moves from a cell c' to another cell c. The perceived cells C_{q,c_q} will be updated due to the movement of q. Some cells will be newly perceived by q and some cells will not be perceived by q any more. This will further trigger the updates of PRs between q and the owners in those cells. The movement of q from cell c' to cell c will also results in the updates of some PRs in which q serves as an owner. Some avatars perceiving the cell c' may not perceive q any more. In contrast, some avatars perceiving the cell c may newly perceive q. To exactly update those PRs, each cell c maintains a list of avatars, $c.p$, that perceive the cell c. The list is updated when an avatar newly perceives the cell c or it releases the perceiving relationship with cell c. Algorithm 1 describes how PRs are dynamically updated when an avatar q moves from one cell to another.

3.2   Continuous Query Processing

A CRL query is different with some other continuous queries in that, the query results are determined by not only the positions of moving objects but also the interest scores of items. Given a query q, items with top-k interest scores need to be continuously monitored from the dynamic item set I_q. To avoid frequently compute top-k items from scratch, we borrow the idea from [20], by maintaining a short list of top items. One way to continuously process a preference query is to maintain a list of the exact top-k items by addressing the insertions and deletions of items over the item set I_q. Let l_q be the list of top-k items maintained for a preference query q, and b_q be the lowest interest score of items in l_q. For an item i whose interest score $s(q, i) < b_q$, it will be ignored whether it is inserted into or deleted from I_p because the l_q will not be affected. When an inserted item whose interest score $s(q, i) > b_q$, item i will substitute the kth item in l_q. When a deleted item i is maintained by l_q, to still keep top-k items in l_q, all

---

Algorithm 1. OnMove(q, c', c)

Input: q, an avatar.
Input: c', the previous cell that q falls in.
Input: c, the new cell that q falls in.
1. **for** each cell $\hat{c} \in C_{q,c}$ and $\hat{c} \notin C_{q,c'}$ **do**
2.   **for** each owner o in cell $\hat{c}$ **do**
3.   **OnDetectPR**(q, o)
4. $\hat{c}.p.add(q)$
5. **for** each cell $\hat{c} \in C_{q,c'}$ and $\hat{c} \notin C_{q,c}$ **do**
6.   **for** each owner o in cell $\hat{c}$ **do**
7.   **OnReleasePR**(q, o)
8. $\hat{c}.p.remove(q)$
9. **for** each avatar $\hat{q} \in c.p$ and $\hat{q} \notin c'.p$ **do**
10.   **OnDetectPR**$(\hat{q}, q)$
11. **for** each avatar $\hat{q} \in c'.p$ and $\hat{q} \notin c.p$ **do**
12.   **OnReleasePR**$(\hat{q}, q)$

---

items in I_q have to be re-scanned so that a new item with the kth interest score substitutes the item i. This is called a *refill* process which is expensive.

It is costly to maintain lists of top-k items when items in I_q frequently update. Instead of caching only top-k items in l_q, Yi et al. [20] propose a solution of maintaining a larger list of top-k' items, where $k \leq k' \leq k_{max}$. A top-k query is processed by retrieving top-k items in list l_q, without the need of scanning and ranking all the items in I_q. A self-maintainable mechanism [20] is designed for handling the frequent updates of items in the dynamic item set I_q. The list l_q of top-k_{max} items is created in the beginning, i.e., k' is initially set as k_{max}. When an item i is deleted from I_q, if it is cached by l_q, l_q has to shrink by removing item i out. When the size of l_q, k', drops to $k - 1$, a *refill* process will be conducted to recharge l_q into a list of top-k_{max} items again. By caching more data items in l_q, the frequency of costly refill process is significantly reduced. When an item i is inserted into I_q, if its score is larger than b_q, l_q will be enlarged by recruiting i in. If $k' > k_{max}$ after the enlargement, the k' item will be removed from l_q so that l_q at most caches top k_{max} items.

The updates of the list l_q are triggered by the change of items in I_q. Therefore, when an update of a PR $q \triangleright o$ happens (triggered in Algorithm 1), all items of o may trigger the updates of l_q. They are consequently processed in batch. Based on the techniques of the CRL problem discussed above, we design the update algorithms of PRs in Algorithms 2 and 3. When a PR $q \triangleright o$ is newly detected, a straightforward way of updating l_q is to scan and compute interest scores of all items in I_o, which will be the major cost of the PR update operation. However, if an effective upper bound ($\hat{b}(q, o)$ in the first step in Algorithm 2) of interest scores of items in I_o can be given, we will save the cost of scanning and computing interest scores for all items in I_o if $\hat{b}(q, o) < b_q$. During the refill process, owners in O_q are ranked based on their upper bound score $\hat{b}(q, o)$. The owners are scanned in the order of $\hat{b}(q, o)$. The scanning terminates if $k' = k_{max}$ and $b_q > \hat{b}(q, o)$. The $\hat{b}(q, o)$ is important in determining the cost of processing CRL queries.

3.3   Upper Bound the Interest Scores

In the above solution of CRL query processing, the major computing cost comes from scanning and aggregating interest scores of the items of the perceived owners. In the Algorithms 2 and 3, before scanning the items of an owner, the upper bound score $\hat{b}(q, o)$ of items I_o is checked so that we can save the cost of computing interest scores for items in I_o if I_o can be directly pruned out. The $\hat{b}(q, o)$ upper bounds the maximal interest score $b(q, o)$ of all items in I_o,

---

Algorithm 2. OnDetectPR(q, o)

Input: $r = q \triangleright o$, a PR newly detected.
 1. $\hat{b}(q, o)$, the upper bound of interest scores of items in I_o
 2. **if** $\hat{b}(q, o) \geq b_q$ **then**
 3.   **for** each item $i \in I_o$ **do**
 4.   compute $s(q, i)$
 5.   **if** $s(q, i) \geq b_q$ **then**
 6.   insert i into l_q
 7.   **if** $k' > k_{max}$ **then**
 8.   remove the k'th item from l_q
 9.   update b_q

Algorithm 3. OnReleasePR(q, o)

Input: $r = q \triangleright o$, a PR to be released.
 1. **for** each item $i \in (I_o \cap l_q)$ **do**
 2.   remove i from l_q
 3. **if** l_q is shrunk **then**
 4.   **if** $k' < k$ **then**
 5.   **Refill**(l_q)
 6.   update b_q

Refill(l_q):
 1. $heap_1$, to rank the upper bound scores for owners in O_q
 2. **for** each owner o in O_q **do**
 3.   push $(o, \hat{b}(q, o))$ into $heap_1$
 4. $heap_2$, to extract top k_{max} items, of size k_{max}
 5. **while** $heap_1.size() > 0$ **do**
 6. $o = heap_1.retrieveTop()$
 7.   **if** $heap_2.getMinScore() \geq \hat{b}(q, o)$ **then**
 8.   break
 9.   **for** each item i in I_o **do**
10.   **if** $s(q, i) > heap_2.getMinScore()$ **then**
11.   push $(i, s(q, i))$ into $heap_2$
12. push items in $heap_1$ into l_q

$\hat{b}(q,o) \geq b(q,o) = \sup_{i \in I_o} \mathbf{q} \cdot \mathbf{i}$, where $b(q,o)$ is the exact maximal interest score of all items in I_o to the given query q. Because the upper bound score $\hat{b}(q,o)$ is used for pruning the scanning and aggregation of interest scores, the tightness of $\hat{b}(q,o)$ is therefore very important. We can evaluate the tightness of an upper bound score $\hat{b}(q,o)$ as a bounding ratio $t = \frac{\hat{b}(q,o)}{b(q,o)}$. The ideal bounding ratio is 1, i.e., $\hat{b}(q,o) = b(q,o)$. The lower the t, the tighter the $\hat{b}(q,o)$, the larger probability that $\hat{b}(q,o) < b_q$ in Algorithm 2, and the larger chance that I_o can be pruned out.

We borrow the Lemma 1 from [5], and upper bound the interest score a query q by some views:

Lemma 1. *Given a query q and k views $v_1, \ldots, v_k$, if $\mathbf{q} = \Sigma_{i=1}^{k}(w_i \mathbf{v_i})$ for a set of w_i, where $0 \leq w_i \leq 1$ and $\Sigma_{i=1}^{k} w_i = 1$, then we have $\hat{b}(q,o) = \Sigma_{i=1}^{k}(w_i b(v_i, o)) \geq b(q,o)$.*

By applying techniques proposed in [5], we are able to create lightweight views for tightly bounding $b(q,o)$ of any query q. This will further speed up the performance of the proposed CRL algorithm.

3.4   Effective Size of Materialized Item Lists

On processing a CRL query, the size of l_q, k', drops from k_{max} to $k-1$ within the period of two consecutive refill processes. During this period, although k' can be sometimes enlarged (by recruiting new items in), the interest score b_q, however, monotonously increases. Parameter k_{max} affects the frequency of refill processes and the probability of a new PR intervening the item list l_q, and therefore affects the cost of CRL query processing. A cost model of conducting continuous top-k queries is given in [20]:

$$C = C_{update} \times f_{update} + C_{refill} \times f_{refill}. \tag{1}$$

In the CRL problem, C_{update} and f_{update} are the cost of updating a PR and the frequency of updating PRs respectively. C_{refill} and f_{refill} are the cost of a refill process and the frequency of refill processes respectively. An insertion of a PR (Algorithm 2) may require to scan and compute the interest scores of all items in I_o. Comparatively, a release of a PR (Algorithm 3) is much more efficient because it only needs to remove related items from l_q (if no refill process occurs). Therefore, we simply count the **OnDetectPR** operations for update cost.

The larger the k_{max}, the more top items cached in l_q, and therefore the less the f_{refill}. However, larger k_{max} drops b_q, and therefore enlarges f_{update}. An effective k_{max} is required to minimize the overall cost C. In our problem, we assume that the average rate of detecting and releasing PRs for an avatar are both λ. That is, an average of λ owners are newly perceived by q every second, and an average of λ PRs generated by q are released every second. Let $z = |O_q|$. f_{update} can be modeled as $f_{update} = \frac{\lambda k'}{z}$. This is because, for a newly perceived owner, the probability that it has an item whose interest score is no less than b_q is $\frac{k'}{z}$, where $k' \ll z$. Although k' is not fixed, $E(k')$ can be simplified as $E(k') = \frac{(k_{max}+k)}{2}$. Therefore, the expected f_{update} is:

$$f_{update} = \frac{\lambda(k_{max} + k)}{2z} \qquad (2)$$

The expected size of l_q is reduced by one when removing the k'th item from l_q. In our problem, it is caused by the leaving of the owner holding the k'th item from O_q. Given a released PR $q \triangleright o$, the probability that the owner o holds the k'th item is $\frac{1}{z}$. The refill query occurs when k' drops from k_{max} to $k - 1$. Therefore, f_{refill} can be computed as:

$$f_{refill} = \frac{\lambda \cdot \frac{1}{z}}{(k_{max} - k + 1)} = \frac{\lambda}{z(k_{max} - k + 1)} \qquad (3)$$

The cost of scanning and computing interest scores of items of an owner, C_{update}, is treated as one unit. For the refill process described in Algorithm 3, the cost C_{refill} can be measured on the fly, as the number of owners scanned in the last refill process of avatar q, z_q. The overall cost is then modeled as $C = 1 \times \frac{\lambda(k_{max}+k)}{2z} + z_q \times \frac{\lambda}{z(k_{max}-k+1)}$.

$$C' = \frac{\lambda}{z}(\frac{1}{2} - \frac{z_q}{(k_{max} - k + 1)^2})$$
$$C' = 0 \Longrightarrow k_{max} = k - 1 + \sqrt{2z_q}$$
$$C'' = \frac{\lambda z_q}{2z \cdot (k_{max} - k + 1)^3} > 0$$

The cost C is minimized when $k_{max} = k - 1 + \lfloor\sqrt{2z_q}\rfloor$. We therefore dynamically adjust k_{max} as $k - 1 + \lfloor\sqrt{2z_q}\rfloor$ during each refill process.

4   Performance Evaluation

In this section, we compare three methods discussed in this paper on addressing the CRL problem. The first one (shorted as **BF** in our experiments) is to periodically scan all items perceived by an avatar in a brute force way, which is discussed in Sect. 2.2; The second method (shorted as **NV**) applies the CRL algorithm discussed in Sects. 3.1 and 3.2, whereas no upper bound scores $\hat{b}(q, o)$ are applied in Algorithms 2 and 3. The items I_o will have to be scanned every time it may affect the results of the CRL queries; In the third method (shorted as **EV**), lightweight views are applied to compute $\hat{b}(q, o)$ in processing the CRL queries.

4.1 Experimental Settings

To our knowledge, there are no public data sets on items held by game avatars. We therefore synthesize a data set of items with 10 features. Each owner holds no more than 500 items. In VE applications, players are more interested in items of higher interest scores, which are very scarce. We therefore generate the interest scores of features of all the items following a long tail distribution between 0 and 1. We randomly generate queries of 2 to 4 constituent features. Considering players' interests are often biased on some common features, we therefore generate queries by using a skewed distribution over different features.

We generate a number of static owners and moving avatars (which are also counted as owners). They are randomly distributed in the whole game space. The direction of the movements of avatars is totally random. The velocity of an avatar is updated in average of 20 seconds. Each avatar updates its position and its perceiving owners every 5 seconds. Therefore, the game is simulated 5 seconds a tick. Although the real games are usually simulated several ticks per second, in the CRL problem, an update of interesting items for an avatar every 5 seconds is reasonable. Within each tick, the updates of avatars are conducted one by one. We measure the time cost of conducting updates for all avatars within a tick, to evaluate the performance of the CRL query processing methods. The perceiving radius of avatars are fixed as 5 in our experiments. The width of a cell is fixed as 80. The update intensity is controlled by the maximal speed of avatars, the number of avatars (m), and the number of owners (n).

Our experiments are conducted on a machine with 4 Quad- Core AMD Opteron (1150MHz) processors and 64GBytes RAM, running Red Hat Linux kernel version 2.6.9. We implement all the algorithms in C++. There are several parameters used in our experiments. The settings and default values (in bold) of these parameters are shown in Table 1.

Table 1. Experimental parameters and settings

Parameter	Setting
Number of avatars m	1k, 2k, 3k, 4k, **5k**, 6k, 7k, 8k, 9k, 10k
Number of owners n	10k, 20k, 30k, 40k, **50k**, 60k,..., 100k
Top k	1, 3, **5**, 10, 20
Maximal speed	10, 20, 30, 40, **50**, 60, 70, 80
Number of views u	25, 50, 100, **200**, 400, 800

4.2 Results

In the experimental results, to effectively evaluate whether a CRL query processing method can simulate the updates of avatars in time, we draw the time line of a tick (5 seconds) in the figures of scalability tests. We first test the scalability of the competitors in terms of the number of avatars they can support.

As shown in Fig. 2(a), the BF method is the most expensive one. It cannot support even 1000 avatars. The NV method has better performance, which can support to simulate around 2000 avatars when fully loaded. Comparatively, the EV method is much more efficient than the other two. It can support up to 10000 avatars. The simulating time of a tick by the EV method is around 1/5 of that by the NV, and 1/30 of that by the BF. The simulating time is basically proportion to the number of avatars for each method. We also measure the simulating time of only supporting the movements of avatars (without CRL query processing), it only takes around 1/10 of the simulating time of the EV method.

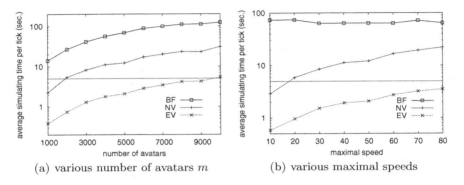

(a) various number of avatars m (b) various maximal speeds

Fig. 2. The scalability comparison of different CRL methods

The results of scalability tests of various methods over various maximal speeds are shown in Fig. 2(b). The simulating time of the BF method is almost not affected by the enlargement of speeds. Comparatively, the cost of the other methods are enlarged with the enlargement of speeds. This is because the other methods are all based on continuous processing of CRL queries. They are therefore affected by the enlargement of PR updates caused by the increasing of the speed of moving avatars.

We then test the scalability of these methods by varying the number of owners. The simulating time of various methods is shown in Fig. 3(a). Obviously, the performance of the EV method is still much better than that of the others, especially when n is large. The EV method can easily support more than 100,000 owners. However, NV can at most support 13, 000 owners. The pruning power of the EV method when adjusting n is shown in Fig. 3(b). The pruning power slightly increases when n is enlarged. This is because, with n increasing, the average number owners perceived by an avatar is enlarged. As a result, b_q (in Algorithms 2 and 3) of avatars increases. More item lists can then be pruned out. In our experimental settings, the average number of owners perceived by an avatar is around 50.

We adjust the parameter k in CRL queries, the simulating time and pruning power of various methods are shown in Fig. 3(c) and (d) respectively. As shown, when parameter k is enlarged, the pruning power of the EV drops slightly.

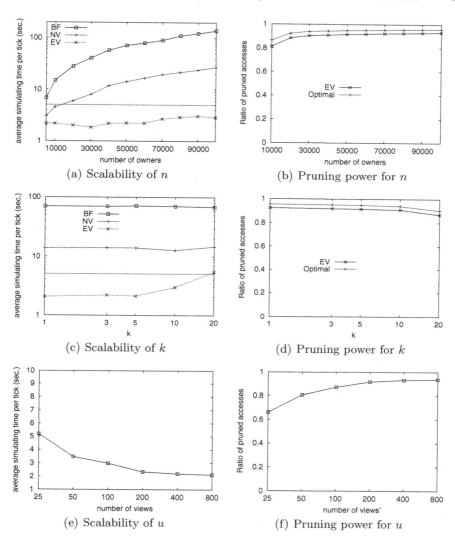

Fig. 3. The scalability comparison under various factors

Consequently, the simulating time increases slightly. This is because the enlargement of k drops b_q for each query.

The performance of the EV method will be affected by the number of u views created. We test the simulating time and pruning power of the EV method by varying u. As shown in Fig. 3(e) and (f), with the enlargement of u, higher pruning power is achieved. The computation cost consequently drops. In these tests, the number of avatars is the default value, 5000. We can find from

Fig. 3(e) and (f) that, when u increases from 200 to 800, the pruning power increases slightly. We therefore set 200 as the default value of u.

5   Related Work

Top-k queries have been widely studied [4,8,10,13,18,19]. A survey of top-k query is given in [12], with most top-k query algorithms assuming that the data are stored in external devices. Therefore, reducing the cost of sequential and random I/Os is the goal of many top-k algorithms. In TA and its variants [8], for retrieving top-k items, the ranking lists (one for each feature) are sequentially scanned in parallel. Since items in the ranking lists are ranked, the upper bound of interest scores of the unseen items can be easily computed. Once the score of the kth item detected is no less than the upper bound, all the unseen items can be directly pruned out. However, these top-k algorithms assume that the ranking lists are almost static once created, which is not the case in the problems of CRL for VE applications. Answering top-k queries by materialized views has also been tried in [5,7]. In these studies [5,7], a view is an ranking list created from the results of a specific top-k query. The TA algorithm [8] is modified to scan high scoring tuples over the materialized views. However, how to create effective views is not addressed in these studies.

Efficient continuous top-k queries has been studied in [20]. The authors [20] propose to continuously maintain top-k tuples by materializing a dynamic top-k' view, so that the frequency of refill processes can be reduced. Such an approach avoids the usage of ranking lists for top-k queries. Continuous top-k queries on streaming data has been recently studied in [6,15]. However, as stated, these methods assume that all the queries are conducted over the same evolving data set, which is not the case in our CRL problem. If applying these methods, maintaining complex data structure for each dynamic I_q, to support only one query q, will be extremely expensive in our CRL problem.

With the widely prevalence of online games [1,2], game applications become more and more data intensive [16]. Adapting database techniques to computer game applications have been tried in [16]. The CRL queries can be applied in data intensive VE applications. However, achieving efficiency and scalability for the expensive CRL query processing has not been addressed.

6   Conclusion

This paper addresses the problem of continuous preference queries in VE applications. We show that CRL queries can be processed by a marriage of continuous range queries and continuous top-k queries. To further achieve efficiency for processing CRL queries, we apply lightweight views to derive tight upper bound scores of items. Our experiments over the synthesis data sets and query sets demonstrates that the proposed solution on efficient processing of CRL queries achieves much desired efficiency, compared with some other alternative solutions. It can be applied to support CRL queries for a large number of avatars and owners in VE applications.

References

1. http://www.secondlife.com
2. http://www.worldofwarcraft.com/index.xml
3. Bruno, N., Gravano, L., Marian, A.: Evaluating top-k queries over web-accessible databases. In: ICDE, pp. 369–380 (2002)
4. Cai, Y., Li, Q., Xie, H., Min, H.: Exploring personalized searches using tag-based user profiles and resource profiles in folksonomy. Neural Netw. **58**, 98–110 (2014)
5. Chen, Y., Cui, B., Du, X., Tung, A.K.H.: Efficient approximation of the maximal preference scores by lightweight cubic views. In: EDBT, pp. 240–251 (2012)
6. Das, G., Gunopulos, D., Koudas, N., Sarkas, N.: Ad-hoc top-k query answering for data streams. In: VLDB, pp. 183–194 (2007)
7. Das, G., Gunopulos, D., Koudas, N., Tsirogiannis, D.: Answering top-k queries using views. In: VLDB, pp. 451–462 (2006)
8. Fagin, R., Lotem, A., Naor, M.: Optimal aggregation algorithms for middleware. J. Comput. Syst. Sci. **66**(4), 614–656 (2003)
9. Gedik, B., Liu, L.: MobiEyes: distributed processing of continuously moving queries on moving objects in a mobile system. In: Bertino, E., Christodoulakis, S., Plexousakis, D., Christophides, V., Koubarakis, M., Böhm, K. (eds.) EDBT 2004. LNCS, vol. 2992, pp. 67–87. Springer, Heidelberg (2004)
10. Hristidis, V., Koudas, N., Papakonstantinou, Y.: Prefer: a system for the efficient execution of multi-parametric ranked queries. In: SIGMOD Conference, pp. 259–270 (2001)
11. Hu, H., Xu, J., Lee, D.L.: A generic framework for monitoring continuous spatial queries over moving objects. In: SIGMOD Conference, pp. 479–490 (2005)
12. Ilyas, I.F., Beskales, G., Soliman, M.A.: A survey of top- query processing techniques in relational database systems. ACM Comput. Surv. **40**(4), 1–58 (2008)
13. Jin, C., Zhang, J., Zhou, A.: Continuous ranking on uncertain streams. Front. Comput. Sci. **6**(6), 686–699 (2012)
14. Mokbel, M.F., Xiong, X., Aref, W.G.: Sina: Scalable incremental processing of continuous queries in spatio-temporal databases. In: SIGMOD Conference, pp. 623–634 (2004)
15. Mouratidis, K., Bakiras, S., Papadias, D.: Continuous monitoring of top-k queries over sliding windows. In: SIGMOD Conference, pp. 635–646 (2006)
16. White, W.M., Demers, A.J., Koch, C., Gehrke, J., Rajagopalan, R.: Scaling games to epic proportion. In: SIGMOD Conference, pp. 31–42 (2007)
17. Wu, W., Guo, W., Tan, K.-L.: Distributed processing of moving k-nearest-neighbor query on moving objects. In: ICDE, pp. 1116–1125 (2007)
18. Xie, H., Li, Q., Mao, X., Li, X., Cai, Y., Zheng, Q.: Mining latent user community for tag-based and content-based search in social media. Comput. J. **57**(9), 1415–1430 (2014)
19. Xie, M., Lakshmanan, L.V.S., Wood, P.T.: Composite recommendations: from items to packages. Front. Comput. Sci. **6**(3), 264–277 (2012)
20. Yi, K., Yu, H., Yang, J., Xia, G., Chen, Y.: Efficient maintenance of materialized top-k views. In: ICDE, pp. 189–200 (2003)

Knowledge Communication Analysis Based on Clustering and Association Rules Mining

Qingyuan Wu[1], Qi Wu[1], Sidi Zhao[1(✉)], Mingxue Wei[1], and Fu Lee Wang[2]

[1] School of Management, Beijing Normal University, Zhuhai, China
wuqingyuan@bnuz.edu.cn,
{mynameis577,rongruo6626}@163.com, 464660015@qq.com
[2] Caritas Institute of Higher Education, Tseung Kwan O, Hong Kong
pwang@cihe.edu.hk

Abstract. With the growth of knowledge sharing, an increasingly large amount of Open-Access academic resources are being stored online. This paper systematically studies the method of mining knowledge communication via Open-Access Journals. We first designed a new framework of knowledge communication analysis based on clustering and association rule mining. Then, we proposed two improved indexes named cited frequency and weighted cited frequency. Extensive evaluations using real-world data validate the effectiveness of the proposed framework of knowledge communication analysis.

Keywords: Open access · Knowledge communication · Knowledge source · Knowledge diffusion · Weighted cited frequency

1 Introduction

The rapid development of Internet has been a great boon for the Open-Access of academic resources. Open-Access academic resources can be divided into two categories: Open-Access Journals (OAJs) and Open-Access Repository. OAJ means that authors pay for the peer review, manuscript editing, electronic journal publishing, and so forth, but readers access resources on the Internet for free. The typical examples of OAJs include BioMed Central (BMC) and Public Library of Science (PLOS). Open-Access Repository refers to storing the published work on document servers, disciplinary archive or institutional repository by authors or entities such as arXiv.

In this paper, we aim to study the knowledge communication among different disciplines by analyzing the knowledge source and knowledge diffusion. By exploiting the citation relationship (i.e., the knowledge source and knowledge diffusion) between the selected OAJ and its external related journals, we compared

The research work described in this article has been supported by a project granted from the National Social Science Foundation of China (Project No. 14CTQ041), and a grant from MOE (Ministry of Education of China) Project of Humanity and Social Science Youth Fund (Project No. 12YJC870012).

A. Liu et al. (Eds.): DASFAA 2015 Workshops, LNCS 9052, pp. 66–75, 2015.
DOI: 10.1007/978-3-319-22324-7_6

the two improved indexes named cited frequency and weighted cited frequency to the traditional ones, and provided a framework of knowledge communication analysis via OAJs.

The remainder of this paper is organized as follows. In Sect. 2, we introduce the related work in knowledge communication analysis. In Sect. 3, we propose knowledge communication analysis framework and our improved indexes. Then the experimental evaluation is shown in the Sect. 4. In Sect. 5, we present our conclusions and future research directions.

2   Related Work

The traditional knowledge communication analysis used the references (i.e., citations) as data sources, and measured the influence of authors, papers and journals by the cited frequency.

Price [1] first described papers' citations in the form of network, and then it became the prototype of citation network and one of the most typical forms of information network [2,3]. Marion et al. stated that the combination of bibliometrics and citation network is expected to reveal the scientific structure and knowledge diffusion more comprehensively [4]. The important metrics in bibliometrics include h index [5], SJR index [6], and eigenvector Eigenfactor [7]. These metrics have different characteristics, in which SJR index and Eigenfactor are relevant to social network analysis methods, and can be used for further studies on knowledge communication.

As an academic communication mechanism, Open-Access has been developed in recent years. The existing researches on Open-Access resources were mainly focused on their quality and development status, as well as several preliminary studies on mining their knowledge communication network. For instance, Chen [8] designed a structural variation model to predict the citation frequency. Three metrics, i.e., modularity change rate, cluster linkage and centrality divergence were employed in the model. The experimental results found that extrinsic features, e.g., the number of references cited by a paper, the prestige of authors and journals tend to be associated with high citations.

In conclusion, it lacks the research on knowledge communication analysis based on Open-Access paper content, citation relationships, and a systematic framework. In our work, the Open-Access paper content and citation relationships were utilized to build a framework of knowledge communication analysis.

3   Knowledge Communication Analysis Framework

In this section, we propose a framework of mining knowledge communication via OAJs, which includes the clustering analysis and association rule mining modules. For each module, we first briefly describe the traditional index, and then illustrate our new methods.

3.1 Clustering Analysis

Clustering analysis is an unsupervised learning technique. In our knowledge communication framework, clustering analysis is employed to divide all knowledge sources (i.e., references) and the knowledge diffusion (i.e., cited papers) of the given dataset into multiple groups or classes, so as to explore the implicit knowledge or information in the dataset.

The vector space model (VSM) is important to many clustering algorithms, here we first introduce the Boolean frequency (BF) used for generating the VSM in traditional knowledge communication analysis methods. Then, two new metrics named cited frequency (CF) and weighted cited frequency (WCF) are proposed for Open-Access academic resources.

Given a paper $p1$ with six references in journals $J1$, $J2$, $J1$, $J3$, $J1$, $J2$, respectively, we can get the total cites of each reference in the full-text of $p1$, e.g., 1, 3, 1, 3, 2, 2. Given a paper $p2$ with four references in journals $J2$, $J4$, $J2$, $J3$, respectively, we can also get the total cites of each reference in the full-text of $p2$, e.g., 1, 4, 2, 2. According to the BF index, the vectors of four source journals are as follows: $J1 =< 1, 0 >, J2 =< 1, 1 >, J3 =< 1, 1 >, J4 =< 0, 1 >$. Due to the fact of limited information in traditional non-Open-Access academic resources, we can only use the Boolean index to indicate the citation relationship.

To generate a fine-grained VSM, we first design the CF index exploiting the number of references in the same journal. The vectors of four source journals are given below: $J1 =< 3, 0 >, J2 =< 2, 2 >, J3 =< 1, 1 >, J4 =< 0, 1 >$. We also design the WCF index by estimating the total cites of each reference in the full-text of $p1$ or $p2$. The vectors of four source journals are as follows: $J1 =< 4, 0 >, J2 =< 5, 3 >, J3 =< 3, 2 >, J4 =< 0, 4 >$.

After transforming the knowledge communication network of OAJs to each VSM, we can use clustering algorithms to divide all knowledge sources and the knowledge diffusion. In the experiment, we employ K-Means algorithm as a demonstration.

3.2 Association Rule Mining

In the knowledge communication network of OAJs, association rule mining can be used to identify the frequent co-occurrence value of variable X in knowledge sources or the knowledge diffusion. The examples of variable X can be journals and papers.

For instance, we denote three Open-Access papers in a OAJ as $\{p1, p2, p3\}$. The collection of all references and cited papers in these three papers is $R = \{r1, r2, ..., r7\}$ and $C = \{c1, c2, c3, c4\}$, respectively. In particular, the collection of references in $p1$ is $\{r1, r2, r3, r4\}$, the first two of which are from journal $J1$, $r3$ is from journal $J2$, while $r4$ is from journal $J3$. The references in $p2$ are $r5$ and $r6$, which are from $J1$ and $J3$, respectively. The reference of $p3$ is $r7$, which is from $J1$. The cited paper of $p1$ is $c1$, which is from $J4$. The cited papers of $p2$ are $c2$ and $c3$, which are from $J4$ and $J6$, respectively. The collection of cited papers for $p3$ is $\{c1, c2, c4\}$, in which $c1$ and $c2$ are from $J4$, and $c4$ is from $J5$.

Thus, the collection of source and diffusion journals is $RJ = \{J1, J2, J3\}$, and $CJ = \{J4, J5, J6\}$, respectively.

According to the theory of association rule mining, we can generate the knowledge communication matrix of each paper based on the traditional BF index, as shown in Table 1.

Table 1. Knowledge communication matrix based on BF

Paper	J1	J2	J3	J4	J5	J6
p1	1	1	1	1	0	0
p2	1	0	1	1	0	1
p3	1	0	0	1	1	0

To measure the detailed strength of knowledge communication, we further construct the knowledge communication matrix of each paper based on CF, as shown in Table 2. For example, two source journals $J1$ and $J3$ are all cited by $p1$. They are equal weight in terms of BF. However, there are two papers in $J1$ are cited by $p1$ (i.e., $r1$ and $r2$), while only one paper $r4$ in $J3$ is cited by $p1$. It is more reasonable to use CF to mine the association strength of $p1$ and $J1$, which is larger than that of $p1$ and $J3$.

Table 2. Knowledge communication matrix based on CF

Paper	J1	J2	J3	J4	J5	J6
p1	2	1	1	1	0	0
p2	1	0	1	1	0	1
p3	1	0	0	2	1	0

The WCF index is proposed by the depth analysis of the full-text of Open-Access papers, which can also improve the representativeness of knowledge communication matrix on OAJs' knowledge communication strength. Taking the above data as an example, we assume that the total cites of $r1$, $r2$, $r3$, $r4$ in the full-text of $p1$ are 3, 2, 1, 6, respectively. Since $r1$, $r2$, $r3$ and $r4$ are from $J1$, $J1$, $J2$ and $J3$, the WCF values of $J1$, $J2$, $J3$ are 5, 1, 6 for $p1$. Similarly, if the total cites of $r5$ and $r6$ in the full-text of $p2$ are all 3, then the WCF value of $J1$ or $J3$ is 3 for $p2$. If the total cites of $r7$ in the full-text of $p3$ is 8, then the WCF value of $J1$ is 8 for $p3$. For the cited papers, we assume that the total cites of $p1$ in the full-text of $c1$ is 3. Since $c1$ is from $J4$, the WCF value of $J4$ is 3 for $p1$. If the total cites of $p2$ in the full-text of $c2$ and $c3$ are 2 and 4, then the WCF values of $J4$ and $J6$ are 2 and 4 for $p2$. If the total cites of $p3$ in the full-text of $c1$, $c2$, $c4$ are 1, 5, 9, then the WCF values of $J4$ and $J5$ are 6 and 9, respectively. Table 3 shows the generated knowledge communication matrix based on WCF index.

Table 3. Knowledge communication matrix based on WCF

Paper	J1	J2	J3	J4	J5	J6
p1	5	1	6	3	0	0
p2	3	0	3	2	0	4
p3	8	0	0	6	9	0

The knowledge communication matrix based on WCF can quantify the strength of knowledge communication in more detail. For example, $J1$ and $J3$ in RJ are both cited by $p1$. Although there are two papers ($r1$ and $r2$) in $J1$ and only one paper ($r4$) in $J3$ are cited by $p1$, the association strength of $p1$ and $J3$ is larger than that of $p1$ and $J1$ according to WCF. The reason is that six positions in the full-text of $p1$ have cited $r4$, while five positions in the full-text of $p1$ have cited $r1$ and $r2$.

4   Experiments

To evaluate the proposed framework, experiments are designed to observe the influence of BF, CF and WCF respectively, on the performance of knowledge communication analysis.

4.1   Data Set

The PubMed Central (PMC, http://www.ncbi.nlm.nih.gov/pmc/) website was employed as the OAJ source for our knowledge communication analysis. For all OAJs in PMC, we compared their impact factors to the average impact factor of biological medical journals included in the journal citation report (JCR), and used one of the OAJs named "BMC Medical Research Methodology" for analysis. There were 500 papers published from January 1, 2001 to December 31, 2010 in the above journal. For these 500 papers, the total number of references is 14499, and there were 2649 cited papers till January 2013.

4.2   Clustering Analysis for Knowledge Sources

We used K-Means algorithm based on BF, CF and WCF to analyze the 92 highly cited source journals, and extracted a specific number of knowledge source clusters. The parameter K was set to 4, i.e., the number of classes is 4.

The distribution of source journal clusters based on BF is as follows: the "Class-4" contains 86 kinds of source journals; the size of "Class-2" is 4, and the source journals are "Biometrics", "Statistics in Medicine", "Biometrika", "Journal of the American Statistical Association"; the "Class-1" and "Class-3" both contain only one source journal. The results indicate that the performance of clustering is poor based on the traditional BF index, where only one meaningful cluster of source journals had been identified, i.e., "Class-2".

The size of "Class-1" to "Class-4" is respectively 1, 83, 2 and 6 based on CF. The six source journals of "Class-4" are "J Clin Epidemiol", "Stat Med", "Am J Epidemiol", "N Engl J Med", "Ann Intern Med" and "BMC Med Res Methodol". The two source journals of "Class-3" are "Lancet" and "JAMA". Thus, we can extract two meaningful source journal clusters, i.e., "Class-4" and "Class-3" by CF index.

The size of "Class-1" to "Class-4" is respectively 1, 80, 2 and 9 based on WCF. The nine source journals of "Class-4" are "J Clin Epidemiol", "Stat Med", "Am J Epidemiol", "N Engl J Med", "Ann Intern Med", "BMC Med Res Methodol", "Int J Epidemiol", "Control Clin Trials" and "Health Technol Assess". The two source journals of "Class-3" are "Lancet" and "JAMA". Through WCF index, we can also identify two meaningful source journal clusters, i.e., "Class-4" and "Class-3".

4.3   Clustering Analysis for Knowledge Diffusion

In this section, we respectively used K-Means algorithm based on BF, CF and WCF to analyze the 102 diffusion journals which cited no less than 4 papers from "BMC Medical Research Methodology", and extracted a specific number of knowledge diffusion clusters. The number of classes was set to 10.

The distribution of diffusion journal clusters based on BF is as follows: the "Class-8" contains 67 diffusion journals, and the size of "Class-10" is 27. The size of "Class-1" to "Class-7", and "Class-9" is all 1, which means that these clusters only contain one diffusion journal. The results suggest that the performance of clustering is poor based on the traditional BF index, where only two meaningful clusters of diffusion journals had been identified, i.e., "Class-8" and "Class-10".

Through the CF index, we can extract four meaningful diffusion journal clusters, i.e., "Class-10", "Class-7", "Class-6" and "Class-9". The size of them is respectively 68, 15, 10, and 3. We can also identify four meaningful diffusion journal clusters by WCF, i.e., "Class-10", "Class-9", "Class-8", "Class-3", and their size is respectively 56, 26, 10, and 4.

4.4   Association Rule Mining for Knowledge Sources

In this section, association rule mining was used to extract the frequent candidates of source journals and rules for 92 source journals. To refine the core source journals, the proposed CF and WCF were applied to filter the generated candidates and rules.

Firstly, we set the threshold values of minimum support and confidence to 1 % in Apriori algorithm, which generated 35199 rules. Based on the descending order of support (i.e., the co-citation ratio), the high frequency source journal collections based on BF are shown in Table 4. The results indicate the strong relationship between epidemiology and statistical medicine [9]. However, for that the weight of any journal in the same collection is 1 according to the BF index, we can not identify the core source journals within each collection.

Table 4. High frequency source journal collections based on BF

Size	High frequency source journal collection	Co-citation ratio
2	{BMJ, JAMA}	14.0 %
2	{BMJ, Lancet}	13.6 %
2	{Lancet, JAMA}	11.4 %
3	{BMJ, Lancet, JAMA}	7.6 %
3	{Lancet, JAMA, J Clin Epidemiol}	6.2 %
3	{BMJ, Lancet, J Clin Epidemiol}	5.8 %
3	{BMJ, JAMA, J Clin Epidemiol}	5.8 %
3	{BMJ, J Clin Epidemiol, Stat Med}	5.0 %
4	{BMJ, Lancet, JAMA, J Clin Epidemiol}	4.6 %

Secondly, we applied CF and WCF to extract the core source journals for each collection (as shown in Table 5). It is worth noting that "Lancet" and "J Clin Epidemiol" in the collection {Lancet, JAMA, J Clin Epidemiol} have the same degree of core if using CF, while the total cites of "Lancet" is larger than that of "J Clin Epidemiol", which can be captured by WCF.

Table 5. CF and WCF of journals within high frequency source journal collections

High frequency source journal collection	CF value	WCF value
{BMJ, JAMA}	190, 151	278, 234
{BMJ, Lancet}	189, 121	309, 181
{Lancet, JAMA}	107, 151	154, 234
{BMJ, Lancet, JAMA}	118, 73, 104	177, 108, 164
{Lancet, JAMA, J Clin Epidemiol}	56, 93, 56	89, 160, 86
{BMJ, Lancet, J Clin Epidemiol}	85, 51, 54	153, 82, 91
{BMJ, JAMA, J Clin Epidemiol}	81, 71, 51	139, 127, 89
{BMJ, J Clin Epidemiol, Stat Med}	73, 60, 89	125, 114, 182
{BMJ, Lancet, JAMA, J Clin Epi-demiol}	69, 43, 60, 40	120, 71, 107, 64

Finally, we generated 163 strong rules with 100 % confidence. The results indicate that for any item in the 500 papers from "BMC Medical Research Methodology", the one cited papers from "Stat the Methods Med Res" also cited papers from "Stat Med". The one cited papers from both "CMAJ" and "Ann Intern Med" also cited papers from "BMJ".

4.5   Association Rule Mining for Knowledge Diffusion

The process of conducting association rule mining for knowledge diffusion is similar to that of the previous section. We also used Apriori algorithm based on

Table 6. High frequency diffusion journal collections based on BF

Size	High frequency diffusion journal collection	Coupling ratio
2	{BMC Med Res Methodol, PLoS One}	10.7%
2	{BMC Med Res Methodol, BMC Public Health}	6.8%
2	{BMC Med Res Methodol, BMJ}	6.8%
3	{BMC Med Res Methodol, PLoS One, BMJ}	3.4%
3	{BMC Med Res Methodol, PLoS One, CMAJ}	2.4%
3	{BMC Med Res Methodol, PLoS One, PLoS Med}	2.4%
3	{BMC Med Res Methodol, BMJ, PLoS Med}	2.4%
3	{BMC Med Res Methodol, PLoS One, BMC Public Health}	2.4%
4	{BMC Med Res Methodol, PLoS One, BMJ, PLoS Med}	1.7%

BF, CF and WCF to analyze the 102 diffusion journals which cited no less than 4 papers from "BMC Medical Research Methodology".

Firstly, we set the threshold values of minimum support and confidence to 1%, which generated 22006 rules. Based on the descending order of support (i.e., the coupling ratio), the high frequency diffusion journal collections based on BF are shown in Table 6. However, for that the weight of any journal in the same collection is 1 according to the BF index, we can not identify the core diffusion journals within each collection.

Secondly, we applied CF and WCF to extract the core diffusion journals for each collection (as shown in Table 7). The results show that "PLoS One" is more important than "BMC Med Res Methodol" for each collection in terms of CF, while "BMC Med Res Methodol" is more important than "PLoS One" for several collections in terms of WCF. This is because the number of cited papers is larger for "PLoS One" but "BMC Med Res Methodol" gains more total cites.

Table 7. CF and WCF of journals within high frequency diffusion journal collections

High frequency diffusion journal collection	CF value	WCF value
{BMC Med Res Methodol, PLoS One}	83, 89	133, 108
{BMC Med Res Methodol, BMC Public Health}	61, 54	103, 60
{BMC Med Res Methodol, BMJ}	69, 57	124, 74
{BMC Med Res Methodol, PLoS One, BMJ}	36, 45, 38	74, 56, 54
{BMC Med Res Methodol, PLoS One, CMAJ}	28, 45, 18	43, 55, 20
{BMC Med Res Methodol, PLoS One, PLoS Med}	29, 42, 14	48, 52, 20
{BMC Med Res Methodol, BMJ, PLoS Med}	35, 34, 16	55, 48, 23
{BMC Med Res Methodol, PLoS One, BMC Public Health}	28, 28, 30	42, 33, 36
{BMC Med Res Methodol, PLoS One, BMJ, PLoS Med}	25, 32, 27, 11	43, 42, 41, 17

Finally, we generated 92 strong rules with 100 % confidence. The results suggest that the papers from "J Headache Pain" and "BMJ" would cite the same item in the 500 papers from "BMC Medical Research Methodology".

5 Conclusion

In this paper, we carried out a systematic study on knowledge communication analysis, and proposed two indexes named cited frequency and weighted cited frequency for OAJs. Extensive evaluations using real-world data validate the effectiveness of them. Our main contributions are as follows:

1. There are a series of theories and methods of knowledge communication analysis based on citation analysis, co-word analysis and network analysis. In this study, we first employed clustering analysis and association rule mining to enrich the theory and method of knowledge communication analysis. Clustering analysis can reveal the information and knowledge hidden in the dataset. Association rules mining can reveal the relationship between knowledge sources and knowledge diffusion networks.
2. In terms of clustering analysis, the cited frequency can reflect the strength of knowledge communication by considering the specific number of cited papers. The weighted cited frequency estimated by the full-text can also improve the performance of clustering analysis for OAJs' knowledge communication. Furthermore, the degree of core source and diffusion journals can be quantified by the improved indexes for association rule mining.

For future work, we plan to extend the work in the following directions.

1. The process of knowledge communication contains a complex mechanism including social influence and sentiments. Thus, we can explore the theories and methods in user community detection [10], user profile mining [11], and sentiment analysis [12–15] for knowledge communication development.
2. The mutual reference of journals reflects the knowledge communication among subjects. This paper systematically analyzed knowledge communication based on OAJs on the journal dimension. The authors in the knowledge communication network can improve the sketch of scientific knowledge map. We will use author co-citation and coupling analysis to reveal authors' activeness and influence, so as to provide a more comprehensive perspective of OAJs' knowledge communication.
3. The comparative analysis of knowledge communication behavior between OAJs and non-OAJs can also help us to understand different knowledge communication networks. Comparative analysis is able to reveal the knowledge communication relationship between OAJs and non-OAJs, as well as their different perspectives such as disciplines, Countries, publishing entities and application platforms.

References

1. Price, D.J.S.: Networks of scientific papers. Science **149**(3683), 510–515 (1965)
2. Newman, M.E.J.: The structure and function of complex networks. SIAM Rev. **45**(2), 167–256 (2003)
3. Newman, M.E.J.: Networks: An introduction. Oxford University Press, New York (2010)
4. Marion, L.S., Garfield, E., Hargens, L.L., Lievrouw, L.A., White, H.D., Wilson, C.S.: Social network analysis and citation network analysis: Complementary approaches to the study of scientific communication. Proc. Am. Soc. Inform. Sci. Technol. **40**(1), 486–487 (2003)
5. Hirsch, J.E.: An index to quantify an individual's scientific research output. Proc. Nat. Acad. Sci. U.S.A. **102**(46), 1–5 (2005)
6. Butler, D.: Free journal-ranking tool enters citation market. Nature **451**, 6 (2008)
7. Bergstrom, C.: Eigenfactor: measuring the value and prestige of scholarly journals. Coll. Res. Libr. News **68**(5), 314–316 (2007)
8. Chen, C.M.: Predictive effects of structural variation on citation counts. J. Am. Soc. Inform. Sci. Technol. **63**(3), 431–449 (2012)
9. Salsburg, D.: The Lady Testing Tea: How Statistics Revolutionized Science in the Twentieth Century. W. H. Freeman and Company, New York (2001)
10. Xie, H.R., Li, Q., Mao, X.D., Li, X.D., Cai, Y., Zheng, Q.R.: Mining latent user community for tag-based and content-based search in social media. Comput. J. **57**(9), 1415–1430 (2014)
11. Xie, H.R., Li, Q., Mao, X.D., Li, X.D., Cai, Y., Rao, Y.H.: Community-aware user profile enrichment in folksonomy. Neural Netw. **58**, 111–121 (2014)
12. Li, X.D., Xie, H.R., Chen, L., Wang, J.P., Deng, X.T.: New impact on stock price return via sentiment analysis. Knowl.-Based Syst. **69**, 14–23 (2014)
13. Rao, Y.H., Lei, J.S., Wenyin, L., Li, Q., Chen, M.L.: Building emotional dictionary for sentiment analysis of online news. World Wide Web: Internet Web Inf. Syst. **17**, 723–742 (2014)
14. Rao, Y.H., Li, Q., Mao, X.D., Wenyin, L.: Sentiment topic models for social emotion mining. Inform. Sci. **266**, 90–100 (2014)
15. Rao, Y.H., Lei, J.S., Wenyin, L., Wu, Q.Y., Quan, X.J.: Affective topic model for social emotion detection. Neural Netw. **58**, 29–37 (2014)

Sentiment Detection of Short Text via Probabilistic Topic Modeling

Zewei Wu[1], Yanghui Rao[1]([✉]), Xin Li[1], Jun Li[1], Haoran Xie[2],
and Fu Lee Wang[2]

[1] School of Mobile Information Engineering,
Sun Yat-sen University, Zhuhai, China
{wuzw3,lixin77,lijun223}@mail2.sysu.edu.cn, raoyangh@mail.sysu.edu.cn
[2] Caritas Institute of Higher Education, Tseung Kwan O, Hong Kong
hrxie2@gmail.com, pwang@cihe.edu.hk

Abstract. As an important medium used to describe events, the short text is effective to convey emotions and communicate affective states. In this paper, we proposed a classification method based on probabilistic topic model, which greatly improve the performance of sentimental categorization methods on short text. To solve the problems of sparsity and context-dependency, we extract hidden topics behind the text and associate different words by the same topic. Evaluation on sentiment detection of short text verified the effectiveness of the proposed method.

Keywords: Short text classification · Sentiment detection · Topic-based similarity

1 Introduction

With the broad availability of portable devices such as tablets, online users can now conveniently express their opinions through various channels. Facing the large-scale user-generated content, it becomes useful and necessary to detect sentiments automatically. Given a set of text with sentimental labels T, it is straightforward to predict the sentiment of new unlabeled text t based on the similar text of t in T. Traditional models of text similarity estimation were mainly based on term frequency, such as TF-IDF, which measured the similarity by the number of words co-occurred int two text. These models can usually achieve good results on long text, because long text share many common words. However, short text have limited words, these models often fail to achieve desired accuracy because of sparse features and strong context dependency.

To tackle this issue, topic model is employed in our work, which takes the latent semantic into consideration. The similarity of two short texts depends on their topic relevancy rather than their common words. Our classification

---

The research work described in this article has been substantially supported by "the Fundamental Research Funds for the Central Universities"(Project Number: 46000-31121401).

A. Liu et al. (Eds.): DASFAA 2015 Workshops, LNCS 9052, pp. 76–85, 2015.
DOI: 10.1007/978-3-319-22324-7_7

method of short text lessened the impact of the limitation of short text and the association of different words on topic improved the performance of short text classification.

The remainder of this paper is organized as follows. In Sect. 2, we will introduce the related work in sentiment detection, short text classification and topic model. In Sect. 3 we will propose probabilistic topic model and our classification algorithm. Then the experimental evaluation will be shown in the Sect. 4. In Sect. 5, we present our conclusions.

2 Related Work

2.1 Sentiment Detection

Sentiment detection aims to identify and extract to attitude of a subject, i.e., an opinion holder, towards either a topic or the overall tore of a document [16]. The existing methods of sentiment detection were mainly based on lexicons, supervised learning and unsupervised learning algorithms. The lexicon-based methods [1–3] tagged sentiments by constructing word- or topic-level emotional dictionaries. Methods based on supervised learning used the traditional categorization algorithms, which included naïve Bayes [6], maximum entropy and support vector machines [4]. The unsupervised learning method first computed the overall polarity documents by counting the occurrence of positive and negative terms, then detected the corresponding emotional orientation [5]. However, the methods mentioned above were mainly proposed for long text.

2.2 Short Text Classification

Most classification algorithms such as naïve Bayes [6], support vector machine [7] are based on term frequency and common words exist in both documents. Due to the fact of feature sparsity and context dependency, these methods may not achieve good results on short text. Short texts have limited words and the information that can be distilled are little. In some cases, short texts are strongly related to each other although they do not share any common words. To overcome this weakness, Sahami and Heilman utilized the external texts from the web to expand the short text [8]. Another method [9] applied the existing knowledge bases such as WordNet or Wikipedia to mine the semantic association between words. In most cases, the method performed well, but when words which do not exist in the database occurred in the short text, the performance was poor.

2.3 Latent Semantic Analysis

To estimate the similarity of two short texts, we should not only focus on the common words they both have. Take the process of composing one short text as an example. We first select some topics for it, and then choose some words for each topic. Thus, a short text is often comprised of a couple of topics, and for each topic, words are chosen from the distribution over the topic. Probabilistic latent semantic analysis (PLSA) [11] and latent Dirichlet allocation (LDA) [10] are two popular models of simulating the above processes. In the PLSA model, the word-by-document matrix is projected to a lower rank space, i.e., the document-topic distribution θ and each topic can be represented by the topic-word distribution ϕ. The LDA model adopted the Dirichlet distribution, which is the conjugate distribution of polynomial distribution, as the prior distribution of θ and ϕ. Both PLSA and LDA are proposed to capture the latent semantic of documents, and can be used to mine the latent user community [17], community-aware resource profiling [18] and user profiling [19]. In our work, we employ the LDA model to capture latent topics, which will be described is Sect. 3.

3   Short Text Sentiment Detection

In this section, we first introduce the algorithm of generating probabilistic topics, and then describe our method of sentiment detection with topic-based similarity. Given a collection of short text and a certain number of topics, words inside the collection would in some extent relate to these topics. Then, some kinds of similarity exist between any two words, based on the topics. With similarities among words, the similarity between two pieces of short text could be calculated [15]; and with similarities among a short text collection containing sentiment labels, the sentiment of new unlabeled short text could be classified by some specific text classification techniques.

3.1   Probabilistic Topic Modeling

Probabilistic topic model (PTM) is a statistical model encoding some abstract "topics" exist in a collection of documents. The collection of topics could be customized since they are "virtual", and it depends on the content of the corpus.

Intuitively, given that a document is about a particular topic, one would expect particular words to appear in the document more or less frequently. For instance, the words "dinosaur" and "bone" will appear more often in documents about dinosaur, "teacher" and "blackboard" will appear in documents about teacher, and "the" and "is" might appear equally in them both. A document typically contains several topics in different extent; thus, in a document that is 10 % about teacher and 90 % about dinosaur, there would probably be about 9 times more words on dinosaur than teacher. The PTM captures this intuition in a mathematical framework, which allows examining a set of documents and discovering, based on the statistics of the words in each document, what topics might be and what each document's topics distribution is.

In the PTM, the process of generating each document is shown as follows [12, 15]:

1. For each document d, pick a topic from the distribution of d over topics;
2. Sample a word from the distribution over the words associated with the chosen topic;
3. Repeat process 1, 2 until all of the words are traversed.

Two parameters can be deducted from the data, i.e., the topic distribution $p(t|d = d_i) = \Theta$ for each document d_i and the word distribution $p(w|t = t_j) = \Phi$ for every topic t_j. In this study, Gibbs Sampling, which is a special case of Markov-chain Monte Carlo [13,14], is applied to estimate the above two parameters.

The whole process of generating the PTM is described in Algorithm 1.

Algorithm 1. Probabilistic Topic Modeling

Input:

1: D: A collection of documents for training;

2: T: Collection of topics, each of which is marked as $0 \ldots T - 1$

3: N_{iter}: Number of iteration of Gibbs-Sampling;

4: λ: Determines how w_{prev} effects on the probability of w_{cur};

5: α: The hyper-parameter associates with the probability $P(t_i)$;

6: β: The hyper-parameter associates with the probability $P(w_{cur}|t_k)$;

Output:

7: $P(t_i, w_j)$: A joint probability of the ith topics and the jth words in D;

8:

9: **procedure** BUILD-PTM

10:      // Gibbs-Sampling

11:      Randomly distribute topics on every word of D: $t_{w_j} = rand() * T$;

12:      Randomly reorder the documents of D;

13:      **repeat**

14:          **for all** $w_{cur} \in d_{cur}, d_{cur} \subset D$ **do**

15:              Remove $t_{w_{cur}}$: $t_{w_{cur}} = -1$;

16:              **for all** $t_k \in T$ **do**

$$P(w_{cur}|t_k) = \frac{(\#(w_{cur}, t_k) + \beta)}{\#(t_k) + \#(w) * \beta}, P(w_{prev}|t_k) = \frac{(\#(w_{prev}, t_k) + \beta)}{\#(t_k) + \#(w) * \beta}$$

$$p_k = \left(\#(d_{cur}, t_k) + \alpha \right) * \left(\lambda * P(w_{prev}|t_k) + P(w_{cur}|t_k) \right)$$

17:              **end for**

18:              Assign t_{w_j}, so that:

$$\sum_{t_k=t_0}^{t_{w_j}} p_k > \left(\sum_{t_k} p_k \right) * rand()$$

19:          **end for**

20:      **until** N_{iter} times

21:      // Probability of t_i

$$P(t_i) = \frac{\#(t_i) + \alpha}{\sum(\#(t_i) + \alpha)}$$

22: **end procedure**

3.2 Sentiment Detection with Topic-Based Similarity

As mentioned in the previous section, the PTM assumes that a document is composed of multiple topics, and a word is, more or less, related to every topic. The joint probabilities of every word and every topic, $P(w_i, t_k)$, are estimated in the PTM. Therefore, two pieces of short text, each containing limited words, could be compared to each other, based on the topics they associate with. The similarity is calculated as follow:

$$Similarity(d_1, d_2) = \sum_{t_i \in T} similarity_{t_i}(d_1, d_2) * P(t_i) \tag{1}$$

Here, d_1, d_2 are the two pieces of short text, T is the total collection of topics, t_i is a certain topic in T, $P(t_i)$ is the prior probability of t_i, $similarity_{t_i}(d_1, d_2)$ is the similarity between d_1 and d_2 based on t_i.

In the equation above, $P(t_i)$ is estimated by the PTM. Given a topic t_i, the similarity between d_1 and d_2, $similarity_{t_i}(d_1, d_2)$, can be calculated by the following cosine function:

$$similarity_{t_i}(d_1, d_2) = \frac{\sum_{w_j \in d_1 \cap d_2} p_{t_i}(w_j)}{\sqrt{\sum_{w_j \in d_1} p_{t_i}(w_j)^2} * \sqrt{\sum_{w_j \in d_2} p_{t_i}(w_j)^2}} \tag{2}$$

After estimating the similarity between two pieces of short text, it is straightforward to detect the latent sentiment of unlabeled text d, based on a training set D with sentimental labels. In our work, we first retrieve k pieces of short text and the corresponding sentimental labels in D that are most similar referenced to d, and then use the most frequent sentimental label as the category of d.

The process of short text sentiment detection (STSD) with the proposed method is shown in Algorithm 2.

4   Experiments

To evaluate the proposed method, experiments are designed to observe the influence of parameters k and T respectively, on the performance of short text sentiment classification.

4.1 Data Set

SemEval. This is an English data set used in the 14th task of the 4th International Workshop on Semantic Evaluations (SemEval-2007). The attributes include the news headline, the user scores over emotions of anger, disgust, fear, joy, sad and surprise, which are normalized from 0 to 100. The data set contains 1,246 valid news headlines with the total score of the 6 emotions larger than 0.

The information of this data set is summarized in Table 1. The number of articles of each emotion label represents the sum of documents having the most ratings over that emotion. For instance, the largest category is "joy". There are

---

Algorithm 2. Short Text Sentiment Detection

Input:
 1: D: A collection of short text for training;
 2: d: A new unlabeled short text; k: The number of referenced short text
Output:
 3: $similarityList$: An ordered list containing the similarities between d and each short text in D;
 4: $result$: Sentimental category of d;
 5:
 6: **procedure** SHORT-TEXT-SENTIMENT-DETECTION
 7:     Build-PTM$(D, T, N_{iter}, \lambda, \alpha, \beta)$;
 8:     **for all** $d_i \in D$ **do**
 9:         $similarityList[i] = $ Similarity(d, d_i); (refer to Eq. 1 & 2)
10:     **end for**
11:     Sort$(similarityList)$;
12:     $result = $ the category that most of the k most similar short text belong to;
13: **end procedure**

---

441 news headlines having the most user ratings over that emotion. The smallest category is "disgust", which only has 42 documents. Among the 1,246 pieces of short text, the first 1,000 of them are collected to be a training set, and the rest 246 to be a testing set.

Table 1. Statistics of *Semeval*

Data set	Emotion label	# of article	# of ratings
SemEval	anger	87	12,042
	disgust	42	7,634
	fear	194	20,306
	joy	441	23,613
	sad	265	24,039
	surprise	217	21,495

4.2   Parameters

The parameters we use for the PTM are shown as follows (Table 2):

4.3   Results and Analysis

The average accuracy of our method with different parameters are shown in Table 3, and depicted in Fig. 1.

In the STSD_k, we evaluate the influence of the parameter k on the accuracy of short text sentiment detection.

Table 2. Parameters of PTM

Parameter	Value
N_{iter}	100
λ	0.5
α	0.5
β	0.01

Table 3. Average accuracy with different parameters

(a) Influence of k (T=100)   (b) Influence of T (k=15)

k	Accuracy(%)
8	38.2
9	35.4
10	37.4
11	39.0
12	37.0
13	37.4
14	37.8
15	39.4
16	39.2
17	38.6
18	38.2
19	37.4
Average value	37.92

T	Accuracy(%)
50	36.6
60	40.2
70	41.1
80	37.0
90	38.2
100	39.4
110	37.8
120	39.4
130	39.8
140	38.6
150	40.2
160	38.6
Average value	38.91

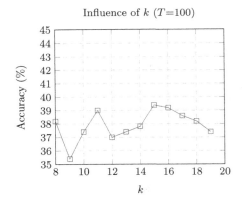
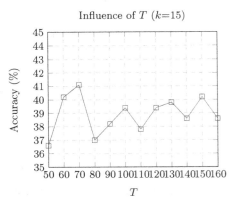

Fig. 1. Average accuracy with different parameters

Given the number of topics equals to 100, the classification accuracy has a maximal value (39.4 %) when $k = 15$, and its average value is 37.98 %. Intuitively, the curve tells that bigger k does not always bring a better accuracy.

Table 4. Comparison with baselines

Method	Accuracy(%)
STSD_T	39.0
STSD_k	38.0
SWAT	33.5
ET	31.0
ETM	25.6

In the STSD_T, we evaluate the influence of the parameter T, i.e., number of topics.

Given the number of referenced short text equals to 15, the classification accuracy has a maximal value (41.1 %) when $T = 70$, and its average value is 38.91 %. The curve fluctuates wildly when T is small, and as the number of topics is increasing, the magnitude of fluctuation becomes smaller and smaller, with a rising trend of accuracy.

We also compare the performance of STSD with various existing mode, as shown in Table 4. Compared to the baselines of SWAT [20], ET and ETM [21], the accuracy of STSD_k improved 13.4 %, 22.5 %, 48.4 %, respectively, and the accuracy of STSD_T improved 16.1 %, 25.5 %, 52.0 %, respectively.

5   Conclusion

In this paper, we proposed a method of sentiment detection for short text. In this method, the probabilistic topic model is generated, which provides the relationship between every word of a collection and every topic. With this model, the similarity of two pieces of short text, each of which is considered as a collection of limited words, could be calculated.

To test the proposed method, experiments with varied parameters are designed. One is designed to evaluate the influence of parameter k, the other is of parameter T. The conclusions can be summarized as follows:

1. Given a certain number of topics, there is an "optimal" k that achieves the best accuracy.
2. The accuracy is also effected by the number of topics, and as the number of topics is increasing, the accuracy is increased with a rising trend.

For future work, we plan to evaluate the performance of all approaches with different ratio of training set and test set, compare the effectiveness of our method with more baselines [22], and estimate the parameters of PTM by a validation set.

Acknowledgements. The authors are thankful to the anonymous reviewers for their constructive comments and suggestions on an earlier version of this paper. The research described in this paper has been supported by "the Fundamental Research Funds for the Central Universities" (46000-31610009), and a grant from the Research Grants Council of the Hong Kong Special Administrative Region, China (UGC/FDS11/E06/14).

References

1. Rao, Y.H., Lei, J.S., Wenyin, L., Li, Q., Chen, M.L.: Building emotional dictionary for sentiment analysis of online news. World Wide Web: Internet Web Inf. Syst. **17**, 723–742 (2014)
2. Rao, Y.H., Li, Q., Mao, X.D., Wenyin, L.: Sentiment topic models for social emotion mining. Inf. Sci. **266**, 90–100 (2014)
3. Rao, Y.H., Lei, J.S., Wenyin, L., Wu, Q.Y., Quan, X.J.: Affective topic model for social emotion detection. Neural Netw. **58**, 29–37 (2014)
4. Pang, B., Lee, L., Vaithyanathan, S.: Thumbs up? sentiment classification using machine learning techniques. In: Empirical Methods in Natural Language Processing, ACL, pp. 79–86 (2002)
5. Turney, P.D.: Thumbs up or thumbs down? semantic orientation applied to unsupervised classication of reviews. In: Proceedings of the 40th Annual Meeting of the Association for Computational Linguistics (ACL), pp. 417–424 (2002)
6. Kim, S.-B., Han, K.-S., Rim, H.-C., Myaeng, S.H.: Some effective techniques for naïve bayes text classification. IEEE Trans. Knowl. Data Eng. **18**(11), 1457–1466 (2006)
7. Wang, J., Yao, Y., Liu, Z.J.: A new text classification method based on HMM-SVM. In: International Symposium on Communications and Information Technologies (2007)
8. Sahami, M., Heilman, T.D.: A Web - based kernel function for measuring the similarity of short text snippets. In: Proceedings of the 15th Conference on World Wide Web, pp. 377–386 (2006)
9. Banerjee, S., Ramanathan, K., Gupta, A.: Clustering short texts using Wikipedia. In: Proceedings of the 30th Annual International ACM SIGIR Conference on on Research and Development in Information Retrieval, pp. 787–788 (2007)
10. David, M.B., Andrew, Y.Ng., Micheal, I.J.: Latent dirichlet allocation. In: Advances in Neural Information Processing Systems, vol. 14 (2002)
11. Thomas, H.: Probabilistic latent semantic analysis. In: Proceedings of Uncertainty in Artificial Intelligence (1999)
12. Weng, J.S., Lim, E.P., Jiang, J., He, Q.: TwitterRank: finding topic-sensitive influential twitters. In: Proceedings of the Third International Conference on Web Search and Web Data Mining (2010)
13. David, J.M.: Information Theory, Inference, and Learning Algorithms. Cambridge University Press, Cambridge (2003)
14. Liu, J.S.: Monte Carlo Strategies in Scientific Computing. Springer, New York (2001)
15. Quan, X.J., Liu, G., Lu, Z., Ni, X.L., Wenyin, L.: Short text similarity based on probabilistic topics. Knowl. Inf. Syst. **25**(3), 473–491 (2010)
16. Gangemi, A., Presutti, V., Recupero, D.R.: Frame-based detection of opinion hodlers and topics: a model and a tool. IEEE Comput. Intell. Mag. **9**(1), 20–30 (2014)
17. Xie, H.R., Li, Q., Mao, X.D., Cai, Y., Zheng, Q.R.: Mining latent user community for tag based and content-based search in social media. Comput. J. **57**(9), 1415–1430 (2014)

18. Xie, H.R., Li, Q., Cai, Y.: Community-aware resource profiling for personalized search in folksonomy. J. Comput. Sci. Technol. **27**(3), 599–610 (2012)
19. Xie, H.R., Li, Q., Mao, X.D., Li, X.D., Cai, Y., Zheng, Q.R., Rao, Y.H.: Community-aware user profile enrichment in folksonomy. Neural Netw. **58**, 111–121 (2014)
20. Katz, P., Singleton, M., Wicentowski, R.: Swat-mp: The semeval-2007 systems for task 5 and task 14. In: The 4th International Workshop on Semantic Evaluations, pp. 308–313. Association for Computational Linguistics (2007)
21. Bao, S., Xu, S., Zhang, L., Yan, R., Su, Z., Han, D., Yu, Y.: Mining social emotions from affective text. IEEE Trans. Knowl. Data Eng. **24**(9), 1658–1670 (2012)
22. Huang, Z., Zhao, Z., Liu, Q., Wang, Z.: An unsupervised method for short-text sentiment analysis based on analysis of massive data. In: Wang, H., Qi, H., Che, W., Qiu, Z., Kong, L., Han, Z., Lin, J., Lu, Z. (eds.) ICYCSEE 2015. CCIS, vol. 503, pp. 169–176. Springer, Heidelberg (2015)

Schema Matching Based on Source Codes

Guohui Ding[1](✉), Guoren Wang[2], Chunlong Fan[1], and Shuo Chen[3]

[1] College of Computer Shenyang Aerospace University, Shenyang, China
dinggh.sau@gmail.com, fanchunlong@sau.edu.cn
[2] College of Information Science and Engineering,
Northeastern University, Shenyang, China
wanggr@mail.neu.edu.cn
[3] Information and Communication Branch,
Liaoning Electric Power Company Limited, Shenyang, China
chenshuo@sina.com.cn

Abstract. Schema matching is a critical step in numerous database applications, such as web data sources integrating, data warehouse loading and information exchanging among several authorities. Existing techniques for schema matching are classified as either schema-based, instance-based, or a combination of both. In this paper, we propose a new class of techniques, called schema matching based on source codes. The idea is to exploit the *exterior schema* extracted from the source codes to find semantic correspondences between attributes in the schemas to be matched. Essentially, the *exterior schema* is a schema that is used to be exposed to final users and is in the outermost shell of applications. Thus, it typically contains complete semantics of data, which is very helpful in the solution of schema matching. We present a framework for schema matching based on source codes, which includes three key components: extracting the *exterior schema*, evaluating the quality of matching and finding the optimal mapping. We also present some helpful features and rules of the source codes for the implementation of each component, and address the corresponding challenges in details.

1 Introduction

Schema matching is an essential building block in many data-management applications that involve data sharing and data transformation across multiple data sources. With the huge increase of structured data sources on the Internet, the need for schema matching solutions is growing. A large number of flight databases of different airlines need to be integrated for a unified access interface which enables users easily to pose queries, further, compare price, itinerary efficiently and make the final decision. Lots of e-commerce companies depend on e-commerce information consolidated for the development of strategies. Thus, the research on schema matching is valuable in practice.

In essence, the problem of schema matching is to find the semantic correspondences between the attributes of two given schemas. These correspondences are also called *matches* traditionally. The task of finding matches is inherently

© Springer International Publishing Switzerland 2015
A. Liu et al. (Eds.): DASFAA 2015 Workshops, LNCS 9052, pp. 86–97, 2015.
DOI: 10.1007/978-3-319-22324-7_8

difficult because the exact semantics of data is not completely captured by the schema itself. The schemas to be matched usually have different structures and representations, because they are typically developed by different persons who have different experiences, habits and characteristics. More efforts should be paid on this area.

A large number of techniques, also called *matchers*, have been proposed, e.g., [3, 6, 10, 12, 16, 17]. These works almost explore all of sources that imply the semantics of schemas from different dimensions, such as data instances, query logs, etc. However, there is no perfect matcher that can return absolutely correct matches. Thus, more sources containing schema semantics should be continually probed for schema matching solutions. In this paper, we propose a new class of techniques, called schema matching based on source codes. The idea is to exploit the *exterior schema* extracted from the source codes to find the matches. The advantage is that the proposed method will be still available when the schema information is incomplete or the size of the data instances is too small to collect statistics.

We propose the new concept *exterior schema* for the first time. It is essentially a schema that is used to be exposed to final users, and contains complete semantics of original data. We will use an example to show the exterior schema. We browsed two webpages demonstrating information about laptops in the famous e-commerce websites, Amazon [23] and Newegg [24], and show them in Fig. 1. For lack of space, we do not show the whole pages. They demonstrate the technical details about two laptop products, such as CPU, memory, hard disk, etc. The layout of the webpages is like a table with two columns. The first column is labels, while the second column is the data described by the labels in the first column. Actually, the labels {"Screen Size", "Processor", "RAM", "Series", "Item model number", ...} compose the exterior schema within Amazon application while {"Brand", "Series", "Model", "CPU Type", "Screen", ...} constitute the one within Newegg application. Obviously, the exterior schema is used to describe data, and the labels should be called attributes. It can be seen that the attributes in the exterior schema are basically made up of full words or very common abbreviations (like "RAM"), so are able to transmit the complete semantic information of data to final users. In what follows, we will discuss why the exterior schema has full semantics and how to exploit the semantics in schema matching.

We focus on database schemas in this paper, which are the most common case of schema matching. Schemas are used by developers or experts to describe, structuralize data in databases. They usually consist of abbreviations, several words concatenated by underline, or combination of incomplete words, etc. Thus, they are just mnemonic symbols with less semantics for developers. However, applications based on databases not only use databases to store data, but also show these data on front-end interfaces to final users who have no development experiences or expertise. Thus, the labels describing these data on front-end interfaces must have explicit and complete semantics. These labels compose the

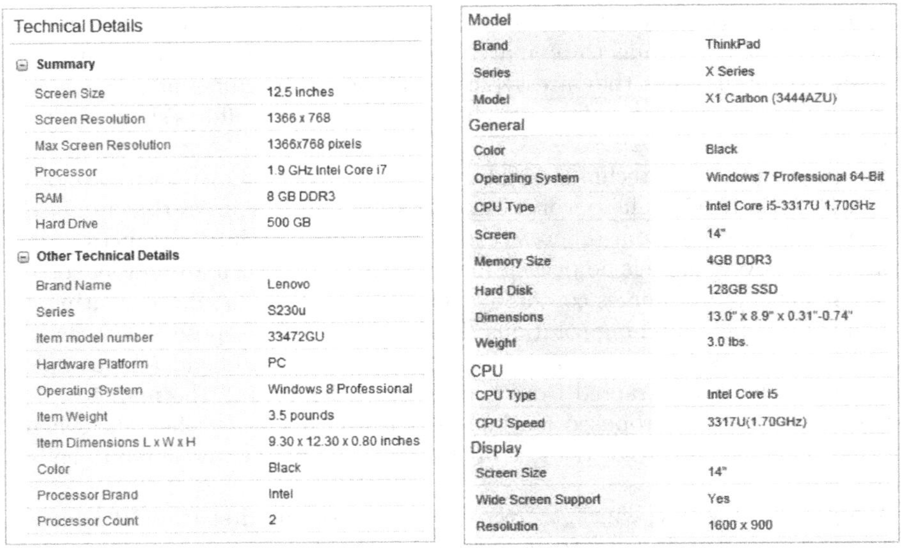

Technical Details	
⊟ **Summary**	
Screen Size	12.5 inches
Screen Resolution	1366 x 768
Max Screen Resolution	1366x768 pixels
Processor	1.9 GHz Intel Core i7
RAM	8 GB DDR3
Hard Drive	500 GB
⊟ **Other Technical Details**	
Brand Name	Lenovo
Series	S230u
Item model number	33472GU
Hardware Platform	PC
Operating System	Windows 8 Professional
Item Weight	3.5 pounds
Item Dimensions L x W x H	9.30 x 12.30 x 0.80 inches
Color	Black
Processor Brand	Intel
Processor Count	2

Model	
Brand	ThinkPad
Series	X Series
Model	X1 Carbon (3444AZU)
General	
Color	Black
Operating System	Windows 7 Professional 64-Bit
CPU Type	Intel Core i5-3317U 1.70GHz
Screen	14"
Memory Size	4GB DDR3
Hard Disk	128GB SSD
Dimensions	13.0" x 8.9" x 0.31"-0.74"
Weight	3.0 lbs.
CPU	
CPU Type	Intel Core i5
CPU Speed	3317U(1.70GHz)
Display	
Screen Size	14"
Wide Screen Support	Yes
Resolution	1600 x 900

(a) A Partial Webpage of Laptop Information, Amazon

(b) A Partial Webpage of Laptop Information, Newegg

Fig. 1. Examples of exterior schemas from E-commerce websites, amazon and newegg

exterior schemas as shown in Fig. 1. That is the reason why e-schemas have complete semantics.

For simplicity, the database schema is denoted as the i-schema ("i" means internal) while the exterior schema denoted as the e-schema. The corresponding match and mapping are also prefixed with "i-" or "e-" respectively. There exist correspondences between the e-schema and the i-schema in the same application, for example, the label "screen size" in Fig. 1(a) may correspond to the field "sz" of table "T_Laptop" in the database of Amazon website. It can be seen that the e-schema is actually another version of the i-schema with full semantics or the external display of the i-schema. Thus, we transform the problem of matching i-schemas with less semantics into the problem of matching e-schemas. Given two i-schemas, we first find the e-matches between their respective e-schemas, then, use the correspondences between the i-schema and the e-schema to generate the final real i-matches. For example, if we know the attribute "Brand Name" in Fig. 1(a) is similar to "Brand" in Fig. 1(b) by comparing strings, and we also discover that "Brand Name" corresponds to the field "BName" in the database of Amazon while "Brand" corresponds to "brand_name" in the database of Newegg, then we can conclude that fields "BName" and "brand_name" have the same semantics, outputted as the final match.

The i-schemas to be matched, the e-schemas and the correspondences between them all exist in the source codes, thus, we propose schema matching based on source codes. It should be noted that it is not difficult to obtain the source codes. For example, if Amazon acquired Newegg, Amazon can easily obtain the source codes of Newegg. Then, the technical staffs of Amazon can use our approach to complete the task of schema matching between their databases. We present a general framework solving schema matching via the source codes, which includes three key components: extracting the exterior schema, evaluating the quality of matching and finding the optimal mapping. We also present some helpful features and rules of the source codes for the implementation of each component, and address the corresponding challenges in details. This paper makes the following contributions:

1. We define a new class of techniques, called schema matching based on source codes.
2. We present a general framework for this variety of matching techniques.
3. We present some helpful features and rules of the source codes for the implementation of each component.

2 A General Framework for Schema Matching Based on Source Codes

In this section, we first summarize the general framework, then discuss each of its components in details.

The intuition underlying schema matching based on source codes is that the e-schema has complete semantics, and we transform schema matching into the problem of matching e-schemas depending on the fact that the i-schema, the e-schema and the relation between them all exist in the source codes. The framework should consider how to extract the e-schema, how to find the correspondence between the e-schema and the i-schema (this could be a difficult problem), how to evaluating the quality of matching and how to generate the final real matches (i-matches). The solution for the former problem is the base of the latter. The first two problems will be the core of the framework and be more difficult than the latter ones which can be solved by leveraging the ideas from existing matching techniques.

2.1 Extracting Exterior Schemas

The source codes actually consist of a number of text files. It might be difficult to directly search e-schemas from lots of text without any clues. The e-schema is related to the i-schema which is given a priori. That is, the attributes in the e-schema correspond to the fields in the i-schema as shown in Sect. 1, and we denote this correspondence as the ei-correspondence briefly. A feasible idea is to first find the code snippet including the given i-schema in the source codes, then, start with the clues provided by the snippet and follow the ei-correspondences to extract the e-schema. We discover that the ei-correspondences will occur in the following situations:

– Demonstrating Data (*DD*): Developers pose SQL queries over the i-schema to the database to read data, then show them to users via the e-schema in the frontend interfaces.
– Adding Data (*AD*): The e-schema is used to guide users to fill in each input field of forms in interfaces. Then, the data inputted is embedded into queries over the i-schema, and these queries will be posed to databases to insert data. This is a very common situation encountered when we register members of some applications.
– Updating Data (*UD*): This is the combination of *DD* and *AD*. First, the e-schema is used to demonstrate data, then to guide users to enter data in the fields which he or she want to modify.

The discovery above is very important for finding the ei-correspondences, because the clues implying the ei-correspondences exist in these situations in different forms. Actually, the clues are the data from the database or users. For the first situation *DD*, the data flows from the i-schema to the frontend e-schema, while for *AD*, the process is converse. For *UD*, the data firstly flows from the database (i-schema) to frontend interfaces (e-schema), then the modified data from users flows back to databases. Whether *DD*, *AD* or *UD*, the data is the key to relating the e-schema to the i-schema. Thus, finding the data flowing between the i-schema and the e-schema is an important step. However, it is a challenge, which is discussed in the following.

We all know that the data is represented by variables in programs. Both the data in databases and the data from frontend users are denoted as variables in the source codes. However, a single field value is generally denoted as multiple variables in the source codes because of the consideration of readability, scalability, flexibility of codes or the requirement of some techniques. Intuitively, given a field value (like "Lenovo" in Fig. 1(a)), there is a chain between the i-schema and the e-schema, and each node of the chain is a variable used to represent the given value. We will use an example to demonstrate the chain of variables. Mvnforum [25] is a very famous discussion board software (forum) developed in Java language and Jsp technology, which is an Open Source project. We download their source codes and choose some code snippets in it. These snippets are used to search member data in databases and show returned results to users. To be consistent with our motivation example in Fig. 1, the attributes of members are replaced with the attributes of the laptop in the code snippets. For lack of space, the statements about some insignificant attributes (like "Hardware Platform", Series, etc.) are omitted in the code snippets. These modifications do not effect the truth, the structure and the logic of the original code snippets.

The changed version of the original snippet is shown in Fig. 2, which includes three parts. The first part and the second part are two functions written in Java and shown in Fig. 2(a) and (b) respectively. The third part is extracted from a Jsp page shown in Fig. 2(c), which is made up of Html tags and Java codes (between the Jsp tag <% and %>). The first function is used to pose a query to the database to read a laptop instance whose ID is given as an argument *laptopID* of this function. The returned laptop instance is represented as a Java

Bean. The second function makes use of the first function to prepare data in the *request* object. Then, the laptop instance is shown in the final Jsp page.

In order to highlight the e-schema and the i-schema, we put them into the red rectangles in Fig. 2(a) and (c) respectively. The i-schema includes eight fields: "LID", "SSize", "Pro", "RAM", "HDrives", "Bra", "Ser", "Col", in Fig. 2(a). The e-schema includes seven attributes: "Screen Size", "Processor", "RAM", "Hard Drive", "Brand Name", "Series", "Color", in Fig. 2(c). These attributes correspond to each other (except for field "LID" which is not shown to users), for example, "SSize" corresponding to "Screen Size". From a whole perspective, the laptop instance first starts from Fig. 2(a), then passes Fig. 2(b), finally arrives at the Jsp page in Fig. 2(c), which falls into the first situation DD we discuss above. There is a chain of variables between the codes in Fig. 2(a) and (c). For intuition, we use red arrow lines to draw out the chain in Fig. 2, while the variables in the chain are also marked with red bold underlines. There are five variables in the chain, which are numbered from 1 to 5. The chain starts from number 1 and ends up with number 5. These variables are all used to represent the same laptop instance in different cases. In Fig. 2(a), the returned laptop record is first saved in the variable *resultset* which is numbered 1. Then, the record is transferred to a JavaBean *bean*, which is numbered 2, because of requirements of J2EE technical specifications. The assignment between the *resultset* and the *bean* is implemented by the one-by-one assignment between their attributes, where the variables for the attribute values of the laptop record are numbered from 1.1 to 1.8, namely the field values of the first variable, *resultset*. In Fig. 2(b), the function *getLaptop()* is first used to obtain the laptop bean, then the variable *laptopBean*, which is numbered 3, is used to represent it. For preparation of the following demonstration, the variable *laptopBean* is inserted into the *request* object as its attribute *LaptopBean* numbered 4. In Fig. 2(c), the Java codes between < % and % > are first used to get out the attribute *LaptopBean* from the *request* object, then another variable *laptopBean*, which is numbered 5, is used to save it. Finally, each field value of *laptopBean*, which is numbered from 5.1 to 5.7, is shown over the e-schema on the webpage.

As mentioned, the i-schemas to be matched are generally given a priori. Thus, the first variable in the chain can be found via the input i-schemas. Next, it seems to be easy to start from the first variable, go along the chain, and finally find the e-schema and the ei-correspondences. However, tracking each variable in the chain is actually a challenge we will discuss in the following points.

Finding the first variable close to the i-schema in the chain is the first challenge. The i-schema may appear for many times even in the same source code file. Although the i-schema is given a priori, we do not know which one is used to provide data for variables in the chain. In the example in Fig. 2(a), the similar queries appear for a number of times in different functions. These functions are all related to the laptop instances, so have similar names, such as *getLaptop()*, *getLaptops()*, *getLaptopCollection()*, etc. However, their purposes are all different, more than just demonstration of data. Thus, it is difficult to

(a) Function used to Pose Query for a Laptop

(b) Function of Preparing Data

(c) Demonstrating the Laptop on Webpage

Fig. 2. Examples of the variable chain between I-Schema and E-Schema in real source codes

automatically differentiate these functions, and find out the function used to save data in the first variable in the chain.

Given a variable, tracking the next adjacent variable in the chain is the second challenge. There exist three difficulties for this problem. First, the mode of value transmission between variables is various. It can be implemented by assignment operators (like =, :=), functions or argument matching. It is difficult to depend on a certain way to track variables. Even if we know that the function is used to transmit values, finding the right function might be a very complex process. Second, the variables in the chain are generally located in different files, which increases the difficulty of finding them. Third, there are lots of variables in the chain. For our example in Fig. 2, there are only five variables in the chain because the laptop instance we consider is a complex variable (a record), not a single value. If we consider each value of the laptop bean, the amount of corresponding variables will be increased twice. The increase of variables in the chain will lead to the increase of opportunities that the previous two difficulties occur.

Next, we will discuss the third challenge. In the source codes, it is common that variables in different scopes have the same name. However, their purposes or functions are generally different. In Fig. 2(a), the string like *strSql, resultSet* and *bean* also appear in other functions as variable names. For lack of space, codes of these functions are not shown. Similarly, different functions are also allowed to have the same name, which is very common because of the introduction of override techniques in programming languages. Thus, it is difficult to find out the variables in the chain because of the interference of these variables or functions with the same name. The last challenge is the redundant source files in projects. In the development process, some source files abandoned are not removed from projects. However, these abandoned files and normal files may overlap within most of their source codes. Thus, these abandoned files will seriously interfere with tracking the variables in the chain.

Except for these challenges, we also discover some valuable rules from the source codes for finding the e-schema. First, the fields of the i-schema are more likely to appear together when the situations (DD, AD or UD) accessing databases occur. This is because it is impossible that only one or several attributes are shown when showing an object to users. For a given code snippet, this rule can be used to judge whether the ei-correspondence exists. Similarly, the attributes in the e-schema often appear together as well, which helps to position the location of the e-schema. Second, the data and its schemas are more likely to appear together and close to each other, for example, the laptop instance and its e-schema in Fig. 2(c). This rule helps to find the first variable in the chain from the i-schema or find the e-schema according to the last variable in the chain. Third, the attributes in the e-schema are all represented by string constants. As mentioned, the e-schema is shown to users, so these string constants must consist of words in the language that users can understand, such as Chinese, French, German, English etc. We all know that most elements of the source codes are made up of numbers or letters in English alphabet. Thus, the e-schema can be easily found by this rule together with the second rule when the e-schema is not English.

Except for these rules, the knowledge provided by function libraries or compilers also can be used to find the e-schema. The advantage of tracking the variable chain is that the ei-correspondences can be obtained while finding the e-schema. In the following subsections, we will discuss other parts of our framework.

2.2 Evaluating the Quality of Matching

In this section, we will discuss the second part of our framework. That is how to choose metrics to evaluate the quality of matching.

Given two i-schemas to be matched, their respective e-schemas can be obtained via the discussion above. We will show how to choose metrics, also called scoring functions, for the e-mapping between the attributes of e-schemas. As mentioned, the attribute names in the e-schema have full semantics, so the similarity between strings can be used as the metric. There are lots of techniques for string comparison, such as edit-distance like functions, token-based similarity measures, and hybrid similarity measures [7]. All these methods can be used for the evaluation of the e-mapping. Here, we just give two problems that should be considered for the evaluation. First, the attribute names to be compared do not usually consist of English words. So, the string comparison method used should adjust itself according to the specific language. Second, there exist synonyms with different representation in e-mappings, such as "RAM" and "Memory" in Fig. 1. The thesaurus in [8] can be used to solve this problem.

2.3   Finding the Optimal Mapping

The last component of our framework will be presented in this subsection. We will discuss how to find the optimal e-mapping, further, generate the optimal mapping (i-mapping).

A similarity score is assigned to each instance of e-mapping, and the goal is to find an e-mapping that optimize the similarity, i.e., maximize it. Given two e-schemas S_a and S_b, S_a includes n_1 attributes $\{a_1, a_2, ..., a_{n_1}\}$ while S_b includes n_2 attributes $\{b_1, b_2, ..., b_{n_2}\}$. We first consider the case $n_1 = n_2 = n$. If the attributes of S_a are regarded as a fixed sequence $a_1 a_2 ... a_n$, any instance of the permutation of all attributes of S_b corresponds to an e-mapping instance. The number of the permutations of all attributes of S_b is $n!$. It can be seen that the space of all e-mapping instances exponentially grows with the increase of the schema size. Thus, an exhaustive search is not feasible. Traditionally, the common approaches are to employ approximate search algorithms to solve this problem, such as Ant Colony Optimization, Simulated Annealing, Genetic Algorithm, etc. The optimal i-mapping can be obtained via the optimal e-mapping together with the ei-correspondences in Subsect. 2.1. The growth trend of the space when $n_1 \neq n_2$ is similar to the case $n_1 = n_2$, so the discussion is omitted.

The e-mapping is basically equal to the traditional i-mapping. All existing techniques for i-mappings can be leveraged by e-mappings. Thus, we do not present much discussion relative to the previous problem in Subsect. 2.1.

3   Related Work

Schema matching has been an active research field for a long time, and a significant amount of approaches have been proposed, e.g., [3,6,8,12–14,17,22]. Work [3] presents a survey of approaches to automatic schema matching. They classify them as schema-based and instance-based. The available information for schema-based methods includes the usual properties of schema elements, such as the attribute name, description, data type of the field, relationship types between tables, etc. The instance-based techniques exploit the statistical information or data characteristics of instances saved in the schemas to finish the matching task.

As far as we know, there are no related works that use the source codes to find matches between attributes of schemas. We just review works related to the topic in this paper, namely schema matching. Work [5] proposes a similarity flooding approach. They first transform the schema information to a graph, then exploit a fixpoint computation algorithm to match schema graphs. Work [8] proposes a new paradigm: statistical schema matching. An approach discovering the complex matchings with the cardinality $m : n$ is proposed in [9]. The context is introduced into schema matching in work [13] in order to improve the quality of matching results. Work [12] proposes the problem of multi-column substring matching for database schema translation. Work [16] develops a method for validating $1 : 1$ and compositional schema matchings.

In work [1], a tool called SEMINT is developed to identify attribute correspondences in heterogeneous databases using neural networks. The machine-learning techniques are used by work [2] to reconcile schemas of disparate data sources. Work [6] considers a special matching situation where the column names in the schemas and the data in the columns are "opaque" or very difficult to interpret. The duplicates of the instances is used to discover *matches* in work [10]. Corpus-based schema matching is proposed in work [11], where the corpus is actually a collection of a large number of schemas and mappings. Work [18] proposes a new category of techniques, called usage-based schema matching. They exploit the features, called co-occurrence patterns, extracted from the query logs, to find semantic correspondences between attributes.

Recently, a new concept named *probabilistic mapping* is introduced into data integration in work [14]. A probabilistic mapping is actually a common mapping with a probability. The probabilistic mapping can be used to capture the uncertainty generated in the process of integrating lots of schemas. Work [19] proposes multilingual schema matching for Wikipedia Infoboxes. Work [20] proposes a self-configuring schema matching system that can automatically adapt to the given mapping problem at hand. Work [21] proposes to automatically construct schema mappings, in real time, from user-input sample target instances.

All methods described above require schema information or instance information to perform schema matching. The semantics conveyed by these information is generally obscure or indirect, which forms a bottleneck for improving the quality of schema matching. The e-schema in the source codes has complete and explicit semantics. This is necessary to find more accurate matches. So far, no work considers the source codes for schema matching. Thus, this paper proposes schema matching based on source codes. This is significant for improving the quality of matching results.

4   Conclusions

In this paper, we propose a new class of techniques, called schema matching based on source codes. The important role which the source codes play in schema matching is to provide the e-schema and the correspondences between the e-schema and the i-schema (ei-correspondence). The attributes in the e-schema have full and explicit semantics. Meanwhile, they can be mapped into the fields in the i-schema. Thus, we transform common schema (i-schema) matching into the problem of e-schema matching. The advantage is that the full semantics of attributes can improve the quality of schema matching. We present a framework for schema matching based on the source codes, which includes three key components: extracting the exterior schema, evaluating the quality of matching and finding the optimal mapping. We also present some helpful features and rules of the source codes for the implementation of each component, and address the corresponding challenges in details. We just present a new research idea for schema matching without developing any schema matching system.

Acknowledgments. This research was supported by the National Natural Science Foundation of China (Grant No. 61303016) and the Normal Project Foundation of Education Department of LiaoNing Province (Grant No. L2012045).

References

1. Li, W.-S., Clifton, C.: SEMINT: A tool for identifying attribute correspondences in heterogeneous databases using neural networks. Data Knowl. Eng. **33**(1), 49–84 (2000)
2. Doan, A., Domingos, P., Halevy, A.: Reconciling schemas of disparate data sources: a machine-learning approach. In: Proceedings of Special Interest Group on Management Of Data (SIGMOD), pp. 509–520 (2001)
3. Rahm, E., Bernstein, P.A.: A survey of approaches to automatic schema matching. J. Very Large Data Bases (VLDB) **10**(4), 334–350 (2001)
4. Do, H.-H., Rahm, E.: COMA - A system for flexible combination of schema matching approaches. In: Proceedings of Very Large Data Bases (VLDB), pp. 610–621 (2002)
5. Melnik, S., Garcia-Molina, H., Rahm, E.: Similarity flooding: a versatile graph matching algorithm and its application to schema matching. In: Proceedings of the International Conference on Data Engineering (ICDE), pp. 117–128 (2002)
6. Kang, J., Naughton, J.F.: On schema matching with opaque column names and data values. In: Proceedings of the Special Interest Group on Management Of Data (SIGMOD), pp. 205–216 (2003)
7. Cohen, W. W., Ravikumar, P., Fienberg, S.E.: A comparison of string distance metrics for name-matching tasks. In: Proceedings of the IJCAI Workshop on Information Integration on the Web (IIWeb), pp. 73–78 (2003)
8. He, B., Chang, K.C.: Statistical schema matching across web query interfaces. In: Proceedings of Special Interest Group on Management Of Data (SIGMOD), pp. 217–228 (2003)
9. He, B., Chang, K.C.-C., Han, J.: Discovering complex matchings across web query interfaces: a correlation mining approach. In: Proceedings of Knowledge Discovery and Data Mining (KDD), pp. 148–157 (2004)
10. Bilke, A., Naumann, F.: Schema Matching using Duplicates. In: Proceedings of International Conference on Data Engineering (ICDE), pp. 69–80 (2005)

11. Madhavan, J., Bernstein, P.A., Doan, A., Halevy, A.Y.: Corpus-based schema matching. In: Proceedings of International Conference on Data Engineering (ICDE), pp. 57–68 (2005)
12. Warren, R.H., Tompa, F.: Multicolumn substring matching for database schema translation. In: Proceedings of Very Large Data Bases (VLDB), pp. 331–342 (2006)
13. Bohannon, P., Elnahrawy, E., Fan, W., Flaster, M.: Putting context into schema matching. In: Proceedings of Very Large Data Bases (VLDB), pp. 307–318 (2006)
14. Dong, X., Halevy, A.Y., Yu, C.: Data integration with uncertainty. In: Proceedings of Very Large Data Bases (VLDB), pp. 687–698 (2007)
15. An, Y., Borgid, A., Miller, R.J.: A semantic approach to discovering schema mapping expressions. In: Proceedings of International Conference on Data Engineering (ICDE), pp. 206–215 (2007)
16. Dai, B.T., Koudas, N., Srivastavat, D., Tung, A.K.H., Venkatasubramaniant, S.: Validating Multi-column Schema Matchings by Type. In: Proceedings of International Conference on Data Engineering (ICDE), pp. 120–129 (2008)
17. Sarma, A.D., Dong, X., Halevy, A.: Bootstrapping pay-as-you-go data integration systems. In: Proceedings of Special Interest Group on Management Of Data (SIGMOD), pp. 861–874 (2008)
18. Chan, C., Elmeleegy, H.V.J.H., Ouzzani, M., Elmagarmid, A.: Usage-based schema matching. In: Proceedings of International Conference on Data Engineering (ICDE), pp. 20–29 (2008)
19. Nguyen, T., Moreira, V., Nguyen, H., Nguyen, H., Freire, J.: Multilingual schema matching for wikipedia infoboxes. In: Proceedings of Very Large Data Bases (VLDB), pp. 133–144 (2011)
20. Peukert, E., Eberius, J., Rahm, E.: A self-configuring schema matching system. In: Proceedings of International Conference on Data Engineering (ICDE), pp. 306–317 (2012)
21. Qian, L., Cafarella, M.J., Jagadish, H.V.: Sample-driven schema mapping. In: Proceedings of Special Interest Group on Management Of Data (SIGMOD), pp. 73–84 (2012)
22. Zhang, M., Chakrabarti, K.: Infogather+: semantic matching and annotation of numeric and time-varying attributes in web tables. In: Proceedings of Special Interest Group on Management Of Data (SIGMOD), pp. 145–156 (2013)
23. http://www.amazon.com
24. http://www.newegg.com/
25. http://www.mvnforum.com/

A Quota-Based Energy Consumption Management Method for Organizations Using Nash Bargaining Solution

Renting Liu[1]([✉]) and Xuesong Jonathan Tan[2]

[1] Information Center, University of Electronic Science
and Technology of China, Chengdu, China
liurt@uestc.edu.cn
[2] National Key Laboratory on Science and Technology of Communications,
University of Electronic Science and Technology of China, Chengdu, China
xstan@uestc.edu.cn

Abstract. The increasing development of energy consumption monitoring system at public buildings, hospitals, campus, and factories enables more and more organizations to measure and coordinate the energy consumption activity of individual key energy users within them. To enhance the overall performance of energy consumption under limited energy budget, the present paper proposes a new criterion, i.e., energy consumption satisfaction degree (ECSD), for an organization to evaluate the satisfaction of each key energy user in the consumption of its allocated energy quota. Inspired by the classical Nash bargaining solution (NBS) in cooperative game theory, we further develop a quota-based energy consumption management method to effectively guarantee the annual energy-saving target of the organization. Numerical simulation shows that, compared with the equal and priority-based energy allocation schemes, the proposed method can maximize the overall satisfaction of all key energy users and, meanwhile, maintain a reasonable fairness among them.

Keywords: Energy consumption · Nash bargaining solution · Quota · Energy allocation

1 Introduction

The issue for monitoring energy consumption among consumers has drawn more and more attentions from national policy makers and energy management operators. In order to satisfy the growing requirement of energy consumption, each energy user has to better utilize its rationed energy to support its production and tasks. Millions of dollars are spent to build energy consumption monitoring systems nationwide for providing real-time energy consumption monitoring and displaying statistical and real-time energy-related information. This provides data support for improving energy efficiency and developing optimized

© Springer International Publishing Switzerland 2015
A. Liu et al. (Eds.): DASFAA 2015 Workshops, LNCS 9052, pp. 98–108, 2015.
DOI: 10.1007/978-3-319-22324-7_9

energy management strategies [1–3] for various organizations. Many works have been done to better understand the energy consumption characteristics of different consumers [4]. In residential environment, as the energy monitoring systems show the detailed pictures of energy consumption to end users, they can influence energy consumption behavior of households [5,6]. Based on this information, cost sensitive consumers may be able to adjust their demand according to energy price and independently make appropriate consumption scheduling decisions for maximizing their utilities in energy consumption [7,8]. In [9–11], the energy consumption monitoring technology is utilized in demand-side management [12,13] in residential smart grid to minimize the energy cost and inconvenience of energy consumption customers.

While these works can effectively optimize the construction of energy monitoring system as well as the energy scheduling as part of the smart grid, there exists rather limited research work with the focus on how to deal with the annual energy consumption targets and how to set a reasonable and applicable quota for each energy user under a fixed annual energy budget, which has become a commonly mandatory assignment to large organizations and enterprises in many countries. For example, [14] proposes to allocate energy consumption quotas, which consider the energy utilization efficiency, CO_2 emission per capita and GDP as the core factors, among 30 province-level administrative regions in China. However, none of the energy allocation schemes proposed in [14] can be directly adapted to determine energy consumption quota inside large organizations with fixed energy budgets and mandatary energy conservation targets.

In fact, followed by a wide range development and construction of energy consumption monitoring systems for public buildings, hospitals, campus, and factories, more and more organizations are able to deploy smart meters for measuring the energy consumption of individual key energy users within them and implementing quota-based energy management to coordinate the energy consumption among users. Moreover, facing the fixed target of total energy-savings in the national 5-year plan which specifies the annual energy-saving target and total energy consumption budget, the energy managers of large organizations in China are seeking effective energy evaluation and allocation system to enhance the overall performance of energy consumption under limited energy budget, restrain the energy-waste behaviors without affecting normal production activities, and minimize the environmental effect of the whole production processes. In summary, the optimal allocation of energy budget among energy users of an organization under the constraint of a total energy budget becomes an urgent problem for energy managers. First, the energy allocation process should satisfy the basic energy requirement of all energy users while promote the overall energy efficiency. Second, the energy manager should also make energy allocation among different users based on their historical performance of energy consumption so as to stimulate each user to improve their energy efficiency. Moreover, since different energy users may play different roles for the ordinary operation of an organization, the energy allocation scheme should incorporate the priority of energy users into consideration such that the energy requirement of those users with higher priority can be better satisfied than that of those users with lower priority.

In view of these considerations, the present paper first proposes an energy consumption satisfaction degree (ECSD) in Sect. 2 for an organization to evaluate the satisfaction of each key energy user for consuming its allocated energy quota. Inspired by the classical Nash bargaining solution (NBS) in cooperative game theory, Sect. 3 further develops a quota-based energy consumption management method for energy managers to effectively guarantee the annual energy-saving target of the organization with a fixed energy budget. The advantages of the proposed scheme include that it can always meet the basic energy requirement of all energy users, satisfy the preset annual energy budget of the organization, and support the priorities of different energy users in energy allocation. Numerical simulation in Sect. 4 then shows that, compared with the equal and priority-based energy allocation schemes, the proposed energy allocation scheme can maximize the overall satisfaction of all key energy users and, meanwhile, maintain a reasonable fairness among them. The main contribution of the present paper is finally concluded in Sect. 5.

2   Energy Consumption Satisfaction Degree

Consider an organization consisting of N key energy users, each of which has to consume a certain number of energy for implementing its normal production activities. Without the loss of generality, the present paper restricts the energy type consumed by each user to be electricity only. Meanwhile, all energy-related parameters and allocation schemes formulated in this paper can also be applied for other energy types, like gas, coal oil, and gasoline, by using tonne coal equivalent (TCE) as a basic unit.

To optimize the energy allocation among N key users, the organization should first evaluate the performance or satisfaction of each user in energy consumption based on its historical records. This evaluation would serve as the basis for optimizing the ensuing energy allocation in the current year. For this purpose, the energy consumption satisfaction degree (ECSD) of each energy user $i \in [1, N]$ in the past m years, where $m \geq 1$, can be defined as follows:

$$\overline{\text{ECSD}_{i,-m}} = \max \left\{ \text{ECSD}_{\min}, \sum_{k=1}^{m} \alpha_{i,k} p_{i,k} \right\} \tag{1}$$

where $p_{i,k} = \frac{R_{i,k} - C_{i,k}}{R_{i,k}}$ for $k \in [1, m]$ denotes the annual profit rate of user i in the past kth year, while $R_{i,k}$ and $C_{i,k}$, respectively, denote the annual revenue and cost of user i in the past kth year; $\alpha_{i,k}$ for $k \in [1, m]$ denotes the weight of annual profit rate of user i in the past kth year; $\text{ECSD}_{\min}(\geq 0)$ denotes the minimum ECSD that each energy user should be guaranteed by the organization so as to obtain a minimum energy quota for meeting its basic operation requirement. Normally, the coefficient $\text{ECSD}_{\min}(\geq 0)$ is preset by the energy manager of the organization and should satisfy the constraint $\sum_{i=1}^{N} Q_i(\text{ECSD}_{\min}) \leq Q_{\text{total}}$, where Q_i represents the annual amount of energy allocated to the energy user

i and can be expressed as a monotonically increasing function of its obtained ECSD, while Q_{total} the annual energy budget of the organization. Since the ECSD of each user i in a more recent year can better reflect its performance of energy consumption than that in an older year, the weights $\alpha_{i,1}, \alpha_{i,2}, \cdots, \alpha_{i,m}$ for each energy user i should satisfy $\sum_{k=1}^{m} \alpha_{i,k} = 1$ and $\alpha_{i,1} \geq \alpha_{i,2} \geq \cdots \geq \alpha_{i,m}$.

In general, when the total amount of energy consumed by each energy user i becomes higher, its total profit (in terms of number of qualified products, fulfilled tasks, achievements, etc.) will not decrease. Moreover, this increasing should satisfy the following two properties:

(a) If an energy user consumes more energy, it should achieve higher profit;
(b) When the amount of energy consumed by an energy user increases, the increasing speed of its total profit is gradually decreasing.

Under the properties of (a) and (b), we can fit the value $\overline{\text{ECSD}_{i,-m}}$ of each energy user i based on its average ECSD in the past m years for evaluating its historical average performance of energy consumption. One possible fitting for the ECSD in the past m years can be formulated as follows:

$$\overline{\text{ECSD}_{i,-m}} = 1 - e^{-c_i \sum_{k=1}^{m} \alpha_{i,k} q_{i,k} [1 - f_i \cdot U(x_{i,k} - q_{i,k})]} \tag{2}$$

where $c_i \in [0, \infty)$ denotes the coefficient of the energy user i for ECSD fitting; $q_{i,k}, k \in [1, m]$, denotes the annual energy quota allocated to energy user i in the past kth year; $x_{i,k}, k \in [1, m]$, denotes the actual annual energy consumed by the energy user i in the past kth year; f_i denotes the energy punishment coefficient imposed by the energy manager of the organization to the energy user i when the actual annual energy consumed by the user i exceeds its annual allocated energy quota; $U(y)$ denotes the step function, i.e., if $y > 0$, then $U(y) = 1$; otherwise, $U(y) = 0$.

Therefore, as the basis for the quota-based energy consumption management in the current year, the organization can update the coefficient c_i of each energy user i for ECSD fitting in (2) as follows:

$$c_i = -\frac{1}{\sum_{k=1}^{m} \alpha_{i,k} q_{i,k} \left[1 - f_i \cdot U\left(x_{i,k} - q_{i,k}\right)\right]} \ln \overline{\text{ECSD}_{i,-m}} \tag{3}$$

Based on the coefficient c_i so updated, the ECSD of each energy user i in the current year, i.e., ECSD_i, can be defined as follows:

$$\text{ECSD}_i(Q_i) = \begin{cases} 1 - e^{-c_i Q_i}, & \text{when } \overline{\text{ECSD}_{i,-m}} \geq \text{ECSD}_{\min} \\ \text{ECSD}_{\min}, & \text{when } \overline{\text{ECSD}_{i,-m}} < \text{ECSD}_{\min} \end{cases} \tag{4}$$

where $c_i \in [0, \infty)$ denotes the updated coefficient of energy user i for ECSD fitting according to (3) and Q_i denotes the annual energy quota allocated to energy user i in the current year. Obviously, when an energy user satisfies $\overline{\text{ECSD}_{i,-m}} \geq \text{ECSD}_{\min}$ and the ECSD fitting follows (2), the value of ECSD_i can satisfy both (a) and (b) and hence qualify as a convex function.

3   The Proposed NBS-based Energy Allocation Scheme

Once the ECSD coefficient c_i of each energy user i is determined by (3), the organization can further optimize the allocation of its energy quota to all energy users based on their latest ECSDs. This allocation, however, is constrained by the total annual energy budget of the organization, i.e. $\sum_{i=1}^{N} Q_i = Q_{\text{total}}$, where Q_i denotes the energy quota allocated to each energy user i and Q_{total} denotes the total amount of energy quota prescribed by the organization.

First of all, in order to satisfy the basic operation requirement of each energy user i, the organization can classify all the N energy users into two classes based on the average ECSD of user in the past m years, i.e. $\overline{\text{ECSD}}_{i,-m}$. If $\overline{\text{ECSD}}_{i,-m} \geq \text{ECSD}_{\min}$ is satisfied, then the energy user i can be classified as Class-I users; otherwise, the user i is classified as Class-II users. In other words, all Class-I users outperform all Class-II users in terms of the historical ECSD or energy consumption performance. For each Class-II user j, its ECSD in the current year will be set as a fixed value $\text{ECSD}_{\min}$, and its allocated energy quota can be calculated as $Q_j = -\frac{1}{c_j} \ln\left(1 - \text{ECSD}_{\min}\right)$. That is, each Class-II user with less energy efficiency than a Class-I user will be allocated a minimum quota of energy to guarantee its basic energy requirement.

Denote by n the number of Class-I energy users and by Q_{I} and Q_{II} the total amount of energy allocated to all Class-I or -II users, respectively. Thus the number of Class-II energy users is $N - n$ and $Q_{\text{total}} = Q_{\text{I}} + Q_{\text{II}}$. As for Class-I users, the organization needs to further allocate energy quotas for improving their total satisfaction for energy consumption. This problem can be appropriately solved by the Nash bargaining solution [15] in classic cooperative game theory.

More specifically, the allocation of energy quotas for Class-I users can be generalized into the following constrained optimization problem:

$$\begin{aligned} \max_{Q_{e(1)},Q_{e(2)},\ldots,Q_{e(n)}} & \prod_{i=1}^{n} \left[\text{ECSD}_{e(i)}\left(Q_{e(i)}\right)\right]^{\tau_{e(i)}} \\ \text{s.t.} \quad & \sum_{i=1}^{n} Q_{e(i)} \leq Q_{\text{I}} \\ & \sum_{i=1}^{n} \tau_{e(i)} = 1 \\ & \tau_{e(i)} \geq 0 \ \forall \ i \in [1,n] \end{aligned} \tag{5}$$

where e denotes the one-to-one mapping from the label set $\{1, 2, \ldots, n\}$ of all n class-I users to the set $\{1, 2, \ldots, N\}$ of all N energy users, while $\tau_{e(i)} \in [0,1]$ denotes the priority of Class-I $e(i)$ energy user in the energy allocation process, i.e., the larger the value of $\tau_{e(i)}$, the higher priority of the class-I user $e(i)$ in energy allocation.

When each $\text{ECSD}_{e(i)}\left(Q_{e(i)}\right), \forall i \in [1,n]$, is a convex function, e.g., the exponential fitting in (2), this constrained optimization problem (5) can be directly solved by introducing Lagrange multipliers. More specifically, the problem (5) can be transferred into the following equations:

$$\begin{cases} \frac{\partial}{\partial Q_{e(i)}} \left\{ \prod_{k=1}^{n} [\text{ECSD}_{e(k)}(Q_{e(k)})]^{\tau_{e(k)}} - \lambda\left(\sum_{j=1}^{n} Q_{e(j)} - Q_{\text{I}}\right) + \mu_i Q_{e(i)} \right\} = 0, \forall i \\ \lambda\left(\sum_{j=1}^{n} Q_{e(j)} - Q_{\text{I}}\right) = 0 \\ \mu_i = 0, \forall i \end{cases} \tag{6}$$

where λ and μ_i are the Lagrange multipliers. By solving these equations, we can obtain the unique optimal solution of the formula (5) via the interior-point approach [16] in a polynomial computation time.

On the other hand, if each $\text{ECSD}_{e(k)}\left(Q_{e(k)}\right)$ is non-convex, geometric programming (GP) [17] can be applied to solve the constrained problem (5). Under GP, the problem (5) can be transformed into the following convex form:

$$
\begin{aligned}
&\text{minimize} \ln \left(\prod_{j=1}^{n} \frac{1}{\text{ECSD}_{e(j)}(Q_{e(j)})} \right) \\
&\text{s.t.} \ \frac{\sum_{j=1}^{n} Q_{e(j)}}{Q_{\text{I}}} \leq 1 \text{ and } Q_{e(j)} \geq 0 \text{ for } 1 \leq j \leq n
\end{aligned} \tag{7}
$$

Because the optimization target in (7) is equivalent to

$$
\ln \left(\prod_{j=1}^{n} \frac{1}{\text{ECSD}_{e(j)}(Q_{e(j)})} \right) = -\sum_{j=1}^{n} \ln \left(\text{ECSD}_{e(j)}(Q_{e(j)}) \right), \tag{8}
$$

the constrained optimization problem (7) can be transformed into:

$$
\begin{aligned}
&\text{maximize} \sum_{j=1}^{n} \ln \left(\text{ECSD}_{e(j)}(Q_{e(j)}) \right) \\
&\text{s.t.} \ \sum_{j=1}^{n} Q_{e(j)} \leq Q_{\text{I}} \text{ and } Q_{e(j)} \geq 0 \text{ for } 1 \leq j \leq n
\end{aligned} \tag{9}
$$

Because $\ln \left(\text{ECSD}_{e(j)}(Q_{e(j)}) \right)$ is convex, the problem (9) has a unique optimal solution.

Through introducing Lagrange multiplier, we can construct the function $\phi(Q_{e(j)}, \lambda, \mu_j)$ as

$$
\phi(Q_{e(j)}, \lambda, \mu_j) = \sum_{k=1}^{n} \ln \left(\text{ECSD}_{e(k)}(Q_{e(k)}) \right) - \lambda \left(\sum_{k=1}^{n} Q_{e(k)} - Q_{\text{I}} \right) + \mu_j Q_{e(j)} \tag{10}
$$

By differentiating $\phi(Q_{e(j)}, \lambda, \mu_j)$ with respect to $Q_{e(j)}$, we have

$$
\frac{\partial \phi(Q_{e(j)}, \lambda, \mu_j)}{\partial Q_{e(j)}} = \sum_{k=1}^{N} \frac{\partial}{\partial Q_{e(k)}} \ln \left(\text{ECSD}_{e(k)}(Q_{e(k)}) \right) - \lambda + \mu_j \tag{11}
$$

This equation, together with (9), implies:

$$
\begin{cases}
\frac{1}{\text{ECSD}_{e(k)}(Q_{e(k)})} \cdot \frac{\partial}{\partial Q_{e(k)}} \text{ECSD}_{e(k)}(Q_{e(k)}) - \lambda + \mu_j = 0 \\
\lambda \left(\sum_{k=1}^{N} (Q_{e(k)}) - Q_{\text{I}} \right) = 0 \\
\mu_j = 0
\end{cases} \tag{12}
$$

By solving the above equations, we can obtain the unique optimal solution of $Q_{e(j)}$ for all j.

4   Numerical Simulation

This section simulates the performance of the proposed NBS-based energy allo-cation scheme. In this simulation, we consider an organization consisting of 6 key energy users, which are labeled from 1 to 6 and will participate into the whole energy allocation process. Moreover, the parameter setting of this simulation is summarized as follows:

The profit weights of user 1 to user 6 in the past year 1 are $\alpha_{1,1} = \alpha_{2,1} = \alpha_{3,1} = \alpha_{4,1} = \alpha_{5,1} = \alpha_{6,1} = 0.3$; the profit weights of user 1 to user 6 in the past year 2 are $\alpha_{1,2} = \alpha_{2,2} = \alpha_{3,2} = \alpha_{4,2} = \alpha_{5,2} = \alpha_{6,2} = 0.25$; the profit weights of user 1 to user 6 in the past year 3 are $\alpha_{1,3} = \alpha_{2,3} = \alpha_{3,3} = \alpha_{4,3} = \alpha_{5,3} = \alpha_{6,3} = 0.2$; the profit weights of user 1 to user 6 in the past year 4 are $\alpha_{1,4} = \alpha_{2,4} = \alpha_{3,4} = \alpha_{4,4} = \alpha_{5,4} = \alpha_{6,4} = 0.15$; the profit weights of user 1 to user 6 in the past year 5 are $\alpha_{1,5} = \alpha_{2,5} = \alpha_{3,5} = \alpha_{4,5} = \alpha_{5,5} = \alpha_{6,5} = 0.1$.

The profit rates of user 1 in the past 5 years are $p_{1,1} = 0.5$, $p_{1,2} = 0.9$, $p_{1,3} = 0.8$, $p_{1,4} = 0.9$ and $p_{1,5} = 0.3$, respectively; the profit rates of us-er 2 in the past 5 years are $p_{2,1} = 0.9$, $p_{2,2} = 0.9$, $p_{2,3} = 0.6$, $p_{2,4} = 0.2$ and $p_{2,5} = -0.08$, respectively; the profit rates of user 3 in the past 5 years are $p_{3,1} = 0.8$, $p_{3,2} = 0.98$, $p_{3,3} = 0.44$, $p_{3,4} = 0.78$ and $p_{3,5} = 0.23$, respectively; the profit rates of user 4 in the past 5 years are $p_{4,1} = 0.63$, $p_{4,2} = 0.85$, $p_{4,3} = 0.8$, $p_{4,4} = -0.05$ and $p_{4,5} = 0.9$, respectively; the profit rates of user 5 in the past 5 years are $p_{5,1} = 0.84$, $p_{5,2} = 0.6$, $_{5,3} = 0.7$, $p_{5,4} = 0.12$ and $p_{5,5} = -0.33$, respectively; the profit rates of user 6 in the past 5 years are $p_{6,1} = 0.62$, $p_{6,2} = 0.34$, $p_{6,3} = 0.11$, $p_{6,4} = -0.22$ and $p_{6,5} = -0.34$, respectively.

For all the 6 users, the minimum ECSD is $\text{ECSD}_{min} = 0.4$, and the punish-ment coefficient is set as $f_1 = f_2 = f_3 = f_4 = f_5 = 0.1$ when the actual annul energy consumption exceeds its annual energy quota allocated. The total energy budget of the current year is 12000 kwh.

The allocated energy quotas of user 1 to user 6 in the past 5 years are 2000kwh per year. The actual energy consumption of user 1 in the past 5 years are $x_{1,1} = 2000$, $x_{1,2} = 3000$, $x_{1,3} = 1000$, $x_{1,4} = 2000$ and $x_{1,5} = 2800$, respectively; the actual energy consumption of user 2 in the past 5 years are $x_{2,1} = 2000$, $x_{2,2} = 2600$, $x_{2,3} = 2000$, $x_{2,4} = 3000$ and $x_{2,5} = 1000$,, respectively; the actual energy consumption of user 3 in the past 5 years are $x_{3,1} = 1030$, $x_{3,2} = 1300$, $x_{3,3} = 2000$, $x_{3,4} = 1600$ and $x_{3,5} = 2000$, respectively; the actual energy consumption of user 4 in the past 5 years are $x_{4,1} = 2300$, $x_{4,2} = 2080$, $x_{4,3} = 1000$, $x_{4,4} = 2000$ and $x_{4,5} = 4030$, respectively; the actual energy consumption of user 5 in the past 5 years are $x_{5,1} = 1080$, $x_{5,2} = 2600$, $x_{5,3} = 2000$, $x_{5,4} = 2300$ and $x_{5,5} = 2000$, respectively; the actual energy consumption of user 6 in the past 5 years are $x_{6,1} = 2000$, $x_{6,2} = 1000$, $x_{6,3} = 2000$, $x_{6,4} = 2000$ and $x_{6,5} = 1000$, respectively.

According to the definition of ECSD, the ECSD coefficient c_i for each user $i \in [1,6]$ can be calculated according to the formula (3), i.e. $c_1 = 0.0006238$, $c_2 = 0.0005278$, $c_3 = 0.0006241$, $c_4 = 0.0005523$, $c_5 = 0.0003899$ and $c_6 = 0.0002554$. As we can see, the ECSDs of the users 1 to 5 in the past 5 years are higher than ECSD_{min}, while the ECSD of the user 6 in the past 5 years is lower than ECSD_{min}. Therefore, in the energy allocation process of the current year, the users 1 to 5 are classified as Class-I energy users, while the user 6 is classified as Class-II energy users. Thus, for the user 6, the organization will directly set its ECSD of the current year to $ECSD_{min}$, i.e. 0.4. The calculated energy quota for user 6 of the current year is 2000kwh while the remaining total energy budget to be allocated to user 1 to user 5 is $Q_I = 12000 - 2000 = 10000$ kwh.

Figure 1 shows the unique optimal solution of the constrained problem in formula (5), where the priorities of energy users 1 to 5 are set to be equal, i.e., $\tau_1 = \tau_2 = \tau_3 = \tau_4 = \tau_5 = 0.2$. The vertical coordinate of Fig. 1 denotes the ECSDs of all the 6 users, under the 2 different energy allocation methods, one being the proposed method based on the Nash Bargaining Solution (NBS) and the other the equal allocation of electricity energy among all 5 class-I users. As shown by Fig. 1, compared with the equal allocation of the total energy budget among all 6 users, the NBS-based allocation scheme can provide better fairness among energy users and effectively alleviate the large ECSD gap between the 5 class-I users under the equal allocation scheme, e.g., the user 5, of which the ECSD is the lowest among all class-I users, will obtain a relatively lower ECSD under the equal allocation scheme than that under the NBS-based allocation scheme, while each of the remaining Class-I users, i.e., the users 1 to 4, will obtain a relatively larger ECSD under the equal allocation scheme than that under the NBS-based allocation scheme. Thus the proposed NBS-based allocation scheme helps minimize the gap of energy consumption satisfaction between different class-I energy users and hence outperforms the equal allocation scheme in the fairness of energy allocation.

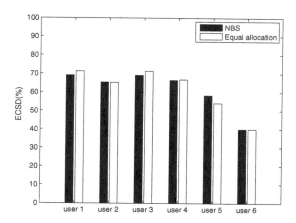

Fig. 1. Comparison of the ECSDs of the energy users 1 to 6 with a same priority.

Figure 2 shows the variation of the ECSD of the energy user 1 in terms of its priority given that the priorities of the users 3 to 5 are fixed as 0.2. Under this priority setting, the priorities of the users 1 and 2 should satisfy the relationship $\tau_1 + \tau_2 = 0.4$ and the variation range of τ_1 or τ_2 is $[0, 0.4]$.

As shown by Fig. 2, compared with the equal energy allocation scheme and the priority-based proportional energy allocation scheme, the proposed NBS-based energy allocation scheme can significantly improve the product of the ECSDs of the 5 energy users. Moreover, the larger the difference between τ_1 and the priorities of the remaining energy users, the more benefit the NBS-based allocation scheme can offer over the other two allocation schemes in terms of the ECSD product. Thus the energy manager of the organization would be better motivated for adopting the NBS-based allocation scheme given that the priority difference among energy users increases.

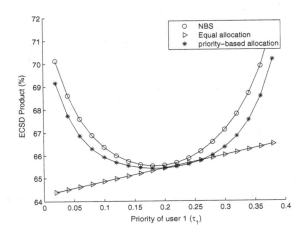

Fig. 2. The ECSD product of the energy users 1 to 5 under different energy allocation schemes is depicted in terms of the possible priority of user 1 given that $\tau_3 = \tau_4 = \tau_5 = 0.2$.

5   Conclusion

In view of the growing deployment of smart meters for monitoring energy consumption, the present paper is devoted to solving the urgent requirement on the quota-based energy consumption management within various organizations. To facilitate this management, we first formulate a basic coefficient, i.e., energy consumption satisfaction degree (ECSD), for an organization to evaluate the performance or satisfaction of each key energy user in energy consumption. The main advantage of this coefficient is its simplicity and easy for the organization to fit it as a convex or non-convex function. Based on ECSD, we further utilize the classical Nash bargaining solution (NBS) in cooperative game theory to develop a quota-based energy consumption management scheme, of which the optimal solution can be easily derived by Lagrange multiplier or geometric programming. This scheme enables energy managers to effectively guarantee the annual energy-saving target of the organization with a fixed energy budget.

Numerical simulation shows that, compared with the equal and priority-based energy allocation schemes, the proposed energy allocation scheme can maximize the overall satisfaction of all key energy users within the organization and, meanwhile, maintain a reasonable fairness among them.

Acknowledgments. This work was supported by grants from the National Natural Science Foundation of China (no. 61471104) and the Science and Technology Program for Public Wellbeing of Chengdu (no. M11036001HMGC201301).

References

1. Nicolae, P.-M., Nicolae, I.-D.: The monitoring of energy consumptions in a urban transportation system. In: International Conference on Electrical and Electronics Engineering (ELECO), pp. I-108–I-112 (2009)
2. Sun, Q., Yang, H., Wang, C., Zang, H.: Energy consumption monitoring systemin discrete manufacturing plants. In: Fourth International Conference on Digital Manufacturing and Automation (ICDMA), pp. 541–544 (2013)
3. Yimin, S., Zhengli, W.: The energy consumption monitoring platform design for large-scale industry users based on the GPRS. In: Second International Conference on Mechanic Automation and Control Engineering (MACE), pp. 7827–7830 (2011)
4. Neagu, B.C., Georgescu, G., Elges, A.: Monitoring system of electric energy consumption to users. In: International Conference and Exposition on Electrical and Power Engineering(EPE), pp. 265–270 (2012)
5. Ultra, C., Boso, A., Espluga, J.: A qualitative study of users' engagement with real-time feedback from in-house energy consumption displays. Energy Policy **61**, 788–792 (2013)
6. Kuo-Ming, C., Shah, N., Farmer, R., Matei, A., Ding-Yuan, C., Schuster-James, H., Tedd, R.: A profile based energy management system for domestic electrical appliances. In: IEEE 7th International Conference on e-Business Engineering(ICEBE), pp. 415–420 (2010)
7. Hassan, N., Pasha, M.A., Yuen, C., Huang, S., Wang, X.: Impact of scheduling flexibility on demand profile flatness and user inconvenience in residential smart grid system. Energies **6**(12), 6608–6635 (2013)
8. Yushar, W., Chai, B., Yuen, C., Smith, D.B., Poor, H.V.: Energy management for a user interactive smart community: a Stackelberg game approach. In: IEEE Proceedings IEEE Innovative Smart Grid Technology (ISGT) Asia (2014)
9. Liu, Y., Yuen, C., Huang, S.: Peak-to-average ratio constrained demand-side management with consumer's preference in residential smart grid. IEEE J. Sel. Top. Signal Process. **8**(6), 1084–1097 (2014)
10. Zhu, Z., Tang, J., Lambotharan, S., Chin, W.H., Fan, Z.: An integer linear programming and game theory based optimization for demand-side management in smart grid. In: GLOBECOM Workshops, pp. 1205–1210 (2011)
11. Mohsenian-Rad, A.-H., Wong, V.W.S., Jatskevich, J., Schober, R., Leon-Garcia, A.: Autonomous demand-side management based on game-theoretic energy consumption scheduling for the future smart grid. IEEE Trans. Smart Grid **1**(3), 320–331 (2010)
12. Cai, Y., Li, Q., Xie, H., Min, H.: Exploring personalized searches using tag-based user profiles and resource profiles in folksonomy. Neural Netw. **58**, 98–110 (2014)

13. Xie, H., Li, Q., Mao, X., Li, X., Cai, Y., Rao, Y.: Community-aware user profile enrichment in folksonomy. Neural Netw. **58**, 111–121 (2014)
14. Li, N., Shi, M., Shang, Z., Yuan, Y.: Impacts of total energy consumption control and energy quota allocation on China's regional economy based on a 30-region computable general equilibrium analysis. Chin. Geogr. Sci. **25**(109), 1–15 (2015)
15. Nash, J.: The bargaining problem. Econometrica **18**, 155–162 (1950)
16. Byrd, R.H., Gilbert, J.C., Nocedal, J.: A trust region method based on interior point techniques for nonlinear programming. Mathematical Programming **89**(1), 149–185 (2000)
17. Chiang, M.: Geometric programming for communication systems. Commun. Inf. Theory **2**(1–2), 1–154 (2005)

Entity Relation Mining in Large-Scale Data

Jingnan Li, Yi Cai$^{(\boxtimes)}$, Qixuan Wang, Shuyue Hu,
Tao Wang, and Huaqing Min

School of Software Engineering,
South China University of Technology, Guangzhou, China
ycai@scut.edu.cn

Abstract. Currently, the web-based Named-Entity relationship extraction has been a new research field with a tremendous potential. The goal of web-based entity relationship extraction is to explore the relationship between a set of realistic entities. It's a challenging research field and has a widely application value in the related fields of text mining. In this paper, we propose a newly defined framework called Snowball++ based on the traditional entity relationship extraction frameworks. In our Snowball++ framework, we focus on the many-to-many relations more than one-to-one relations. The system is also implemented in the many-to-many manner and it improves the precision and recall. It's worth to notice that Snowball++ will assign a specific relation type to each entity-relationship pair and the whole training process only need a few manual labor. For the sake of building a efficient and scalable system, we implement the Snowball++ framework on the Hadoop platform which is a totally distributed computing system. Eventually, the experiments show that our framework and implementation are efficient and effective.

Keywords: Entity relationship extraction · Hadoop · Distributed system

1 Introduction

In past decades, the rapid development of World Wide Web leads to the explosion of data on the Internet. Everyone in the world is continuously producing data. Therefore, a great change is under way in the technology, business, management and finance fields. The era of "Big Data" is coming which is changing our lifestyle and even our way of thinking.

The extraction of Web-based Entity Relation is a fascinating research issue, which is to extract relationships between entities from a corpus. The Web-based Entity Relation Extraction is meaningful in the Information Retrieval, Question Answering and many other related fields of Natural Language Processing.

The Named-Entity is a person, group or object which exists in the real world (e.g., "*Barack Obama*", "*House of Cards*", "*the Forbidden City*" etc.). This paper focuses on the extraction of relationships (e.g., "is husband of", "is author of" etc.) between entities. Given a context containing two or more entities

A. Liu et al. (Eds.): DASFAA 2015 Workshops, LNCS 9052, pp. 109–121, 2015.
DOI: 10.1007/978-3-319-22324-7_10

which were already annotated, our goal is to extract the relation between these annotated entities. For example, the sentence *"Liu Kaiwei* and *Yang Mi* held a wedding ceremony in the *Tahiti"* contains two entities, i.e., *Liu Kaiwei* and *Yang Mi.* From the context of this sentence, we can detect that *"Liu Kaiwei"* and *"Yang Mi"* are couple.

Currently, there are some methods having been proposed to extract the relationship of entities, e.g., in supervised learning methods [1–4] and bootstrapping systems [5–7]. There are also some object-level search engines, which automatically extract and integrate the semantic information about entities and return a list of ranked entities instead of webpages to answer user queries [8,9]. There are several works which are relevant to the topic of entity relation mining, such as event relationship analysis [10], mining latent user community [11] and personalized search [12,13]. *Agichtei* and *Gravano* [6] focus on the one-to-one relationships, such as, a company and its headquarters. However, the entity-relationship is many-to-many in most cases. Accordingly, *Zhu* et al. [14] improve this method to extract many-to-many relationships. However, their method is too time-consuming to deal with the large amount of data on the Internet. In another work, *Ndapandula* and *Martin* et al. [15] propose a method which can achieve a good efficiency on the large-scale data. A limitation of this method is that it needs a large amount of training data in order to increase the precision.

To address the above limitations, we propose *Snowball++* framework which is based on the *Snowball* [6]. The *Snowball++* is to extract the many-to-many relationships rather than the one-to-one relationships in the original Snowball. It also deals with multi-type relationships simultaneously during each iteration of the extraction process. As a result, the recall of relationships gets a considerable improvement. Each pair of entities that contains relationship will be assigned a specific relation type. Besides, *Snowball++* needs a few manual labels.

The development of distributed computing programming flourishes the research of big data. The well-known MapReduce programming framework [16] and Hadoop platform help us building a flexible and scalable distributed entity relation extraction framework. In *Snowball++*, the whole process executes on the Hadoop platform and follow the MapReduce programming framework. Thus, our *Snowball++* is efficient and scalable.

2   Related Works

Agichtei et al. [6] proposed a heuristic strategy and implement it as Snowball system to generate entity-relationship patterns and pairs. The Snowball include a effective evaluation method to measure a newly generated entity-relationship pattern. This is an online learning system which can continuously improve itself. However, Snowball focus on the one-to-one relationships (e.g., a company and it's headquarters), and there is only one type of relationship is concerned when the system is running, it's not flexible and robust because the entity-relationship is many-to-many in most cases(e.g., a movie and it's actors is not one-to-one, the movie may have more than one actor and an actor may have participated in a lot of movies.).

Zhu et al. [14] improve the Snowball by supplementing a statistical model called *Markov Logic Network* and call their new framework as *StateSnowball*. The learning process of Markov Logic Network improve the evaluation of entity-relationship pairs. It involves the weights into the model to generate entity-relationship pairs. Their *StatSnowball* improve the recall of the extraction process but cannot assign a relation type to the extracted entity pair. Besides, they proposed a popular websites called "Renlifang" based on the research. Overall, the StatSnowball is too time-consuming to deal with the large amount of data on the Internet since it's a deep parsing process.

Ndapandula et al. [15] proposed the PROSPERA framework since they recognized the difficulty in balancing recall and precision. Considering the heuristic strategy may lead to lower recall, and the deep parsing method may lead to high consuming, they proposed a method with the n-gram manner. The PROSPERA improves the recall and efficiency. They also elaborated their implementation in the Hadoop platform which is a scalable and flexible distributed system. However, the n-gram method need large amount of training data.

3   The Proposed Framework

The Snowball++ framework contains the following steps:

1. Match the "Entity-Relationship Patterns" in the documents by using the "seeds". The "seeds" are acquired by collecting the Entity-Relationship pairs from the structure web resources (e.g., *Douban.com*). The "Entity-Relationship Pattern" is a short text which contains "seed" pair.
2. Extract more new Entity-Relationship pairs from the documents by employing the "Entity-Relationship Patterns". The newly extracted pairs will be evaluated by a similarity measurement and the low similar pairs will be neglected.
3. The high-credible new pairs will be added to the "seeds" and using in the next extraction iteration.

which is shown in Fig. 1.

3.1   Acquiring the Entity-Relationship Patterns

The "Entity-Relationship Pattern" is a short text which contains "seed" pair. We firstly extract the texts which are matched by the "seeds". The text can be present by segmented to three vectors:

$$p = < left, tag_1, middle, tag_2, right, relation_{type} > \tag{1}$$

where the tag_1 and tag_2 are the feature tags of two entities from one seed pair. In the experiments, our entities are mainly gathered from four classes (i.e., P-people, M-movie/drama, S-music, B-book). The *left*, *middle* and *right* are vectors of three part of texts which are separated by the two tags. The $relation_{type}$

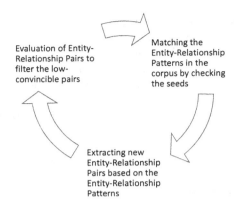

Fig. 1. Framework of Snowball++

denotes the concrete relation type (e.g., "is father of") of the pattern. Besides, the vectors are regularized after extracted. Here is a typical pattern sample

$$p = < \{\}, P, \{sing\_v\_0.049823112\}, S,$$
$$\{at\_p\_0.02461, concert\_n\_0.03371\}, singer > \tag{2}$$

where in each vector, a word contains three parts which are word, POS tag and its term frequency. During the extraction process, two patterns is merged if their similarity exceed a empirical threshold. The new merged pattern is

$$p = < \bar{l}, t_1, \bar{m}, t_2, \bar{r}, relation_{type} > \tag{3}$$

where the $\bar{l}$, $\bar{m}$ and $\bar{r}$ are the average values of the original vectors. Note that only the two patterns belong to the same relation type can be merged.

3.2   Extracting the Entity-Relationship Pairs

After getting considerable "Entity-Relationship Patterns", we employ them to extract new "Entity-Relationship Pairs" in the corpus. The pairs can also be formulated as the vector form of patterns. The newly extracted pairs are then measured by counting the similarity between the existing patterns. The pairs with higher similarity (i.e., exceed a empirical threshold) are remained and the others are neglected. The pairs will be assigned a relation type by voting algorithm (i.e., voted by the most similar patterns). The Algorithm 1 shows the concrete steps.

The *CreatePattern* function in the Algorithm 1 is used to extract a entity-relationship pattern without relation type from the text which contains a pair of entities. The *Sim* function is used to count the similariy between patterns and pairs. The *AddPatternToTuple* function is used to collect the most similar patterns (i.e., similarity exceeding a empirical threshold) to the extracted pair.

Algorithm 1. The algorithm used to extract Entity-Relationship Pairs

Require:
    Entity-Relationship Patterns
Ensure:
    Candidate Tuple Set
1: **for** each $text\_segment \in corpus$ **do**
2:    $\{< e1, e2 >, < l, t1, m, t2, r, () >\} = CreatePattern(text\_segment)$;
3:    $Tc = < e1, e2 >$;
4:    $SimBest = 0$;
5:    **for** each $p \in Patterns$ **do**
6:      $sim = Sim(< l, t1, m, t2, r, () >, p)$;
7:      **if** $(sim \geq T\_sim)$ **then**
8:        $AddPatternToTuple(p, Tc)$;
9:        **if** $(sim \geq SimBest)$ **then**
10:          $SimBest = sim$;
11:          $PBest = p$;
12:        **end if**
13:      **end if**
14:    **end for**
15:    **if** $PBest \neq null$ **then**
16:      $< l, t1, m, t2, r, (PBest.RT) >$;
17:      $Tc.Relation\_type = PBest.RT$;
18:      $CandidateTupleSet.add(Tc)$;
19:    **end if**
20: **end for**
21: **return** $CandidateTupleSet$

These patterns will be used in the next step which is used to assign a relation type to the new pair.

For example, assume that we get a pair like

$$p = < \{\}, Jay, \{sing\_v\_0.043823\}, Juhuatai,$$
$$\{at\_p\_0.021325, concert\_n\_0.034371\}, null > \qquad (4)$$

which contains two entities (i.e., *"Jay"* and *"Juhuatai"*). Its most similar pattern is (2). Therefore, we can assign the "singer" type to the pair and acquire a "Entity-Relationship pair" which is $(Juhuatai, Jay, Singer)$.

3.3   Evaluation of the Candidate Relationship Pairs

The evaluation and filtrate of the entity-relationship pairs is the most important step in the Snowball++. Only the convincible entity-relationship patterns can extract valuable entity-relationship pairs. Furthermore, only the valuable entity-relationship pairs can be the effective seeds which will be used in the next iteration.

We have already collected all the entity-relationship patterns (i.e., by using the AddPatternToTuple function) in the previous step. These patterns are generated by seeds, so their relation types are known. Since every pattern can extract more than one pair and these pairs are assigned possible relation types, we can scan a pattern and pairs which were generated by the pattern, if a pair's relation type is the same with the pattern's, we put the pair into the "positive pattern set", and the others are put into the "negative pattern set". Eventually, patterns are evaluated by the following formula

$$Support(Pattern) = \frac{\mid Positive\_set \mid}{\mid Positive\_set \mid + \mid Negative\_set \mid} \tag{5}$$

in which we can acquire the *Support* value of a pattern. It is based on the intuition that a pattern is more convincible if it can extract more pairs whose relation type is equal to the pattern's.

It should be noted that the Support evaluation is totally different with the original Snowball. Since our Snowball++ is aim to extract the many-to-many relationships, this Support evaluation is more robust in the many-to-many situation.

After counting all the pattern's Support scores, we use them to evaluate the pairs generated in this iteration. Since each pair can be extracted by more than one pattern. We give a possibility to the pair by checking the Support scores of patterns which can extract this pair. Analogously, we still call it as Support

$$Support(Tuple) = 1 - \prod_{i=0}^{P} (1 - Support(P_i)) \tag{6}$$

where the P_i represents each pattern.

Eventually, we set a empirical threshold T_sup to filter the highly convincible pairs as new seeds and add them into the seed set which will take part in the next iteration.

4   Implement of Distributed System

In order to process large amount of data and build the system more scalable, we implement our Snowball++ system in the MapReduce model.

The Algorithms 2 and 3 show the pseudo-code of the extraction process of entity-relationship patterns. The "mapper" function matches the entity-relationship patterns in the corpus by checking the seeds. The "getTupples" function acquires the entities of a short text which had been marked in the preprocessing step. If the two entities exist in one seed pair, the "createPattern" function are called to extract the text as new entity-relationship pattern. The "reducer" function only collects the patterns which contains the same key and output them. Eventually, we acquire the entity-relationship patterns which classified by their entity tags.

The Algorithms 4 and 5 show the pseudo-code of the extraction process of entity-relationship pairs. The "mapper" function is used to process all the pending texts. The text contains a pair of entities will be extracted as a new entity-relationship pair. It will traverse all the known patterns and count the

---

Algorithm 2. The mapper function for extracting entity-relationship patterns

Require:

  i;

  $Text_i$;

1: **if** $Text_i$ has $Entitys$ **then**
2:    $(e1, e2) = getTuples(Text_i)$;
3:    **for each** $seed \in Seeds$ **do**
4:       $p_i = < l, t1, m, t2, r, (RT) > = CreatePattern(Text_i)$;
5:       $tag_i = t1 * t2$;
6:       $write(tag_i, p_i)$;
7:    **end for**
8: **end if**

Algorithm 3. The reducer function for extracting entity-relationship patterns

Require:

  tag_i;

  $[p_1, p_2, p_3, ...]$;

1: **for each** $p_i \in [p_1, p_2, p_3, ...]$ **do**
2:    $write(tag_i, p_i)$;
3: **end for**

Algorithm 4. The mapper function for extracting entity-relationship pairs

Require:

  i;

  $Text_i$;

1: **if** $Text_i$ has $Entitys$ **then**
2:    $\{< e1, e2 >, < l, t1, m, t2, r, () >\} = CreatePattern(Text_i)$;
3:    $T_i = < e1, e2 >$;
4:    $SimBest = 0$;
5:    $pattern_set = \{\}$;
6:    **for each** $p_i \in Patterns$ **do**
7:       $sim = Sim(< l, t1, m, t2, r, () >, p)$;
8:       **if** $(sim \geq T\_sim)$ **then**
9:          $pattern\_set.add(p_i)$
10:          **if** $(sim \geq SimBest)$ **then**
11:             $SimBest = sim$;
12:             $PBest = p$;
13:          **end if**
14:       **end if**
15:    **end for**
16:    **if** $PBest \neq null$ **then**
17:       $T_i.Relation\_type = PBest.RT$;
18:    **end if**
19:    **for each** $pattern \in Pattern\_set$ **do**
20:       $write(pattern, T_i)$;
21:    **end for**
22: **end if**

similarity between each on them with the new extracted pair. The patterns with relatively high similarity (i.e., exceeding a empirical threshold) will be collected in the pattern set of the pair. The "reducer" function traverses all the pairs extracted by a specific entity-relationship pattern and computers the support score for this pattern.

The Algorithms 6 and 7 show the pseudo-code of the evaluation process of entity-relationship pairs. The "mapper" and "reducer" functions collaboratively computer the support scores for each entity-relationship pairs by employing the support scores of entity-relationship patterns acquired before. Eventually, the entity-relationship pairs with support score exceeding a empirical threshold will be retained.

Algorithm 5. The reducer function for extracting entity-relationship pairs

Require:
  p_i;
  $[T_1, T_2, T_3, ...]$;
  1: **for** each $T_i \in [T_1, T_2, T_3, ...]$ **do**
  2:    **if** $T_i.RT = p_i.RT$ **then**
  3:       $PositiveSet.add(T_i)$;
  4:    **else**
  5:       $NegativeSet.add(T_i)$;
  6:    **end if**
  7: **end for**
  8: **for** each $T_i \in [T_1, T_2, T_3, ...]$ **do**
  9:    $write(T_i, Support(p_i))$;
 10: **end for**

Algorithm 6. The mapper function for evaluating entity-relationship pairs

Require:
  T_i;
  sup_{pi};
  1: $t_i = 1 - sup_p i$;
  2: $write(T_i, t_i)$;

Algorithm 7. The reducer function for evaluating entity-relationship pairs

Require:
  T_i;
  $[t_1, t_2, t_3, ...]$;
  1: $t\_total = 1$;
  2: **for** each $t_i \in [t_1, t_2, t_3, ...]$ **do**
  3:    $t\_total* = t_i$
  4: **end for**
  5: $Support(T_i) = 1 - t\_total$;
  6: **if** $Support(T_i) > T\_sup$ **then**
  7:    $write(T_i, Support(T_i))$;;
  8: **end if**

5 Experiments

5.1 Dataset

The datasets used in experiments are summarized to structured and unstructured data. The structured data are crawled from the *Douban.com, MTime.com, JD.com* and many other websites which provide the data with explicit relation information. The data from these websites can be the seeds to extract the implicit entity-relationship patterns in the semi-structured or unstructured data.

The unstructured data are crawled from *baike.baidu.com*. The *baike.baidu.com* is a big encyclopedia which contains more than 700 million entries. It has a good timeliness and coverage. We collect 1,543,669 million entries from it and integrate them as a 100 GB-sized corpus. We extract texts contains 100 pairs of entities which involve people, movies, music and books and call this subset as "100-Tuple".

Our entities are provided by the NDBCCUP2014[1]. It contains four kinds of entities which are people(P), movies(M), music(S) and books(B). Each entity has a exclusive ID for distinguishing.

5.2 Results of Comparison

In contrast, we implement the *PROSPERA* proposed by *Ndapandula* et al. and the original *Snowball* as the baseline approach.

The comparison experiments include three parts. Firstly, we extract entity-relationships from the "100-Tuple" with 1000 seeds by the Snowball++, Snowball and PROSPERA. The results are evaluated by the precision and recall.

Secondly, we conduct the extraction process from "100-Tuple" with different amount of seeds. The seeds are selected to 200, 1000 and 5000 three sets.

Eventually, we employ the three frameworks to conduct the extraction process in the 100 GB-sized corpus crawled from *baike.baidu.com*. Since the amount of the data is too large, we implement the three framework in distributed manner and also compare their efficiency.

Table 1. The number of positive pairs extracted from "100-Tuple"

Approaches	Positive pairs
Snowball	53
PROSPERA	60
Snowball++	61

Table 1 shows the result of first experiment. It's the number of positive pairs extracted after three iterations. We can find that our Snowball++ framework acquire the best performance among the three frameworks.

---

[1] http://iir.ruc.edu.cn/ndbccup2014/

Figures 2 and 3 show the precision and recall evaluation of the three frameworks. It obvious that our Snowball++ acquires the best performance and the Snowball is the worst.

Figures 4 and 5 show the result of second experiment. Our Snowball++ can get a higher performance than the PROSPERA when the seeds is less while PROSPERA overwit Snowball++ when the number of seed is growing. It denotes that our Snowball++ need less manual supervising and the PROSPERA is sensitive to the number of prior knowledge.

Finally, we implement three frameworks in the Hadoop platform. We employ four PC servers with 16G-memory and intel i7 processer to form a small cluster. We employ 1000 pair of seeds in this extraction. Figure 6 shows the efficiency of these three frameworks. The PROSPERA is time-consuming because it need to traverse the documents to acquire the frequent words. Since our Snowball++ and PROSPERA can acquire the higher performance and our Snowball++ is more efficient, Snowball++ is a good enough framework.

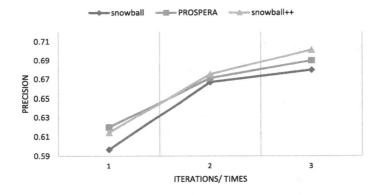

Fig. 2. Precision of three frameworks

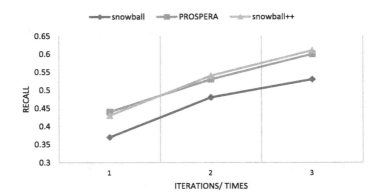

Fig. 3. Recall of three frameworks

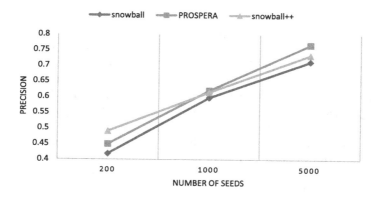

Fig. 4. Precision of three frameworks

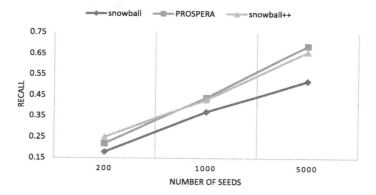

Fig. 5. Recall of three frameworks

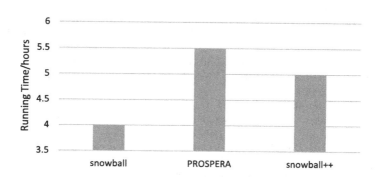

Fig. 6. The running time of three frameworks

6   Conclusion

The Web-based Entity-Relationship Extraction is a new research field with a tremendous potential. We propose a new extraction framework based on the Snowball, StatSnowball and PROSPERA frameworks.Comparing with the original Snowball, the Snowball++ consider the many-to-many multi-relations more than the one-to-one relations. We modify the evaluation method for the entity-relationship pairs and accomplish to improve the recall and precision. Besides, be different from the StatSnowball, we assign a specific relation type to each entity-relationship pair. Comparing with the PROSPERA system, Snowball++ can acquire a better performance with small amount of prior knowledge. Besides, the PROSPERA and our Snowball++ can be implemented in the distributed manner and Snowball++ is more efficient than the PROSPERA.

Acknowledgement. This work is supported by National Natural Science Foundation of China (Grant NO. 61300137), the Guangdong Natural Science Foundation, China (NO. S2013010013836), Science and Technology Planning Project of Guangdong Province China NO. 2013B010406004 the Fundamental Research Funds for the Central Universities, SCUT(NO. 2014ZZ0035).

References

1. Zhou, G., Zhang, M., Ji, D.H., Zhu, Q.: Tree kernel-based relation extraction with context-sensitive structured parse tree information. EMNLP-CoNLL 2007, p. 728 (2007)
2. Giuliano, C., Lavelli, A., Romano, L.: Exploiting shallow linguistic information for relation extraction from biomedical literature. In: EACL, vol. 18, pp. 401–408. Citeseer (2006)
3. Harabagiu, S., Bejan, C.A., Morarescu, P.: Shallow semantics for relation extraction. In: Proceedings of the 19th International Joint Conference on Artificial Intelligence, pp. 1061–1066. Morgan Kaufmann Publishers Inc (2005)
4. Zelenko, D., Aone, C., Richardella, A.: Kernel methods for relation extraction. J. Mach. Learn. Res. **3**, 1083–1106 (2003)
5. Brin, S.: Extracting patterns and relations from the world wide web. In: Atzeni, P., Mendelzon, A.O., Mecca, G. (eds.) WebDB 1998. LNCS, vol. 1590, pp. 172–183. Springer, Heidelberg (1999)
6. Agichtein, E., Gravano, L.: Snowball: extracting relations from large plain-text collections. In: Proceedings of the Fifth ACM Conference on Digital Libraries, pp. 85–94. ACM (2000)
7. Etzioni, O., Cafarella, M., Downey, D., Popescu, A.M., Shaked, T., Soderland, S., Weld, D.S., Yates, A.: Unsupervised named-entity extraction from the web: an experimental study. Artif. Intell. **165**(1), 91–134 (2005)
8. Nie, Z., Wen, J.R., Ma, W.Y.: Object-level vertical search. In: CIDR, pp. 235–246 (2007)
9. Salton, G.: Automatic Text Processing: The Transformation, Analysis, and Retrieval of Reading. Addison-Wesley, New York (1989)

10. Cai, Y., Li, Q., Xie, H., Wang, T., Min, H.: Event relationship analysis for temporal event search. In: Meng, W., Feng, L., Bressan, S., Winiwarter, W., Song, W. (eds.) DASFAA 2013, Part II. LNCS, vol. 7826, pp. 179–193. Springer, Heidelberg (2013)
11. Xie, H., Li, Q., Mao, X., Li, X., Cai, Y., Zheng, Q.: Mining latent user community for tag-based and content-based search in social media. Comput. J. **57**(9), 1415–1430 (2014)
12. Xie, H.R., Li, Q., Cai, Y.: Community-aware resource profiling for personalized search in folksonomy. J. Comput. Sci. Technol. **27**(3), 599–610 (2012)
13. Cai, Y., Li, Q.: Personalized search by tag-based user profile and resource profile in collaborative tagging systems. In: Proceedings of the 19th ACM International Conference on Information and Knowledge Management, pp. 969–978. ACM (2010)
14. Zhu, J., Nie, Z., Liu, X., Zhang, B., Wen, J.R.: Statsnowball: a statistical approach to extracting entity relationships. In: Proceedings of the 18th International Conference on World Wide Web, pp. 101–110. ACM (2009)
15. Nakashole, N., Theobald, M., Weikum, G.: Scalable knowledge harvesting with high precision and high recall. In: Proceedings of the Fourth ACM International Conference on Web Search and Data Mining, pp. 227–236. ACM (2011)
16. Dean, J., Ghemawat, S.: Mapreduce: a flexible data processing tool. Commun. ACM **53**(1), 72–77 (2010)

A Collaborative Filtering Model for Personalized Retweeting Prediction

Jun Li[1], Jiamin Qin[2], Tao Wang[3], Yi Cai[3(✉)], and Huaqing Min[3]

[1] Information Science and Technology School,
Lingnan Normal University, Zhanjiang, China
[2] Department of Engineering and Computer Science,
Syracuse University, Syracuse, NY, USA
[3] School of Software Engineering,
South China University of Technology, Guangzhou, China
ycai@scut.edu.cn

Abstract. As the development of social media, the services in social media have significantly changed people's habits of using Internet. However, as the large amount of information posted by users and the highly frequent updates in social media, users often face the problem of information overload and miss out of content that they may be interested in. Recommender systems, which recommends an item (e.g., a product, a service and a twitter etc.) to users based on their interests, is an effective technique to handle this issue. In this paper, we borrow matrix factorization model from recommender system to predict users' behaviors of retweeting in social media. Compared with previous works, we take the relevance of users' interests, tweets' content, and publishers' influence into account simultaneously. Our experimental results on a real-world dataset show that the proposed model achieves desirable performance in characterizing users' retweeting behaviors and predicting topic diffusion in social media.

1 Introduction

In social media, users are allowed to connect and share information with their friends posting multimedia content, such as text, image, audio and video. Most users update these communication information frequently and a large amount of fresh messages are generated per day. The services of social media have significantly changed the habits of users exploiting Internet. Currently, most of the latest news about critical information in various situations, such as natural disasters and political upheavals, are first made known to the public via social media. Therefore, the abundant social content has become a valuable recourse to analyze users' opinion on social issues.

However, as the rapidly increasing amount of social content, most users cannot effectively and efficiently digest these information, and they generally encounter a serious problem of information overload. Previous studies [1] report that Twitter users follow 80 people on average, and hundreds or even thousands of tweets are posted to each user every day. On the other hand, users usually

© Springer International Publishing Switzerland 2015
A. Liu et al. (Eds.): DASFAA 2015 Workshops, LNCS 9052, pp. 122–134, 2015.
DOI: 10.1007/978-3-319-22324-7_11

want to be informed of the latest updates that they are interested in. However, this kind of information is limited since users usually have the limited and stable social connections in social network [2]. Unfortunately, these information is overwhelmed by the large amount of irrelevant updates. This brings great difficulty for users to find desirable content, and leads to the problem of information shortage. Therefore, it arises the need of techniques to help users filter information based on their interests or preferences.

Recommender systems, which recommends an item (e.g., a product, a service and a twitter etc.) to users based on their interests, is an effective technique to handle this issue [3]. However, different from traditional recommendation tasks, the information that users generate in social media has much diversity, not only including message content but only various behavior data such as following, re-posting (retweeting), replying and commenting. Apart from users' interests which are emphasized in traditional recommender system, the social connections also have a significant impact to influence the behaviors of users in social media. Therefore, recommender systems need to be modified to handle this case.

Nowadays, some tweet recommendation methods have proposed in analyzing users' interests and behaviors in social media, so as to filter content that match users' interests. Most of existing methods exploit tweet content to perform tweet recommendation without considering the collaborative view [4–7]. In addition, these methods recommend tweets to users based on some explicit features. However, the information in social network is so rich and complex that it is insufficient to characterize and analyze users' behaviors only depending on observed features. To better model users' behaviors, some studies employ latent factor model to further find out some factors that are unobserved but have significant influence on users' social interactions, such as [1, 2, 8–10]. These methods consider the relationship between tweets and users and the relationship between users and publishers separately. However, when users face the tweets that have the same content but are posted by different publishers, they may make different decisions. The reason is that different publishers have different influence in the domain that a tweet involves. Therefore, the relationship between users, publishers and twitters should be considered simultaneously in analyzing users' behaviors.

In all behaviors that users make in social media, retweeting is a significant indicator that users are much interested in a tweet, and this behavior is a main way leading to information diffusion. Therefore, in this paper, we study the issue of personalized retweeting prediction to understand users' behaviors in social network. To address this issue, we extent the work in [1] and propose a matrix factorization model to predict users' retweeting behaviors. Different from previous works, in the proposed model, we take the three aspects influencing users' retweeting decisions, i.e., users' interests, content of tweets, the influence of publishers in tweets' domains, into account simultaneously. Our experimental results on a real-world dataset show that the proposed model achieves desirable performance in characterizing users' retweeting behaviors and predicting topic diffusion in social media.

The remainder of this paper is organized as follows. Section 2 introduces the related work. Then, we present the proposed matrix factorization model in

Sect. 3. The experimental evaluation results are discussed in Sect. 4. Finally, we conclude our work in Sect. 5.

2   Related Work

2.1   Collaborative Filtering and Recommender Systems

As the rapidly increasing amount of information online, recommender systems are widely used to address the problem of information overload. Recommender systems exploit the history data that users create online to learn user profiles, and then recommend some content that matches different interests or preferences of users [3]. Currently, recommender systems are classified into content-based methods and collaborative filtering methods. Since content-based methods build user profiles by using the history records, such as the items that users rated, bought or browsed before. These profiles are built using much explicit features. However, in many real-world situations, the content of data that uses leave online is much limited and sparse, so that these explicit terms extracted from history data are insufficient to reflect user's interests. For example, some latent interests or preferences cannot be characterized in content-based methods [11].

In contrast, collaborative filtering (CF) methods build user profiles based on the community data rather [12]. The basic assumption of collaborative filtering methods is that users that have the same opinion on some items trend to hold similar opinions on other items (user-based CF); items that obtain the same opinion from some users trend to receive similar opinions from other users (item-based CF) [3,13]. Generally, CF methods are classified into two categories: neighborhood-based methods [14,15] and model-based methods [11,16]. The former recommends items based on the similarity of user of item neighbors, while the later performs recommendation using matrix factorization model or probabilistic latent factor model. Since model-based methods have be reported to achieve superior performance in recommender system and this kind of methods have better extensibility, we adopt matrix factorization model to handle the problem of personalized retweeting prediction.

2.2   Personalized Retweeting Prediction

In recent years, social network, such as Twitter, Facebook, Sina Weibo, have become increasing popular and have a great impact on people's social connections. Due to the large amount of content and the frequent information updates in social media, personalized recommender systems, which filter and recommend content to users based on their interests, have attracted increasing attentions recently [1,2,17].

Some recommender systems in social media have been proposed. Sun et al. [4] build a diffusion graph to select a small subset of tweets as recommended content for regular users. Chen et al. [5] exploit URL recommendation on Twitter to better direct user attention in information streams. Duan et al. [6] study three types of features and propose a tweet ranking method. Naveed et al. [7] predict the probability of a tweet being retweeted based on content features alone. However, these methods only exploit tweet content to perform tweet recommendation without considering the collaborative view [1]. Furthermore, these methods only exploit explicit features in content but neglect latent factors that influence users' behaviors.

Recently, some methods that employ latent factors analysis have been proposed. Yang et al. [8] employ a factor graph to predict the spread of tweets. Kim and Shim propose a probabilistic model based on Probabilistic Latent Semantic Analysis (PLSA) to recommend potential followers to users in Twitter [9]. However, this method only exploit latent features but neglecting explicit features. Lin et al. [10] propose a generic joint inference framework for topic diffusion and evolution in social network. However, this method is not extendable to integrate with rich features. Chen et al. [1] propose a matrix factorization model including a wide range of features to recommend tweets to users. Hong et al. [2] propose Co-Factorization Machines to predict user decisions and model users' interests. These methods consider the relationship between tweets and users and the relationship between users and publishers separately. However, when users face the tweets that have the same content but are posted by different publishers, they may make different decisions. The reason is that different publishers have different influence in the domain that a tweet involves. In the proposed method, we take the three aspects influencing users' retweeting decisions, i.e., users' interests, content of tweets, the influence of publishers in tweets' domains, into account simultaneously to predict users' retweeting behaviors.

3   Proposed Method

There are many method proposed which is based on finding the explicit factors of tweet text. However, explicit factors are not the whole thing why people will pay attention on one specific tweet. Therefore, we propose our approach which will take implicit factors into account. This is based on our observation that the contain of tweet text involve some implicit factors which can not be said clearly but will certainly affect people's interest of that tweet. We use idea of single value decomposition to track out the implicit factors. In this situation, we build up our matrix like each column represents one user, each row represents tweet.

3.1   Latent Factor Model

Latent factor model kind of technology which is originally used to analyze the implicit syntax of long text in the field of text mining. It can help to find the latent attributes of text. After a long time of using latent factor model, it is found that it can be used in recommender system to cluster user's record. In our tweet recommender system, we use latent factor model to find out those latent factor that will affect people's preference on tweet. To explain what is latent factor, we will take people's behavior on repost a tweet about make up as example. It is likely three users (users A, B, and C) repost the same tweet about make up while they repost it for different factors. Under this situation, user A may repost it because she is concerns about a shop selling that cosmetic on TAOBAO while user B repost it because she cares about the skills of how to use it. And user C repost it because of her favorite actress is in the advertisement of that cosmetic. Obviously, it is difficult to find out the factors that can categorize them into. However, when using latent factor model, factors are not needed to be figured out. The only requirement needed is to construct a matrix like this:

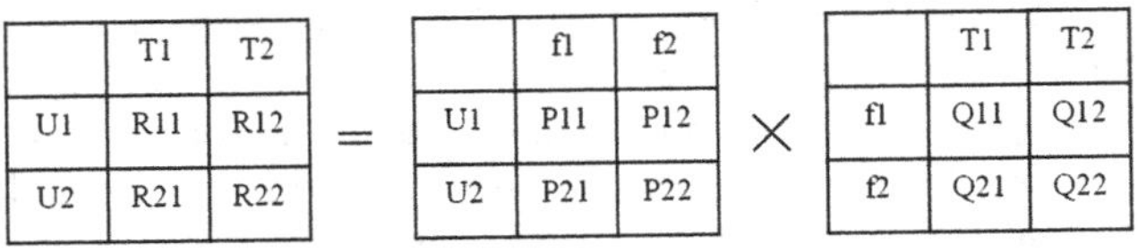

	T1	T2
U1	R11	R12
U2	R21	R22

=

	f1	f2
U1	P11	P12
U2	P21	P22

×

	T1	T2
f1	Q11	Q12
f2	Q21	Q22

Fig. 1. Example of matrix factorization model

In Fig. 1, matrix $R \in \mathbb{R}^{n \times m}$ is a matrix represents the relation between user and tweet. The value of the matrix $R_{i,j}$ represents the preference of user i towards tweet j. This value is the evidence that whether this tweet should be recommended to that user. Matrix $P \in \mathbb{R}^{n \times t}$ and matrix $Q \in \mathbb{R}^{t \times m}$ is the feature matrix for tweet and user which is gotten by decompose. The feature f in feature matrix indicates factors assumed by us which will affect user's preference. The matrix product of P and Q is what we need. The value of preference is calculate by the formula below:

$$R_{i,j} = P_i Q_j = \sum P_{i,k} Q_{k,j}. \tag{1}$$

From the formulas above, it can be found that data can cluster automatically without knowing definitely classification of factors. What's more, the specific granularity is not need to know, too. What we should do is to set a number of how many factors we need, the larger the number, the more the granularity coarse. In this model, what we do is not to classify the tweet, but to calculate the probability that the tweet belongs to a class. In the same way, we are not finding which class a user like but calculating user's favor to classes. To execute latent factor model, we need to settle down things below:

1. The initial value of matrix P and matrix Q. We use the average of words frequency together with some constraint and a random number as initial value.
2. Setting appropriate values for parameters such as the number of feature and the times of iteration in the module.

3.2   Ranking

$P_u \in R^d$ and $Q_u \in R^d$ is used as the representation that indicates user and tweet are in the latent area. Therefore, preference of user u towards tweet i can be defined as follows.

$$y_{u,i} = p_u{}^T q_i, \tag{2}$$

where y is the preference that is calculated for user u to tweet i. For a tweet k which is reposted by user and tweet h which is not reposted by user, we consider user's preference to k is larger than that to h, that is to said $y_{u,k} > y_{u,h}$. So, user's preference model is defined as

$$P(r(k) > r(h)|u) = \frac{1}{1 + e^{-(y_{u,k} - y_{u,h})}}. \tag{3}$$

In this formula, $r(k)$ and $r(h)$ is the rank of tweet k and tweet h among all the tweet. Consequently, it indicates the probability that tweet k rank higher than tweet h in the ranking of user's preference. Hence, it is assumed that people prefer tweet that has been reposted by them than that has not been reposted by them. Therefore, many pairs of data D that can indicate user's preference is need:

$$D = \{< u, k, h > | k \in Re(u), h \notin Re(u)\}. \tag{4}$$

Here k is a positive sample while h is a negative sample. $Re(u)$ is the set of tweet which has been reposted by user u. Due to the assumption that people prefer tweets that have been reposted by them best, among all the tweets can be seen by user u, each of those tweets in set Re k can make up a pair with every tweet of those that are not in set Re. Combining all the formula above, the model is finally represented in the form

$$Min \sum_{<u,k,h> \in D} \ln(1 + e^{-(y_{u,k} - y_{u,h})}) + r. \tag{5}$$

3.3 Decomposing Tweets

In previous section, the problem of sparsity is mentioned many times. However, it can not be solved by a direct way like including more data. Hence, we step up to another level which is the keywords in the tweet. Obviously, it is much likely that user will repost tweets including the same words other than repost the same tweet many times. Finally, a tweet's latent factors becomes the latent factors of some keywords which appear in the tweet.

$$y_{u,i} = p_u{}^T (\frac{1}{Z} \sum_{w \in T_i} q_w). \tag{6}$$

In this formula, p_u is the matrix of users and latent factors. q_w is the matrix of words and latent factors (w are words in tweet i).

3.4 Attributes of Tweets in Latent Factor Model

In the recommendation model, if only one dimension that users' interest to tweet is considered, matrix will become very sparse because of the fact that the number of repost tweet is very small. In order to handle this problem, we take both user, poster and tweet into consideration. Now, the problem of sparsity is solved while more probability can be explained by our model . Being the case that three dimension is considered, the latent factors between users and tweet is still most important. As tweets are already decomposed into words, the formula begin with users matrix and words matrix.

In addition to the content of tweet, the recessive factors model also considers the social relations within users. When using tweeter, an obvious phenomenon occurs, which is that the opportunity of reposting between two socially connected users will increase. More precisely, if the reader has a recessive relationship between the tweet's original publisher, then the reader is more likely to repost the tweet that originally posted by the publisher. Therefore, the potential relations between the publishers and readers are taken into account by the recessive factors model. At this point, a formula is generated

$$y_{u,i} = p_u{}^T d_{p(i)}, \qquad (7)$$

where $d_p(i)$ indicates the publisher of tweet also known as the publisher of $p(i)$ C latent factor matrix. Here, the consideration is that when user u reads about the publishers tweet, what is the probability of u reposting the tweet without taking the tweet's content into account. Such a phenomenon can be seen very often in the real life situation. For instance, during the data gathering process, it appears that one user reposted another user's entire tweets. This sample exactly confirms the reasonability of taking the recessive factors into the model's consideration.

After taking the recessive factors into the model's consideration, the model is extended as the following form:

$$\dot{y}_{u,i} = p_u{}^T \left(\frac{1}{Z} \sum_{w \in T_i} q_w + \alpha d_{p(i)} \right) \qquad (8)$$

In this formula, α is a coefficient that shows the importance of the publisher to the potential influencing factors of a tweet. The determination method to the coefficient α will be introduced in the later sections. The extended model now denotes that when considering a user u's level of interest in a particular tweet i, each keyword w and the publisher $p(i)$ of this tweet will possibly contain recessive factors that can influence user u's level of interest in this particular tweet.

The attributes of tweets' publisher is also probability that will affect the importance of tweets. It is always the situation that when an authority publish a tweet in his own area will attract more attention then other nonsense words. For example, when a famous nutritionist's tweet about how to eat diet may attract many people to read it while a joke publish by that nutritionist will cause no response to other users. The reason of this phenomenon is some people play an important role in some area will make their words about that area also important to other people. Now, the formula can be defined as

$$y_{u,i} = \frac{1}{Z} \sum_{w \in T_i} q_w{}^T d_{p(i)}, \qquad (9)$$

where q is the matrix of words and latent factors, and w is the keywords of tweet i. D is the matrix of publishers $p(i)$ and latent factors. This formula can represent users favor to a tweet when consider the publisher of the tweet. Finally, the formula is defined as:

$$y_{u,i} = p_u{}^T(\frac{1}{Z}\sum_{w\in T_i} q_w + \alpha d_{p(i)}) + \frac{1}{Z}\sum_{w\in T_i} q_w{}^T \beta d_{p(i)}. \tag{10}$$

Like the formula mentioned above, β expresses the importance of latent factors of publisher for tweet. The method to choose a correct value of β will be explain in the next subsection.

3.5 Linear Combination of Latent Factors

In this section, many attributes that will affect users' preference is introduced. To combine them, a special method is need. The method of linear combination will be introduced. In the previous section, a formula about probability of preference is mentioned. To get an optimization, the mathematics method of gradient descent is used. The value of gradient is updated by these ways:

$$-\frac{\partial L}{\partial p_u} = e(\frac{1}{z_{s+}}\sum_{s\in k} q_s - \frac{1}{z_{s-}}\sum_{s\in h} q_s + (d_{p(k)} - d_{p(h)})) - \sigma_1 p_u; \tag{11}$$

$$-\frac{\partial L}{\partial q_{w+}} = e(\frac{1}{Z_{j+}}p_u + \alpha d_{p(k)}) - \sigma_2 q_{w+}; \tag{12}$$

$$-\frac{\partial L}{\partial q_{w-}} = -e(\frac{1}{Z_{j-}}p_u + \alpha d_{p(k)}) - \sigma_2 q_{w-}; \tag{13}$$

$$-\frac{\partial L}{\partial d_{p(k)}} = \alpha e(p_u + \frac{1}{Z_{j+}}q_{w+}) - \sigma_3 d_{p(k)}; \tag{14}$$

$$-\frac{\partial L}{\partial d_{p(h)}} = \alpha e(p_u + \frac{1}{Z_{j-}}q_{w-}) - \sigma_3 d_{p(h)}; \tag{15}$$

$$-\frac{\partial L}{\partial b_j} = e(\gamma_j^+ - \gamma_j^-) - \sigma_4 b_j. \tag{16}$$

In the formula above, e represents the deviation of prediction and truth value.

$$e = 1 - P(r(k) > r(h)|u) = 1 - \frac{1}{1 + e^{-(u_{u,k} - y_{u,h})}}. \tag{17}$$

4 Experiments

In the section, we conduct experiments to evaluate the SVD. The proposed method is based on decomposing matrix to get the implicit factors of tweet. This kind of method is very convenient that it don't need to know exactly what the factors are. To make it more convincing, three intuitive methods are chosen to be baseline methods.

4.1 Data Set

In our method, we pay attention on those users whose number of followers is larger than 15. We regard this kind of users active user. To construct an environment seems like real Tweet, we choose one active, crawl his tweet text. And then crawling all his followers. After collecting those followers, we crawl those followers tweet. After repeating this two steps, we collect 2048 active users and their tweet text and followers and followings. However, some of these users' send tweet very rarely. So we get them rid of. After this steps, we only have 959 users who are very active.

4.2 Data Preprocess

To deal with the original tweet text, we first calculate the length of tweet, and get the tag and URL links out. Secondly, as most of the tweets are written by Chinese which is not structure text, we must separate the text. During the step of separation, we also exclude some words like 'is', 'yes', 'no' and some words always happens in tweet text which don't have any meaning like 'tweet', 'retweet'. After getting all the words we need, we calculate the frequency of words using TF-IDF method. In this step, we exclude some useless words choose only those words with meaning and high frequency. This step is very important in our method. Because the large number of words in tweets makes the matrix very sparse which will affect the effect of SVD. The approach we make to choose words will be declared clearly now. In order to the study of explicit factors in the text, we found that people will pay most attention on nouns and adjectives. So we only keep nouns and adjectives in our data set. What's more, it is clearly that words with higher words frequency value more than words with lower frequency. However, if the a word rarely exists when considering all the text of all the users while it appears a lot of times in a specific user's tweet, this word is an important key word for this user. So, we can't exclude words with very low frequency.

In our method, we consider one tweet which has been reposted by one user as a positive example of that user. And one tweet sent by the following of user which has not been reposted by the user is treated as negative example. It's clear that the number of negative example is much larger than the number of positive example in original data set. That kind of sparsity will make our experiment very hard. In the mean while, people's behavior that they don't repost one tweet do not mean that they are not interested in this tweet. However, we can only know that they are interested in those tweet they repost but we could not find out those tweet they are interested in and they read carefully but not repost. To make our data set much more like the original tweet they will see ranked by time, which may contain 20 % tweet which user will be interested in, we put one positive example together with four negative example in our data set. After doing that, we get a data set with 20 percents of positive example and rank by time for each user. In the data set, there are average 2500 tweet under one user. Finally, we separate the data set into training set and data set proportionally.

4.3 Metrics

When considering the accuracy of recommendation, we tab a tweet as 1 if it is reposted by a user which means it is recommended successfully while tab it as 0 if it is not reposted by the user. Then, mean average precision method is used to value the result of experiment. Using this evaluation can not only consider the percentage of successfully recommended tweets but also take the rank of successfully recommended tweets into account.

$$AP = \frac{\sum_{n-1}^{R} P@n \times retweet(n)}{|R|}, \tag{18}$$

where N is the number of tweets which are reposted by users in test set. $retweet(n)$ is a boolean function. $P@n$ is the accuracy of top n recommendation. After calculating AP for all the users, we can get MAP.

$$MAP = \frac{AP}{N}. \tag{19}$$

4.4 Comparison

In this section, we will qualitatively evaluate the method proposed with baseline method. We have compared our method with several others baseline method. The approaches are list below: Timeline: Tweets are ranked by chronological order without sorting by algorithms. This is the original situation which users facing. Repost times: Repost times is an sign of popularity of tweets. However, this method can not consider users' personality. Similarity: Similarity method is a method which considering the relation of users' favor for a tweet with users' historical tweets.

The same data set which is crawl from Sina Tweet will be used to conduct the proposed method and baseline methods. The results of these four method are show in Table 1.

Table 1. MAP results of all methods

Method	Timeline	Retweet times	Similarity	SVD
MAP	0.163	0.16	0.116	0.304

Collaborative Filtering Model: When setting the number of latent factor to 8, maximum MAP is got. For some of the user, the accuracy of recommendation can be as high as 100 %. Most of the users' accuracy is around 30 % to 60 %. Furthermore, by selecting some users who are extremely active, the accuracy of Collaborative Filtering model can be even higher.

Baseline: Comparing with baseline method, it is clearly that Collaborative Filtering model is accurate. MAP of Collaborative Filtering is nearly twice higher than baseline method. Among three baseline method, repost times method shows lowest effect. It's not hard to explain that. In most situation, a tweet with very high repost times is always tweet sent by famous people or official site. It's like a advertisement instead of normal tweet. Though it can be seen by many users, most of them will not repost it. Therefore, if recommending tweet with highest repost times, it will not meet users' personality need. Similarity method is also not accuracy enough. Through observing the result of similarity method, we found that among all the tweet can be seen by user there only one or two tweet have similarity with user's historical tweet. And tweet with similarity is always not the tweet user will repost. Consequently, the accuracy of similarity method will not be much higher than timeline method. In the timeline method, tweets are ranked by timeline. Since the data set is constructed as one positive example with four negative example, the accuracy of timeline method will be around 20 %.

4.5 Influence of Parameters

In this section, we will discuss the influence of parameters in Collaborative Filtering Model. There are three parameters very important when we deducting the model. They are the number of latent factor, epoch and the number of iteration using in gradient descent. It is illustrated that the larger the number of latent factor, the coarse the granularity. However, it is not the situation that the larger number of latent factor will make our recommendation more accurate. After deducting many times of Collaborative Filtering Model with different number of latent factors, we found that eight latent factors will make the result most accurate.

5 Conclusions

In social media, users are typically overwhelmed with the large amount of information posted by their social friends and miss the content that they may be interested in. Recommender system is an effective technique to filter information for users based on their interests. As the behavior of retweeting is a significant indicator to reflect that a user is much interest in a tweet, in the paper, we study the issue of personalized retweeting prediction to understand users' behaviors in social network. To handle this issue, we propose a matrix factorization model which takes the relevance of users' interests, tweets' content, and publishers' influence into account simultaneously. Experiments are conducted to evaluate our method and show the advantages of the proposed method qualitatively and quantitatively.

As loss function (i.e., object function in optimizing) is an important component in matrix factorization models, in the future, we will explore the different performance of loss functions in recommender systems in social media. Furthermore, we will conduct more extensive experiments to better understand the behaviors of users in social network.

Acknowledgements. This work is supported by National Natural Science Foundation of China (Grant NO. 61300137), the Guangdong Natural Science Foundation, China (NO. S2013010013836), Science and Technology Planning Project of Guangdong Province, China (NO. 2013B010406004), the Fundamental Research Funds for the Central Universities, SCUT(NO. 2014ZZ0035).

References

1. Chen, K., Chen, T., Zheng, G., Jin, O., Yao, E., Yu, Y.: Collaborative personalized tweet recommendation. In: Proceedings of the 35th International ACM SIGIR Conference on Research and Development in Information Retrieval, pp. 661–670. ACM (2012)
2. Hong, L., Doumith, A.S., Davison, B.D.: Co-factorization machines: modeling user interests and predicting individual decisions in twitter. In: Proceedings of the Sixth ACM International Conference on Web Search and Data Mining, pp. 557–566. ACM (2013)
3. Adomavicius, G., Tuzhilin, A.: Toward the next generation of recommender systems: a survey of the state-of-the-art and possible extensions. IEEE Trans. Knowl. Data Eng. **17**(6), 734–749 (2005)
4. Sun, A.R., Cheng, J., Zeng, D.D.: A novel recommendation framework for microblogging based on information diffusion. In: 19th Annual Workshop on Information Technolgies and Systems, (WITS 2009) (2009)
5. Chen, J., Nairn, R., Nelson, L., Bernstein, M., Chi, E.: Short and tweet: experiments on recommending content from information streams. In: Proceedings of the SIGCHI Conference on Human Factors in Computing Systems, pp. 1185–1194. ACM (2010)
6. Duan, Y., Jiang, L., Qin, T., Zhou, M., Shum, H.Y.: An empirical study on learning to rank of tweets. In: Proceedings of the 23rd International Conference on Computational Linguistics, Association for Computational Linguistics, pp. 295–303 (2010)
7. Naveed, N., Gottron, T., Kunegis, J., Alhadi, A.C.: Bad news travel fast: A content-based analysis of interestingness on twitter. In: Proceedings of the 3rd International Web Science Conference, p. 8. ACM (2011)
8. Yang, Z., Guo, J., Cai, K., Tang, J., Li, J., Zhang, L., Su, Z.: Understanding retweeting behaviors in social networks. In: Proceedings of the 19th ACM International Conference on Information and Knowledge Management, pp. 1633–1636. ACM (2010)
9. Kim, Y., Shim, K.: Twitobi: A recommendation system for twitter using probabilistic modeling. In: 2011 IEEE 11th International Conference on Data Mining (ICDM), pp. 340–349. IEEE (2011)
10. Lin, C.X., Mei, Q., Han, J., Jiang, Y., Danilevsky, M.: The joint inference of topic diffusion and evolution in social communities. In: 2011 IEEE 11th International Conference on Data Mining (ICDM), pp. 378–387. IEEE (2011)
11. Koren, Y.: Factorization meets the neighborhood: A multifaceted collaborative filtering model. In: Proceedings of the 14th ACM SIGKDD International Conference on Knowledge Discovery and Data Mining, pp. 426–434. ACM (2008)
12. Su, X., Khoshgoftaar, T.M.: A survey of collaborative filtering techniques. Adv. Artif. Intell. **2009**, 4 (2009)

13. Schafer, J.B., Frankowski, D., Herlocker, J., Sen, S.: Collaborative filtering recommender systems. In: Brusilovsky, P., Kobsa, A., Nejdl, W. (eds.) Adaptive Web 2007. LNCS, vol. 4321, pp. 291–324. Springer, Heidelberg (2007)
14. Sarwar, B., Karypis, G., Konstan, J., Riedl, J.: Item-based collaborative filtering recommendation algorithms. In: Proceedings of the 10th International Conference on World Wide Web, pp. 285–295. ACM (2001)
15. Shi, Y., Larson, M., Hanjalic, A.: Exploiting user similarity based on rated-item pools for improved user-based collaborative filtering. In: Proceedings of the Third ACM Conference on Recommender Systems, pp. 285–295. ACM (2009)
16. Rendle, S.: Factorization machines with libFM. ACM Trans. Intell. Syst. Technol. (TIST) 3(3), 57 (2012)
17. Kywe, S.M., Lim, E.-P., Zhu, F.: A survey of recommender systems in twitter. In: Aberer, K., Flache, A., Jager, W., Liu, L., Tang, J., Guéret, C. (eds.) SocInfo 2012. LNCS, vol. 7710, pp. 420–433. Springer, Heidelberg (2012)

Finding Paraphrase Facts
Based on Coordinate Relationships

Meng Zhao[✉], Hiroaki Ohshima, and Katsumi Tanaka

Graduate School of Informatics, Kyoto University,
Yoshida Honmachi, Kyoto 606–8501, Japan
{zhao,ohshima,tanaka}@dl.kuis.kyoto-u.ac.jp

Abstract. We propose a method to acquire paraphrases from the Web in accordance with a given sentence. For example, consider an input sentence "Lemon is a high vitamin c fruit". Its paraphrases are expressions or sentences that convey the same meaning but are different syntactically, such as "Lemons are rich in vitamin c", or "Lemons contain a lot of vitamin c". We aim at finding sentence-level paraphrases from the noisy Web, instead of domain-specific corpora. By observing search results of paraphrases, users are able to estimate the likelihood of the sentence as a fact. We evaluate the proposed method on five distinct semantic relations. Experiments show our average precision is 60.5 %, compared to TE/ASE method with average precision of 44.15 %. Besides, we can acquire 3 paraphrases more than TE/ASE method per input.

Keywords: Paraphrase acquisition · Coordinate relationship · Web mining · Mutual reinforcement

1 Introduction

Nowadays, it is intuitive to utilize the Web as a huge encyclopedia and trust information on the Web. However, those information is not always correct or true. For example, it has been reported that information on the Wikipedia, which is regarded as the biggest online encyclopedia, is not so credible [9]. Therefore, it is necessary to understand risks of Web information and distinguish facts from it. We assume information, which is often mentioned by people on the Web, is more likely to be correct or true. Consequently, such information is regarded as "fact" with a high possibility. On the contrary, we assume information, which is rarely mentioned by people on the Web, is more likely to be incorrect or untrue, consequently unlikely to be "fact". Based on the assumption, a naive way to estimate the likelihood of a sentence as a fact is to observe its hit count on the Web. However, it always fails since the expression of a user-input sentence may be rarely used on the Web. Suppose a user wants to know whether lemon is a high vitamin c fruit or not. He thinks of a sentence like "Lemon is a high vitamin c fruit" and use it as a query to search on the Web. Neither Google[1] nor

---

[1] http://www.google.com.

© Springer International Publishing Switzerland 2015
A. Liu et al. (Eds.): DASFAA 2015 Workshops, LNCS 9052, pp. 135–151, 2015.
DOI: 10.1007/978-3-319-22324-7_12

Bing[2] return any matches for this query (at the time of writing the document). However, if it is rewritten as "Lemons are rich in vitamin c", or "Lemons contain a lot of vitamin c", adequate number of Web pages can be obtained. Hence, the user can infer that lemon is a high vitamin c fruit. In this paper, we aim at finding sentence-level paraphrases from the noisy Web, instead of domain-specific corpora. By observing search results of paraphrases, especially frequently-used ones, users are able to estimate the likelihood of an input sentence as a fact.

Paraphrases are linguistic expressions that restate the same meaning using different variations. In the most extreme case, they may not be even similar in wording. It has been shown that paraphrases are useful in many applications. For example, paraphrases can help detect fragments of text that convey the same meaning across documents and this can improve the precision of multi-document summarization [6,17]. In the field of machine translation, [8,15,16] show that augmenting the training data with paraphrases generated by pivoting through other languages can alleviate the vocabulary coverage problem. In information extraction, [7,10,26] present approaches incorporating paraphrases to extract semantic relations among entities. In information retrieval, paraphrases have been used for query expansion [2,20,25]. A large proportion of previous work extract and generate paraphrases based on parallel corpora [3,5] or comparable corpora [4,21,23]. However, there are limitations in using those corpora. For example, the quality of obtained paraphrases strongly depends on the quality of the corpus, a high-quality corpus can cost a great deal of manpower and time to construct. Moreover, it may be hard to cover all possible genres. For example, [23] uses a corpus consisted of newswire articles written by six different news agencies.

Entity tuples that describe or are members of the same relationships may be defined as "coordinate tuples" to each other. For example, *(guavas,vitamin c)* and *(tomatoes,potassium)* are coordinate tuples since there is a **highConcentration** relation between *guavas* and *vitamin c*, so is between *tomatoes* and *potassium*. We think it is not easy to find all variations of paraphrases by just one entity tuple, and such variations exist in expressions of its coordinate ones. For example, given the sentence "Guavas are rich in vitamin c", it might be difficult to find part of its paraphrases, such as "Guavas are considered a high vitamin c fruit", since it is seldom used by the entity tuple *(guavas,vitamin c)*. However, such paraphrases can be acquired from the expressions of its coordinate entity tuples, i.e. the former paraphrase can be easily found via *(tomatoes,potassium)*. Thus, we can capture more paraphrases by mining the expressions of coordinate entity tuples.

The distributional hypothesis, attributed to Harris [12], has been the basis for Statistical Semantics. It states that words that occur in the same contexts tend to have similar meanings. Moreover, its extension that if two phrases, or two text units, occur in similar contexts then they may be interchangeable has been extensively tested. Our idea is based on the extended hypothesis: if two templates share more common coordinate entity tuples, then they may be

---

[2] http://www.bing.com.

paraphrase templates; if two entity tuples share more common paraphrase templates, then they may be coordinate entity tuples. Thus, paraphrase templates and coordinate tuples are in a mutually reinforcing relationship, and this relationship can be used to find more paraphrase templates or coordinate tuples.

We assume a sentence is mapped to a template and an entity tuple. Thus, given a sentence query, it can be separated into a template and a corresponding entity tuple. The proposed method first extracts templates that connect that entity tuple and entity tuples mentioned by that template. Several filters and limitations are added to eliminate partial inappropriate templates and entity tuples. A mutually reinforcing approach is proposed to simultaneously identify different templates that convey the same meaning with the given template, and entity tuples which hold the same relation with the given entity tuple. Finally, paraphrase queries can be generated by substituting the given entity tuple into discovered paraphrase templates.

Our contributions can be summarized as follows. First, we propose a method for detecting sentence-level paraphrases and our method does not require deep natural language processing such as dependency parsing. Second, paraphrases are not limited to word-level, or phrase-level. Given a sentence query, our method outputs its paraphrases at the sentence level. Third, instead of using high-quality input data restricted to a particular genre, our method can employ the Web as its data source.

The remainder of the paper is organized as follows. In Sect. 2, we discuss some related work. Section 3 shows some preliminaries and Sect. 4 describes our basic idea. In Sect. 5, we illustrate the method to acquire paraphrases from the Web by a given sentence. We evaluate the proposed paraphrase acquisition method using five semantic relations in Sect. 6. Finally, Sect. 7 concludes the paper and gives an outline of our future work.

2   Related Work

2.1   Semantic Relation Extraction

Snowball [1], KnowItAll [11], TextRunner [26] are famous information extraction systems. All of them extract valuable information from plain-text documents by using lexical-syntactic patterns. Snowball and TextRunner require a handful of training examples from users, while KnowItAll emphasizes its distinctive ability to extract information without any hand-labeled training examples.

In Snowball, given a handful of example tuples, such as organization-location tuple $<o,l>$, Snowball finds segments of text in the document collection where o and l occur close to each other, and analyzes the text that "connects" o and l to generate patterns. It extracts different relationships from the Web by the bootstrap method. Besides, Snowball's patterns include named-entity tags. An example is $<ORGANIZATION>$'s headquarters in $<LOCATION>$. $<ORGANIZATION>$ will only match a string identified by a POS tagger as an entity of type $ORGANIZATION$. So does $<LOCATION>$.

In KnowItAll, its input is a set of predicates that represent classes or relationships of interest. A generic representation of rule templates for binary predicates is *relation(Class1, Class2)*. For example, the predicate *CeoOf(PERSON, COMPANY)* corresponds to the pattern *<PERSON> is the CEO of <COMPANY>*. It learns effective patterns to extract relevant entity names.

In TextRunner, extractions take the form of a tuple $t = (e_i, r_{i,j}, e_j)$, where e_i and e_j are strings meant to denote entities, and $r_{i,j}$ is a string meant to denote a relationship between them. A deep linguistic parser is deployed to obtain dependency graph representations by parsing thousand of sentences. For each pair of noun phrases (e_i, e_j), TextRunner traverses the dependency graph, especially the part connecting e_i and e_j, to find a sequence of words that composes a potential relation $r_{i,j}$ in the tuple t.

2.2   Paraphrase Acquisition

Paraphrase acquisition is a task of acquiring paraphrases of a given text fragment. Some approaches have been proposed for acquiring paraphrases at word, or phrasal level. However, these techniques are designed only suitable for specific types of resources. Both [22] and [24] acquire paraphrases from news article. In [22], Shinyama et al. considered that news articles reported the same event of the same day by different news agents can contain paraphrases. Thus, they proposed an automatic paraphrase acquisition approach based on the assumption that named entities are preserved across paraphrases. Pairs of similar sentences whose similarity is above a certain threshold are chosen. For any pair, if the two sentences share the same number of comparable named entities, then patterns in the two sentences are linked as paraphrases. In [24], news article headlines,which are already grouped by news aggregators such as Google News, are taken for further processing. *k*-means clustering and pairwise similarity are applied to find paraphrases, respectively. These work has explicit access to, and relies strongly on clues such as the news articles that describe the same event.

To acquire paraphrases, some works proposed methods based on deep natural language processing, i.e. dependency parsing. Lin and Pantel introduced an unsupervised method to discover inference rules from text in [14]. Inference rules include not only exact paraphrases, but also related and potentially useful expressions. Their core idea is also based on an extension to the distributional hypothesis: if two paths in dependency trees tend to occur in similar contexts, the meanings of the paths tend to be similar. The words that fill the slots of a path is regarded as a context for the path. Idan et al. [13] took a verb lexicon as the input and for each verb searches the Web for related syntactic entailment templates. Although they did not use the term "coordinate", they used a similar concept called "anchors" referred to lexical elements describing the context of a sentence. Different from our method, they first extract promising anchor sets for the verb lexicon, then extract templates (dependency parse-tree fragments) for which an entailment relation holds with the verb lexicon from sentences containing the promising anchor sets.

Paşca and Dienes proposed a method differed from previous ones in [19]. They use inherently noisy, unreliable Web documents rather than clean, formatted documents so that the paraphrases are not limited to a specific domain or a narrow class. Their proposed method is based on the assumption that if two sentence fragments have common word sequences at both extremities, then the variable word sequences in the middle are potential paraphrases of each other. So actually, their acquired paraphrases are almost word-, or phrase-level ones, while our work aims to get sentential paraphrases.

In [25], Yamamoto and Tanaka also concentrated on improving search results responded by sentence queries. Unlike we focus on paraphrases, they generally collected several types of sentence substitutions, including paraphrases, generalized sentences, detailed sentences and comparative sentences. Based on the criteria that sentence substitutions which appears frequently on the Web and whose context is similar to that of the input sentence query should be ranked higher, a ranking algorithm is also stated.

3   Preliminaries

We assume a sentence consists of a template and an entity tuple. Thus, given a sentence, it can be separated into a template and a corresponding entity tuple. For example, "Google has purchased Nest Labs" consists of the template X *has purchased* Y and the entity tuple *(Google,Nest Labs)*. For further illustration, we borrow the idea about the definition of a relation in [7]. They advocated a relation can be expressed extensionally by enumerating all the instances of that relation. Take the **acquisition** relation[3] for example. An extensional definition of **acquisition** is a set of all pairs of two companies in which one company acquired another, i.e. *(Google,Nest Labs)*, *(Adobe Systems,Macromedia)*. In this paper, entity tuples hold the same relation are defined to be "coordinated" to each other. For simplicity, relations are all binary relations. Thus, in the former example, *(Adobe Systems,Macromedia)* is a coordinate entity tuple of *(Google,Nest Labs)*. Bollegala et al. [7] also introduced an intensional definition of a relation by listing all the paraphrases of that relation. Therefore, finding paraphrases of a template can also be regarded as a way to survey a certain relation. Terminologies used in this paper are listed Table 1.

Let T be the set of all possible templates in the world, E be the set of all possible entity tuples in the world. Three predicates are defined as follows:

fact(e,t). It returns *true* when the statement of sentence mapped by e and t is actually the case or has really occurred, where $e \in E$, $t \in T$. If $fact$ holds for a certain pair of an entity tuple and a template, we call the entity pair is "suitable" for the template and vice-versa.

---

[3] The **acquisition** relation exists between two companies such that one company acquired another.

Table 1. Terminologies

Sentence	Google has purchased Nest Labs
Entity tuple	*(Google,Nest Labs)*
Substitution	**X**=*Google*, **Y**=*Nest Labs*
Template	**X** has purchased **Y**.
Paraphrase templates	**X** buys **Y**, **X** has acquired **Y**, **X** finalizes acquisition of **Y**
Paraphrases	Google buys Nest Labs
	Google has acquired Nest Labs
	Google finalizes acquisition of Nest Labs
Coordinate entity tuples	*(Microsoft,Nokia),(Yahoo,Tumblr),(Amazon,Goodreads)*

$para(t_i, t_j)$. It returns *true* when template t_i and template t_j both convey the same meaning ("paraphrases"), where $t_i, t_j \in T$.

$coord(e_k, e_g)$. It returns *true* when entity tuple e_k and entity tuple e_g hold the coordinate relation, where $e_k, e_g \in E$.

4   Basic Idea

In this paper, in order to find the paraphrases of a sentence query, we aim to find pairs of templates t_i and t_j and coordinates e_k and e_g that make the predicate $para(t_i, t_j)$ and $coord(e_k, e_g)$ are true.

In the ideal world, two templates are paraphrases if every entity tuple that is suitable for one templates is also suitable for the other template. Formally, let $t_i, t_j \in T$, and $E_{t_i} = \{e | fact(e, t_i)\}$, $E_{t_j} = \{e | fact(e, t_j)\}$. If $E_{t_i} = E_{t_j}$, then $para(t_i, t_j) = true$.

Similarly, two entity tuples are coordinates if every template that is suitable for one tuple is also suitable for the other tuple. Formally, let $e_k, e_g \in E$, and $T_{e_k} = \{t | fact(e_k, t)\}$, $T_{e_g} = \{t | fact(e_g, t)\}$. If $T_{e_k} = T_{e_g}$, then $coord(e_k, e_g) = true$.

However, even in the ideal world, we can easily find a counterexample to the above discussion of **para**. Suppose t_i is **X** *and* **Y**, and t_j is **X** *or* **Y**. This is an extreme case where both t_i and t_j are very general templates suitable for almost all entity tuples. Consequently, E_{t_i} might be equal to E_{t_j} so that **X** *and* **Y** and **X** *or* **Y** are misjudged as paraphrases. One may add the following condition to exclude such noisy entity tuples: if $E_{t_i} = E_{t_j}$ and $\forall (e_k, e_g) \in E_{t_i} \times E_{t_i}, coord(e_k, e_g)$, then $para(t_i, t_j)$ is *true*. Soon we find another problem that the newly added condition is too strict and will likely miss many paraphrases. A single template may represent several relations. For example, **X** *direct* **Y** may be

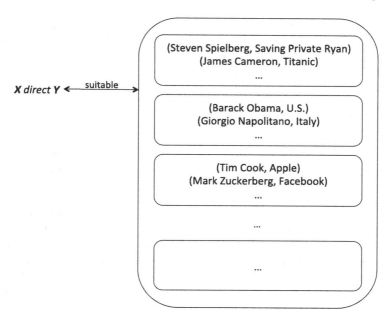

Fig. 1. An example for grouped entity tuples. Entity tuples in big frame are those suitable for the template X *direct* Y, whereas entity tuples in small frame are those held the same relation.

interpreted as the **directorOf** relation[4], the **leaderOf** relation[5], or the **ceoOf** relation[6]. As a result, the entity tuples suitable for X *direct* Y are naturally grouped in accordance with the relation held by each tuple, shown in Fig. 1.

Hence, we moderate the conditions for **para** as follows:

- If $\exists E' \subset E_{t_i} \cap E_{t_j}$ and $|E'| > \alpha$, $\forall (e_k, e_g) \in E' \times E'$, $coord(e_k, e_g)$, then $para(t_i, t_j)$ is *true*.

Here α is a threshold.

Let us look at a single entity tuple. It is easy to find different relationships between the entities of the tuple. Take *(Mark Zuckerberg, Facebook)* as an example. There exists the **founderOf** relation[7] between *Mark Zuckerberg* and *Facebook*. There also exists the **ceoOf** relation between *Mark Zuckerberg* and *Facebook*. Based on our discussion that a relation can be expressed by listing all

---

[4] The **directorOf** relation exists between a director and his works, i.e. *(Steven Spielberg, Saving Private Ryan)*, *(James Cameron, Titanic)*.

[5] The **leaderOf** relation exists between a country and its current leader, i.e. *(Barack Obama, U.S.)*, *(Giorgio Napolitano, Italy)*.

[6] The **ceoOf** relation exists between a company and the chief executive officer of that company, i.e. *(Tim Cook, Apple)*, *(Mark Zuckerberg, Facebook)*.

[7] The **founderOf** relation exists between a person and his founded company, i.e. *(Larry Page, Google)*.

the paraphrases of that relation, we can see the similar phenomenon occurs that the templates suitable for *(Mark Zuckerberg,Facebook)* are naturally grouped in accordance with different relations. Following this, we modify conditions for **coord** as:

- **If** $\exists T' \subset T_{e_k} \cap T_{e_g}$ and $|T'| > \beta$, $\forall(t_i, t_j) \in T' \times T'$, $para(t_i, t_j)$, then $coord(e_k, e_g)$ **is** *true*.

Here β is a threshold.

However, in the real world, it is difficult to find all paraphrases by a single entity tuple perhaps because of idiomatic expressions and personal preferences. For example, consider the sentence "Guavas are rich in vitamin c", where the entity tuple is *(guavas,vitamin c)*, the template is *X are rich in Y*. It might be difficult to find some of its paraphrases, such as *X are considered a high Y fruit*, or *X pop a powerful Y punch*, since people seldom use those expressions to describe the relation between guavas and vitamin c. Similarly, it is difficult to find all coordinate entity tuples by a single template, since the template might be specially used with a subset of entity tuples. Hence, we cannot find the exactly equal sets of entity tuples when considering the value of $para(t_i, t_j)$, and the exactly equal sets of templates when considering the value of $coord(e_k, e_g)$.

In Fig. 2(a), there are two templates t_1 and t_2. Under each template, there is a set of all suitable entity tuples shown in a big oval. Besides, the tuples are further grouped according to the relations they hold, shown in a small oval. If e_3 is coordinated to e_4, then we think they are interchangeable, meaning $\{e_1, e_2, e_4\} = \{e_1, e_2, e_3\}$. In addition, since e_7 is coordinated to e_5, e_6 under the same relation, we think people always use the expression of t_2 to describe e_7 but seldom use the expression of t_1. Therefore, although the sizes of two subsets are different, $\{e_5, e_6\}$ is regarded as equal to $\{e_5, e_6, e_7\}$. Finally, if all pairs of subsets are "equal", t_1 and t_2 are paraphrases, meaning $para(t_1, t_2) = true$. Similarly, in Fig. 2(b), there are two entity tuples e_1 and e_2. Under each tuple, there list all suitable templates in big oval. Besides, they are grouped according to different relations, shown in small oval. If t_3 is paraphrased to t_4, then we think they are interchangeable, meaning $\{t_1, t_2, t_4\} = \{t_1, t_2, t_3\}$. In addition, since t_7 is paraphrased to t_5, t_6 under the same relation, we think people always use the expression of t_7 to describe e_2 but seldom use it to describe e_1. Therefore, although the sizes of two subsets are different, $\{t_5, t_6\}$ is regarded as equal to $\{t_5, t_6, t_7\}$. Finally, if all pairs of subsets are "equal", e_1 and e_2 are coordinate entity tuples, meaning $coord(e_1, e_2) = true$.

5   Our Method

In this paper, the problem to be solved is as follows: given a sentence, its paraphrases are automatically acquired from the Web, and they are ranked in accordance with paraphrase degree. We have stated our basic idea in Sect. 4 that paraphrase relationship and coordinate relationship interdepend and mutually

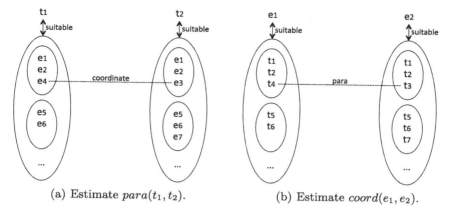

(a) Estimate $para(t_1, t_2)$.   (b) Estimate $coord(e_1, e_2)$.

Fig. 2. A real-world situation

reinforce each other. Hence, at the very beginning, it is necessary to gather templates and entity tuples. Brief introductions of template extraction and entity tuple extraction are given in Sects. 5.1 and 5.2, respectively. Then details of our method are addressed in Sect. 5.3.

5.1   Template Extraction

As we mentioned in Sect. 1, we use the Web as our data source, so we search the Web and extract templates from it. Suppose a given sentence is s which consists of a template t and an entity tuple e. t is actually made by replacing two entities in e respectively with two variables $\mathbf{X}$ and $\mathbf{Y}$ in the sentence s. An example is shown in Table 1. The entity tuple is *(Google,Nest Labs)*. We replace *Goolge* with variable $\mathbf{X}$ and *Nest Labs* with variable $\mathbf{Y}$ and get the template $\mathbf{X}$ *has purchased* $\mathbf{Y}$. An AND query generated from e is issued to the Web, i.e. "Google AND Nest Labs". We gather templates from the top N search results of the query[8] that satisfy the following conditions.

(1) A template must contain exactly one occurrence of each $\mathbf{X}$ and $\mathbf{Y}$ (i.e. exactly one $\mathbf{X}$ and one $\mathbf{Y}$ must exist in a template).
(2) The maximum length of a template is L_{max} times of that of s.
(3) The minimum length of a template is L_{min} times of that of s.
(4) Information such as date, money, quantity, are removed if s doesn't contain such information.
(5) Templates must be consistent of s (if s is a question, gathered templates must limit to questions; if s is a declarative sentence, gathered templates must also be declarative ones).

The values of parameters N, L_{max} and L_{min} are set experimentally, as explained later in Sect. 6. The proposed template extraction algorithm takes all the words

---

[8] Replace entities in e with variables.

in a sentence into account, and is not limited to extract templates only from the portion of a sentence that appears between two entities. Besides, we assume an overlong template is more likely to contain additional information, while a too-short template is more likely to miss some information. Both the situations lead to non-paraphrases. Therefore, we consider two length limitations to exclude some inappropriate templates in advance and reduce the number of templates gathered from the Web. The consideration of (4), (5) is because of similar reasons.

5.2   Entity Tuple Extraction

As we mentioned in Sect. 1, we use the Web as our data source, so we search the Web and extract entity tuples from it. Suppose a given sentence is s which consists of a template t and an entity tuple e. Still use the example presented in Table 1. We first search coordinate terms of two entities in e, respectively, using the bi-directional lexico-syntactic pattern-based algorithm [18]. For example, we get *Yahoo*, *Microsoft*, *Apple* and etc. as coordinate terms of *Google*; *Samsung*, *Dropcam* and etc. as coordinate terms of *Nest Labs*. Next, we issue wildcard queries generated by t and either of the two entities in e or their coordinate terms to the Web and extract the other ones from the top M search results. To detect entities in sentences, we run a POS tagger[9] and only annotate sentences exactly contained the queries with POS tags. Then nouns or noun phrases are selected out. For example, queries, such as "Google has purchased *", or "Yahoo has purchased *", are formed to extract corresponding companions. As a result, entity tuples like *(Google, YouTube)*, or *(Google, Titan Aerospace)* are extracted by the former query, entity tuples like *(Yahoo, Tumblr)*, or *(Yahoo, Blink)* are extracted by the latter query.

We use coordinate terms for the following two reasons. First, there is too massive information on the Web. If we only search by t (i.e. "* has purchased *") and extract entity tuples from corresponding portions of sentences, many irrelevant tuples are gathered, such as *(God, freedom)*. Hence, coordinate terms are used to reduce the number of irrelevant tuples. Second, there might be few entity tuples extracted from the Web if the binary relation in e is one-to-one type. For example, in sentence "The capital of Japan is Tokyo", relation between *Japan* and *Tokyo* belongs to one-to-one type, since we can only find *Tokyo* as the answer for which city the capital of *Japan* is, and vise versa, we can only find *Japan* as the answer for *Tokyo* is the capital of which country. Thus, it is difficult to get other entity tuples from wildcard query "The capital of * is Tokyo" or "The capital of Japan is *". In this case, coordinate terms are used to increase the number of entity tuples extracted from the Web.

5.3   The Mutual Reinforcement Algorithm

Assuming that the set of all extracted templates is T, and the set of all extracted entity tuples is E. Suppose there are m templates in T and n entity tuples in E.

---

[9] http://nlp.stanford.edu/software/tagger.shtml.

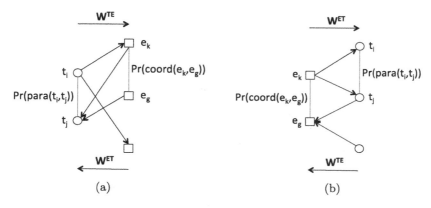

Fig. 3. An example of the mutual reinforcement between $Pr(para(t_i, t_j))$ and $Pr(coord(e_k, e_g))$.

Let $W^{TE} \in \mathbb{R}^{m \times n}$ denote the transition matrix from T to E, whose entry w_{ij}^{te} is the proportion of e_j's occurrence in t_i's top search results. Let $W^{ET} \in \mathbb{R}^{n \times m}$ denote the transition matrix from E to T, whose entry w_{ij}^{et} is the proportion of t_j's occurrence in e_i's top search results.

Since we want to know the quality of a paraphrase rather than treat all paraphrases equally, we introduce paraphrase degree between two templates t_i and t_j as $Pr(para(t_i, t_j))$, which returns a value between 0 and 1. A high value will be returned when t_i and t_j are more likely to be paraphrased to each other. Similarly, we introduce coordinate degree between two entity tuples e_i and e_j as $Pr(coord(e_i, e_j))$, which returns a value between 0 and 1. A high value will be returned when e_i and e_j are more likely to be coordinated to each other.

As we mentioned in Sect. 4, if two templates are paraphrased to each other, they are interchangeable; if two coordinate entity tuples are coordinated to each other, they are interchangeable. In Fig. 3(a), it shows two different situations to consider the paraphrase degree between t_i and t_j. One is exactly equivalence of t_i's suitable entity tuples and t_j's suitable entity tuples, such as e_k. If we can find many such entity tuples, the paraphrase degree between t_i and t_j is high. Another is interchangeability of t_i's suitable entity tuples and t_j's suitable entity tuples, i.e. e_k and e_g are interchangeable with the degree of $Pr(coord(e_k, e_g))$. As a result, the value of $Pr(coord(e_k, e_g))$ is propagated to $Pr(para(t_i, t_j))$ according to the transition probability. Similarly, additional values are propagated from other paris of coordinate entity tuples in E to $Pr(para(t_i, t_j))$, then the value of $Pr(para(t_i, t_j))$ is updated. In Fig. 3(b), it shows the new value is propagated to $Pr(coord(e_k, e_g))$.

Formally, the mutually reinforcing calculations are written as:

$$Pr(para(t_i, t_j)) = \frac{1}{2}(\sum_{e_k, e_g \in E} w_{ik}^{te} w_{gj}^{et} Pr(coord(e_k, e_g)) +$$

$$\sum_{e_k, e_g \in E} w_{jg}^{te} w_{ki}^{et} Pr(coord(e_k, e_g)))$$

$$Pr(coord(e_k, e_g)) = \frac{1}{2}(\sum_{t_i, t_j \in T} w_{ki}^{et} w_{jg}^{te} Pr(para(t_i, t_j)) +$$

$$\sum_{t_i, t_j \in T} w_{gj}^{et} w_{ik}^{te} Pr(para(t_i, t_j)))$$

where $i, j \in [1, m]$, $k, g \in [1, n]$. Especially, when $i = j$, $Pr(para(t_i, t_j)) = 1$, which indicates the exactly equal case. Similarly, when $k = g$, $Pr(coord(e_k, e_g)) = 1$. After values for all pairs of templates are updated, a normalization is taken place. The same for all pairs of entity tuples. Besides, update continues until difference between each new value and old value is smaller than a threshold θ. As a result, the paraphrase degree of two templates will be high if they share many common entity tuples, or have many interchangeable tuples; the coordinate degree of two entity tuples will be high if they share many common templates, or have many interchangeable templates. Finally, we get paraphrases of the given sentence by substituting its entity tuples into discovered paraphrase templates.

6   Evaluation

6.1   Experimental Setting

In this section, we introduce experiments to validate the main claims of the paper.

Given a sentence, it is costly to find all templates and all entity tuples through the whole Web. For our experiments, we set N as 1000, viz. we limit data to the top 1000 search results obtained from Bing Search API[10] for each AND query formed by an entity tuple. Besides, to exclude overlong or too-short templates extracted from the Web, we set $L_{max} = 2$, $L_{min} = 0.5$. We set M as 250, viz. we extract entity tuples by a wildcard query in its top 250 search results. Moreover, since the calculation of W_{TE} requires many accesses to the Web, we only consider 40 most frequently occurring templates. We fix the value of threshold θ to 0.0001 and find values of $Pr(para(t_i, t_j))$ and $Pr(coord(e_k, e_g))$ to converge after $20 \sim 25$ updates.

One claim of this paper is that paraphrase relationship and coordinate relationship mutually reinforce each other, so paraphrase templates can be selected out. To verify this, we evaluate the performance on the following five semantic relations:

---

[10] http://datamarket.azure.com/dataset/bing/search.

1. **highConcentration:** *We define this as a food contains a high amount of a certain nutrient.*
2. **acquisition:** *We define this as the activity between two companies such that one company acquired another.*
3. **founderOf:** *We define this as the relation between a person and his founded company.*
4. **headquarter:** *We define this as the relation between a company and the location of its headquarter.*
5. **field:** *We define this as the relation between a person and his field of expertise.*

Table 2. Input sentences.

Relation	Sentence	Entity tuple
highConcentration	Lemons are rich in vitamin c	*(lemons,vitamin c)*
acquisition	Google has purchased Nest Labs	*(Google,Nest Labs)*
founderOf	Larry Page founded Google	*(Larry Page,Google)*
headquarter	Yahoo is headquartered in Sunnyvale	*(Yahoo,Sunnyvale)*
field	Albert Einstein revolutionized physics	*(Albert Einstein,physics)*

In Table 2, we list five input sentences of the above semantic relations, and the entity tuple extracted from each sentence, respectively. Thus, templates are easily obtained by substituting entity tuples with variables. For example, in the first sentence, let X=*lemons*, Y=*vitamin c*, we have template X *are rich in* Y.

We find paraphrase templates and coordinate entity tuples for each of these inputs by the co-acquisition method described in Sect. 5. Our evaluation will consider only paraphrasing, i.e. given a sentence s, we will assess the quality of its paraphrases we acquire from the Web, whether they convey the same meaning with the given sentence. We do not assess whether it is really a fact.

Table 3. Performance of our method for paraphrase acquisition.

relation	highConcentration	acquisition	founderOf	headquarter	field
# Obtained	16	26	11	10	5
# Paraphrases	9	21	5	4	4
Precision	56.3 %	80.8 %	45.5 %	40 %	80 %
Average Precision	60.5 %				
Average # per input	8.6				

Table 4. An example of some discovered paraphrases.

Sentence	Google has purchased Nest Labs
Correct	Google has acquired Nest Labs
	Google is buying Nest Labs
	Google owned Nest Labs
	Google is buys Nest Labs
	Google has announced their acquisition of Nest Labs
	Google finalizes acquisition of Nest Labs
Incorrect	Google has announced plans to buy **thermostat maker** Nest Labs
	Google has acquired **smart-gadget company** Nest Labs

Table 5. Another example of some discovered paraphrases.

Sentence	Yahoo is headquartered in Sunnyvale
Correct	Yahoo is located in Sunnyvale
	Sunnyvale is home to notable companies such as Yahoo
	Yahoo headquarters in the Sunnyvale area
	Yahoo headquarters in Sunnyvale.
Incorrect	View all Yahoo jobs in Sunnyvale
	Reviews on Yahoo in Sunnyvale

6.2 Results

In this section, we show the results of the experiments and analyze them. Table 3 shows the performance of our proposed method for each of the five semantic relations and their average. We calculate the precision as how many "true" paraphrases are in the paraphrases obtained by our method. From Table 3, we can see the sentence query for the **acquisition** relation achieved the best performance with the precision of 80.8 %, while the sentence query for the **headquarter** relation preforms the worst with the precision of 40 %. As there isn't much work in acquiring sentential-level paraphrases from the Web, it is hard to construct a baseline to compare against. However, we can analyze them in consideration of numbers reported previously for acquiring paraphrases from the Web. TE/ASE method [13] reports obtained precision of 44.15 %, compared to our average precision of 60.5 %. It is difficult to estimate the recall since we do not have a complete set of paraphrases for a given sentence. Instead of evaluating recall, we calculate the average number of correct paraphrases per input sentence. The average number of paraphrases per input is 5.5 of TE/ASE method, compared to our 8.6.

In order to find the reasons why our method succeeds or fails to acquire paraphrases, let us do in-depth analysis especially on the best performance query and the worst performance query, respectively. Table 4 shows some correct and incorrect paraphrases obtained by our method for the query from the **acquisition** relation. As we mentioned before, this query achieves the best performance. Actually, we extract more than 280 templates from the top 1000 search results of the AND query "Google AND Nest Labs". The most frequently occurring templates themselves are good candidates. Therefore, we get more paraphrases with a single input. On the other hand, take the incorrect paraphrase "Google has announced plans to buy thermostat maker Nest Labs." for example. Compared with the given sentence "Google has purchased Nest Labs.", it also contains a further explanation of *Nest Labs* that *Nest Labs* is a thermostat maker, and we think such additional information leads to non-paraphrases. Although its template X *h*as announced plans to buy thermostat maker Y is suitable for few extracted entity tuples, it received the propagated value from the strong coordinate degree between other tuples and *(Google,Nest Labs)*. We surveyed the result of coordinate entity tuples and found that entity tuples such as *(Microsoft,Nokia)*, *(Yahoo,Tumblr)* get higher coordinate values than those of other queries. This leads a misjudgment of paraphrases. Table 5 shows some correct and incorrect paraphrases obtained by our method for the query from the **headquarter** relation. As we mentioned before, this query performs the worst. Actually, we extract even less than 40 templates from the top 1000 search results of the query "Yahoo AND Sunnyvale". The reasons we considered are that firstly, there are not so many search results contained both *Yahoo* and *Sunnyvale* in a single sentence; secondly, even they are in the same sentence, that sentence may be too short, or too long. Besides, advertisements also have an influence. Take the incorrect paraphrase "View all Yahoo jobs in Sunnyvale." for example. Such advertisements are suitable for almost all extracted entity tuples, so they get higher paraphrase values. From the above discussion, we can point out that if the number of extracted templates could increase (i.e. using high-valued coordinate entity tuples to gather more templates), our method's performance would improve to some extent. And we should give a penalty to a too-general template to restrict the value propagation, since it is likely to be an advertisement, or an automatically generated sequence by a website to increase its click rate.

7   Conclusion

Given a sentence, our proposed method aims to find its paraphrases from the noisy Web. Here we incorporate coordinate relationship and take a mutually reinforcing way to calculate paraphrase degree and coordinate degree. Experiments show our average precision is 60.5 %, compared to TE/ASE method with average precision of 44.15 %. Besides, the average number of correct paraphrases is 8.6 of our method, compared to TE/ASE method of 5.5.

As we stated in Sect. 6.2, for some queries, we cannot get enough templates. One way to solve this problem is to use high-valued coordinate entity tuples to gather more templates, and even execute our method in a iterative way. However, it causes too many accesses to the Web, and sometimes, we still cannot find enough templates. Another way to solve this problem is to do syntactic analysis to eliminate some additional information, i.e. "thermostat maker". Furthermore, we will give a penalty to a too-general template to restrict the value propagation.

Acknowledgment. This work was supported in part by the following projects: Grants-in-Aid for Scientific Research (Nos. 24240013, 24680008) from MEXT of Japan.

References

1. Agichtein, E., Gravano, L.: Snowball: extracting relations from large plain-text collections. In: Proceedings of the Fifth ACM Conference on Digital Libraries, DL 2000, pp. 85–94 (2000)
2. Anick, P.G., Tipirneni, S.: The paraphrase search assistant: terminological feedback for iterative information seeking. In: Proceedings of the 22Nd Annual International ACM SIGIR Conference on Research and Development in Information Retrieval, pp. 153–159 (1999)
3. Bannard, C., Callison-Burch, C.: Paraphrasing with bilingual parallel corpora. In: Proceedings of the 43rd Annual Meeting on Association for Computational Linguistics, pp. 597–604 (2005)
4. Barzilay, R., Elhadad, N.: Sentence alignment for monolingual comparable corpora. In: Proceedings of the 2003 Conference on Empirical Methods in Natural Language Processing, pp. 25–32 (2003)
5. Barzilay, R., McKeown, K.R.: Extracting paraphrases from a parallel corpus. In: Proceedings of the 39th Annual Meeting on Association for Computational Linguistics, pp. 50–57 (2001)
6. Barzilay, R., McKeown, K.R., Elhadad, M.: Information fusion in the context of multi-document summarization. In: Proceedings of the 37th Annual Meeting of the Association for Computational Linguistics on Computational Linguistics, pp. 550–557 (1999)
7. Bollegala, D.T., Matsuo, Y., Ishizuka, M.: Relational duality: unsupervised extraction of semantic relations between entities on the web. In: Proceedings of the 19th International Conference on World Wide Web, pp. 151–160 (2010)
8. Callison-Burch, C., Koehn, P., Osborne, M.: Improved statistical machine translation using paraphrases. In: Proceedings of the Main Conference on Human Language Technology Conference of the North American Chapter of the Association of Computational Linguistics, pp. 17–24 (2006)
9. Denning, P., Horning, J., Parnas, D., Weinstein, L.: Wikipedia risks. Commun. ACM **48**(12), 152–152 (2005)
10. Etzioni, O., Banko, M., Soderland, S., Weld, D.S.: Open information extraction from the web. Commun. ACM **51**(12), 68–74 (2008)
11. Etzioni, O., Cafarella, M., Downey, D., Popescu, A.M., Shaked, T., Soderland, S., Weld, D.S., Yates, A.: Unsupervised named-entity extraction from the web: an experimental study. Artif. Intell. **165**, 91–134 (2005)
12. Harris, Z.S.: Distributional structure. Word **10**, 146–162 (1954)
13. Idan, I.S., Tanev, H., Dagan, I.: Scaling web-based acquisition of entailment relations. In: Proceedings of EMNLP, pp. 41–48 (2004)
14. Lin, D., Pantel, P.: Dirt - discovery of inference rules from text. In: Proceedings of the ACM SIGKDD Conference on Knowledge Discovery and Data Mining, pp. 323–328 (2001)

15. Madnani, N., Ayan, N.F., Resnik, P., Dorr, B.J.: Using paraphrases for parameter tuning in statistical machine translation. In: Proceedings of the ACL Workshop on Statistical Machine Translation (2007)
16. Marton, Y., Callison-Burch, C., Resnik, P.: Improved statistical machine translation using monolingually-derived paraphrases. In: Proceedings of the 2009 Conference on Empirical Methods in Natural Language Processing, pp. 381–390 (2009)
17. McKeown, K.R., Barzilay, R., Evans, D., Hatzivassiloglou, V., Klavans, J.L., Nenkova, A., Sable, C., Schiffman, B., Sigelman, S., Summarization, M.: Tracking and summarizing news on a daily basis with columbia's newsblaster. In: Proceedings of the Second International Conference on Human Language Technology Research, pp. 280–285 (2002)
18. Ohshima, H., Oyama, S., Tanaka, K.: Searching coordinate terms with their context from the web. In: Aberer, K., Peng, Z., Rundensteiner, E.A., Zhang, Y., Li, X., Unland, R. (eds.) WISE 2006. LNCS, vol. 4255, pp. 40–47. Springer, Heidelberg (2006)
19. Paşca, M., Dienes, P.: Aligning needles in a haystack: paraphrase acquisition across the web. In: Dale, R., Wong, K.-F., Su, J., Kwong, O.Y., Unland, R. (eds.) IJCNLP 2005. LNCS (LNAI), vol. 3651, pp. 119–130. Springer, Heidelberg (2005)
20. Salton, G., McGill, M.J.: Introduction to Modern Information Retrieval. McGraw-Hill Inc., New York (1986)
21. Shinyama, Y., Sekine, S.: Paraphrase acquisition for information extraction. In: Proceedings of the Second International Workshop on Paraphrasing, vol. 16, 65–71 (2003)
22. Shinyama, Y., Sekine, S., Sudo, K.: Automatic paraphrase acquisition from news articles. In: Proceedings of the Second International Conference on Human Language Technology Research, HLT 2002, pp. 313–318 (2002)
23. Wang, R., Callison-Burch, C.: Paraphrase fragment extraction from monolingual comparable corpora. In: Proceedings of the 4th Workshop on Building and Using Comparable Corpora: Comparable Corpora and the Web, pp. 52–60 (2011)
24. Wubben, S., van den Bosch, A., Krahmer, E., Marsi, E.: Clustering and matching headlines for automatic paraphrase acquisition. In: Proceedings of the 12th European Workshop on Natural Language Generation, ENLG 2009, pp. 122–125 (2009)
25. Yamamoto, Y., Tanaka, K.: Towards web search by sentence queries: asking the web for query substitutions. In: Yu, J.X., Kim, M.H., Unland, R. (eds.) DASFAA 2011, Part II. LNCS, vol. 6588, pp. 83–92. Springer, Heidelberg (2011)
26. Yates, A., Cafarella, M., Banko, M., Etzioni, O., Broadhead, M., Soderland, S.: Textrunner: Open information extraction on the web. In: Proceedings of Human Language Technologies: The Annual Conference of the North American Chapter of the Association for Computational Linguistics: Demonstrations, pp. 25–26 (2007)

The Second International Workshop on Big Data Management and Service (BDMS)

Emergency Situation Awareness During Natural Disasters Using Density-Based Adaptive Spatiotemporal Clustering

Tatsuhiro Sakai$^{(\boxtimes)}$, Keiichi Tamura, and Hajime Kitakami

Graduate School of Information Sciences,
Hiroshima City University, Hiroshima, Japan
my67011@e.hiroshima-cu.ac.jp, {ktamura,kitakami}@hiroshima-cu.ac.jp

Abstract. With the increase in the popularity of social media as well as the emergence of easy-to-use geo-mobile applications on smartphones, a huge amount of geo-annotated data is posted on social media sites. To enhance emergency situation awareness, these geo-annotated data are expected to be used in a new medium. In particular, geotagged tweets on Twitter are used by local governments to determine the situation accurately during natural disasters. Geotagged tweets are referred to as georeferenced documents; they include not only a short text message but also the posting time and location. In this paper, we propose a new spatiotemporal analysis method for emergency situation awareness during natural disasters using (ϵ, τ)-density-based adaptive spatiotemporal clustering. Such clustering can identify bursty local areas by using adaptive spatiotemporal clustering criteria considering local spatiotemporal densities. Extracting (ϵ, τ)-density-based adaptive spatiotemporal clusters allows the proposed method to analyze emergency situations such as natural disasters in real time. The experimental results showed that the proposed method can analyze emergency situations related to the weather in Japan more sensitively compared with our previous method.

Keywords: Emergency situation awareness · Density-based spatiotemporal clustering · Natural disaster · Geotagged tweet · Naive bayes classifier

1 Introduction

In the era of big data, social media are expected to enhance emergency situation awareness during natural disasters such as earthquakes, typhoons, floods, and diseases [17]. During such crises, users usually post text messages on social media about things that they are witnessing [4,8,10]. These posted text messages offer a new information source for local governments to get a better grasp of the situation on the ground. This information provided by the users on social media helps local governments to know what is going on in crisis regions. Therefore, researchers in many application domains have been tackling new challenges related to emergency situation awareness utilizing information available on social media.

© Springer International Publishing Switzerland 2015
A. Liu et al. (Eds.): DASFAA 2015 Workshops, LNCS 9052, pp. 155–169, 2015.
DOI: 10.1007/978-3-319-22324-7_13

Moreover, there has been an emergence of a new type of data on social media: geo-annotated social data, which are often referred to as georeferenced social data [11]. Georeferenced social data usually include location-related topics and events that users are interested in. This trend has been encouraged by the increasing of smartphones equipped with GPS and easy-to-use geo-mobile applications. In particular, geotagged tweets that are georeferenced documents posted on Twitter have received a considerable amount of attention to detect real-time geo-related topics and events. Twitter is one of the most popular social network services where users tweet about what they are thinking and what they are witnessing. For example, Sakaki et al. [13] focused on a method for predicting typhoon trajectories and earthquake epicenters by using geotagged tweets regarding typhoons and earthquakes.

Dense areas related to which many geotagged tweets including keywords related to a natural disaster are posted, are referred to as bursty areas of the crisis. For example, during a natural disaster, the number of tweets including topics about the disaster increases in disaster-stricken areas. Based on this philosophy, we proposed a spatiotemporal analysis method for emergency situation awareness using the (ϵ, τ)-density-based spatiotemporal clustering algorithm [12]. This spatiotemporal analysis method consists of two components: a Naive Bayes classifier and an incremental algorithm for (ϵ, τ)-density-based spatiotemporal clustering. The Naive Bayes classifier extracts georeferenced documents including keywords related to an emergency topic. The incremental algorithm allows one to identify bursty areas whose spatiotemporal densities are higher than those of the other areas in real time.

In this paper, we propose a new spatiotemporal analysis method for emergency situation awareness during natural disasters using novel density-based spatiotemporal clustering called a density-based adaptive spatiotemporal clustering algorithm. The main contributions of our study are as follows:

- To easily handle geotagged tweets, the data model of geotagged tweets is defined. In our model, a geotagged tweet is referred to as a georeferenced document. Moreover, geotagged tweets including an observed emergency topic are referred to as relevant geotagged tweets.
- To identify bursty areas more sensitively, the (ϵ, τ)-density-based adaptive spatiotemporal clustering is proposed. Conventional (ϵ, τ)-density based spatiotemporal clustering [12] cannot identify local high-density areas, even when these areas are high density in local area. The (ϵ, τ)-density-based adaptive spatiotemporal clustering algorithm can identify local high-density areas depending on their neighboring regions.
- To evaluate our new spatiotemporal analysis method for emergency situation awareness during natural disasters, we implemented a web-based application for grasping the emergency situation during natural disasters related to weather in Japan. Moreover, we compared the (ϵ, τ)-density-based adaptive spatiotemporal clustering with our previous method in a weather topic in Japan.

The remainder of this paper is organized as follows: In Sect. 2, we present a brief overview of the related work. In Sect. 3, (ϵ, τ)-density-based adaptive spatiotemporal clusters are introduced. In Sect. 4, we introduce the proposed spatiotemporal analysis method for emergency situation awareness during natural disasters. In Sect. 5, we present the experimental results and the web-based application for grasping the emergency situation during natural disasters related to weather in Japan. In Sect. 6, we conclude this paper.

2 Related Work

In the era of big data, we are witnessing the rapid growth of a new type of information source [11]. In particular, tweets are one of the most widely used microblogging services information available for situation awareness during emergencies [6,17].

Vieweg et al. [16] showed that information related to emergency situations is posted on Twitter during emergencies such as floods and fires. Moreover, Hwang et al. [5] observed flu epidemics by using a spatiotemporal analysis of social media geostreams. Therefore, analysis and situation awareness of emergency topics related to natural disasters, weather, diseases, and other incidents by using georeferenced social data is a major challenge.

Aramaki et al. [1] proposed a novel method for detecting influenza epidemics using tweets; their method utilized classifiers such as support vector machine (SVM) and Naive Bayes to extract tweets that included topics related to influenza. Their geocoding technique was used to map each tweet to a region in Japan. Moreover, their proposed system visualized the increase and decrease in the number of related tweets in each region. As indicated above, their approach is the one most relevant to our work. Thom et al. [15] presented a system that extracts anomalies from geolocated Twitter messages and visualizes them using term clouds on an interactive. Their system can not only detect natural hazards like earthquakes but also a general impression of how severe the situation is by utilizing spatiotemporal clustering and scalable representation. As indicated above, their approach is the one most related to our work; however, their system cannot grasp the details of an emergency situation.

Kim et al. [7] introduced mTrend, which constructs and visualizes spatiotemporal trends of topics, referred to as "topic movements." Avvenuti et al. [2] developed the earthquake alert and report system (EARS), which can identify damage in earthquake-affected areas. Kumar et al. [9] detected road hazards by aggregating hazard-related information posted by Twitter users. Researchers have approached to analyze the situation awareness of various emergencies by using tweets; however, their methods cannot consider the difference between posts of urban areas and those of the countryside.

3 (ϵ, τ)-Density-based Adaptive Spatiotemporal Clusters

This section reviews the definitions of the (ϵ, τ)-density-based spatiotemporal criteria, which were proposed in our previous work. Moreover, the (ϵ, τ)-density-based spatiotemporal criteria are extended for identifying bursty areas affected by natural disasters.

3.1 Data Model

In this study, we focus on geotagged tweets that include location information as well as the posting time and a short text message. Let gt_i denote the i-th geotagged tweet in $GTS = \{gt_1, \cdots, gt_n\}$; then, gt_i consists of three items: $gt_i = < text_i, pt_i, pl_i >$, where $text_i$ is a short text message, pt_i is the time when the geotagged tweet was posted, and pl_i is the location where gt_i was posted or is located (e.g., latitude and longitude).

Let $rgt_j^{(etp)} = gt_{\phi^{(etp)}(j)}$ denote a relevant geotagged tweet in which a message related an emergency topic etp is included in $text_{\phi^{(etp)}(j)}$. A set of relevant geotagged tweets is $RGTS^{(etp)} = \{rgt_1^{(etp)}, \cdots, rgt_m^{(etp)}\}$, where $RGTS^{(etp)} \in GTS$. Function $\phi^{(etp)}(j)$ is an injective function.

$$\phi^{(etp)}(j) : RGTS^{(etp)} \to GTS; \ rgt_j^{(etp)} \mapsto gt_{\phi^{(etp)}(j)} \tag{1}$$

In this study, we cover natural disasters as emergency topics. For example, during heavy rainfall, people often post geotagged tweets including the topic "rain." In this case, the relevant geotagged tweets are geotagged tweets in which each $text_i$ includes the topic "rain."

3.2 Density-Based Spatiotemporal Adaptive Criteria

In our previous work [12], we proposed density-based spatiotemporal criteria for extracting bursty areas as (ϵ, τ)-density-based spatiotemporal clusters, which are a natural extension of the density-based criteria proposed by Ester et al. [3, 14]. In the density-based spatial criteria, spatial clusters are dense areas separated from areas of lower density. The key concept underpinning the use of the density-based spatial clustering algorithm indicates that for each data point within a spatial cluster, the neighborhood of a user-defined radius must contain at least a minimum number of points; that is, the density in the neighborhood must exceed some predefined threshold.

In density-based spatial clustering, the ϵ-density-based neighborhood allows us to recognize areas in which densities are higher than in other areas. However, it does not consider temporal changes. It is important to analyze temporal changes for emergency situation awareness, especially natural disasters. In contrast, the (ϵ, τ)-density-based spatiotemporal clusters cover spatiotemporal clusters that are both temporally and spatially separated from other spatiotemporal clusters. Moreover, bursty areas can be extracted in real time as (ϵ, τ)-density-based spatiotemporal clusters by using an incremental algorithm.

When a natural disaster strikes, the number of relevant geotagged tweets increases in the affected areas; therefore, areas where the density of the posted relevant geotagged tweets is higher than other areas, are the bursty areas of the natural disaster. This concept leads to the idea that bursty areas of the natural disaster can be extracted as density-based spatiotemporal clustering.

For situation awareness of an emergency topic, we must extract information on all areas, namely urban areas and the countryside, because the emergency is felt in all areas irrespective of whether they are urban areas or the countryside. However, it is difficult to seamlessly extract spatiotemporal clusters in both high- and low-density regions. To address this issue, we define density-based adaptive spatiotemporal criteria. Density-based adaptive spatiotemporal criteria consider local densities of geotagged tweets while extracting density-based spatiotemporal clusters.

3.3   Definitions

There are several density-based adaptive spatiotemporal criteria to define (ϵ, τ)-density-based adaptive spatiotemporal clusters.

Definition 1 ((ϵ, τ)-Density-based Neighborhood). Suppose that the observed emergency topic and the set of geotagged tweets related to the topic are etp and $RGTS^{(etp)} \in GTS$, respectively. The (ϵ, τ)-density-based neighborhood of a relevant geotagged tweet $rgt_p^{(etp)}$, which is denoted by $STN_{(\epsilon, \tau)}(rgt_p^{(etp)})$, is defined as

$$STN_{(\epsilon, \tau)}(rgt_p^{(etp)}) = \{rgt_q^{(etp)} \in RGTS^{(etp)} \mid dist(rgt_p^{(etp)}, rgt_q^{(etp)}) \leq \epsilon$$
$$\text{and } iat(rgt_p^{(etp)}, rgt_q^{(etp)}) \leq \tau\}, \quad (2)$$

where the function $dist$ returns the distance between the relevant geotagged tweets $rgt_p^{(etp)}$ and $rgt_q^{(etp)}$, and the function iat returns the interarrival time between them.

Definition 2 (Local Spatiotemporal Density/Adaptive Threshold). The local spatiotemporal density of a relevant geotagged tweet $rgt_p^{(etp)}$ is denoted as $lstd(rgt_p^{(etp)})$. The minimum number of relevant geotagged tweets $MinRGT$, is adjusted in accordance with its local spatiotemporal density. This adjusted minimum number of relevant geotagged tweets is called the adaptive threshold AT:

$$AT(rgt_p^{(etp)}, MinRGT) = (MinRGT - 1) \times lstd(rgt_p^{(etp)}) + 1, \quad (3)$$

where the function $lstd$ returns the degree of local spatiotemporal density ($0 \leq lstd(rgt_p^{(etp)}) \leq 1.0$).

To estimate the local spatiotemporal densities of geotagged tweets, the number of geotagged tweets posted in the past is utilized. The spatiotemporal space is divided into several spatiotemporal grids in three-dimentional space. The number of spatiotemporal grids is $div_{lng} \times div_{lat} \times div_{time}$. lng, lat and $time$ are longitude, latitude and posted time, respectively. For each spatiotemporal grid, the number of geotagged tweets posted in the past is calculated. The degree of local spatiotemporal density of a geotagged tweet is the normalization value of the number of geotagged tweets:

$$lstd(gt_i) = \frac{stnum(geo\_gid(gt_i)) - stnum_{min}}{stnum_{max} - stnum_{min}}, \quad (4)$$

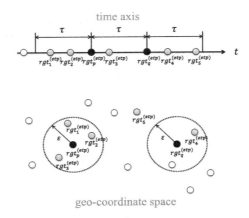

Fig. 1. Example of Definition 3.

where $stnum(i)$ returns the number of geotagged tweets in the i-th grid. Function $geo\_gid(gt_i)$ returns the grid ID where gt_i is located. Furthermore, $stnum_{min}$ and $stnum_{max}$ are the minimum and maximum value, respectively.

Definition 3 (Core-and Border-relevant Geotagged Tweets). A relevant geotagged tweet $rgt_p^{(etp)}$ is called a core-relevant geotagged tweet for an emergency topic etp if there is at least an adaptive threshold, $AT(rgt_p^{(etp)}, MinRGT)$, in $STN_{(\epsilon,\tau)}(rgt_p^{(etp)})$ ($|STN_{(\epsilon,\tau)}(rgt_p^{(etp)})| \geq AT(rgt_p^{(etp)}, MinRGT)$). Otherwise (i.e., $|STN_{(\epsilon,\tau)}(rgt_p^{(etp)})| < AT(rgt_p^{(etp)}, MinRGT)$), $rgt_p^{(etp)}$ is called a border-relevant geotagged tweet for the emergency topic etp.

Suppose that $MinRGT$ is set to 3 and, $lstd(rgt_p^{(etp)})$ and $lstd(rgt_q^{(etp)})$ returns 0.8 ($AT(rgt_p^{(etp)}, MinRGT)$ and $AT(rgt_q^{(etp)}, MinRGT) = 2.6$.) In the Fig. 1, $rgt_p^{(etp)}$ is a core-relevant geotagged tweet, because $|STN_{(\epsilon,\tau)}(rgt_p^{(etp)})|$ is three. On the other hand, $rgt_q^{(etp)}$ is a border-relevant geotagged tweet, because $|STN_{(\epsilon,\tau)}(rgt_p^{(etp)})|$ is one.

Definition 4 ((ϵ, τ)-Density-based Directly Adaptive Reachable). Suppose that a relevant geotagged tweet $rgt_q^{(etp)}$ is in the (ϵ, τ)-density-based neighborhood of $rgt_p^{(etp)}$. If the number of relevant geotagged tweets in the (ϵ, τ)-density-based neighborhood of $rgt_p^{(etp)}$ is greater than or equal to $AT(rgt_p^{(etp)}, MinRGT)$, i.e., if $|STN_{(\epsilon,\tau)}(rgt_p^{(etp)})| \geq AT(rgt_p^{(etp)}, MinRGT)$, $rgt_q^{(etp)}$ is (ϵ, τ)-density-based directly adaptive reachable from $rgt_p^{(etp)}$. In other words, the relevant geotagged tweets in the (ϵ, τ)-density-based neighborhood of a core-relevant geotagged tweet are (ϵ, τ)-density-based directly adaptive reachable.

Definition 5 ((ϵ, τ)-Density-based Adaptive Reachable). Suppose that there is a sequence of relevant geotagged tweets $(rgt_p^{(etp)}, rgt_{(p+1)}^{(etp)}, \cdots, rgt_{(p+l)}^{(etp)})$ and the $(p + i)$-th relevant geotagged tweet $rgt_{(p+i+1)}^{(etp)}$ is (ϵ, τ)-density-based directly adaptive reachable from the $(p + i)$-th relevant geotagged tweet $rgt_{(p+i)}^{(etp)}$. The relevant geotagged tweet $rgt_{(p+l)}^{(etp)}$ is (ϵ, τ)-density-based adaptive reachable from $rgt_p^{(etp)}$.

Definition 6 ((ϵ, τ)-Density-based Adaptive Connected). Suppose that the relevant geotagged tweets $rgt_p^{(etp)}$ and $rgt_q^{(etp)}$ are (ϵ, τ)-density-based adaptive reachable from an arbitrary relevant geotagged tweet $rgt_o^{(etp)}$. If $|STN_{(\epsilon,\tau)}(rgt_o^{(etp)})| \geq AT(rgt_o^{(etp)}, MinRGT)$, $rgt_p^{(etp)}$ is (ϵ, τ)-density-based adaptive connected to $rgt_q^{(etp)}$.

3.4 (ϵ, τ)-Density-based Adaptive Spatiotemporal Cluster

A (ϵ, τ)-density-based adaptive spatiotemporal cluster consists of two types of relevant geotagged tweets: core-relevant geotagged tweets, which are mutually (ϵ, τ)-density-based adaptive reachable, and border-relevant geotagged tweets, which are (ϵ, τ)-density-based directly adaptive reachable from the core-relevant geotagged tweets. A (ϵ, τ)-density-based adaptive spatiotemporal cluster is defined as follows:

Definition 7 ((ϵ, τ)-Density-based Adaptive Spatiotemporal Cluster). A (ϵ, τ)-density-based adaptive spatiotemporal cluster for an emergency topic etp ($ASTC^{(etp)}$) in $RGTS^{(etp)}$ satisfies the following restrictions:

i1j $\forall rgt_p^{(etp)}$, $rgt_q^{(etp)} \in RGTS^{(etp)}$, if and only if $rgt_p^{(etp)} \in ASTC^{(etp)}$ and $rgt_q^{(etp)}$ is (ϵ, τ)-density-based adaptive reachable from $rgt_p^{(etp)}$, and $rgt_q^{(etp)}$ is also in $ASTC^{(etp)}$.

i2j $\forall rgt_p^{(etp)}$, $rgt_q^{(etp)} \in ASTC^{(etp)}$, $rgt_p^{(etp)}$ is (ϵ, τ)-density-based adaptive connected to $rgt_q^{(etp)}$.

Even if $rgt_p^{(etp)}$ and $rgt_q^{(etp)}$ are border-relevant geotagged tweets, $rgt_p^{(etp)}$ and $rgt_q^{(etp)}$ are in the same (ϵ, τ)-density-based adaptive spatiotemporal cluster if $rgt_p^{(etp)}$ is (ϵ, τ)-density-based adaptive connected to $rgt_q^{(etp)}$.

4   Proposed Method

In this section, we propose a new spatiotemporal analysis method for situation awareness during natural disasters.

4.1   Concept and System Overview

When natural disasters occur, especially like emergency weather topics, the struck areas are changed temporally. Therefore, we have to extract bursty areas in real time. Moreover, the struck areas may include not only urban areas but also the countryside. The density-based adaptive spatiotemporal criteria allow us to identify bursty areas irrespective of whether these areas are countryside areas or urban areas. The (ϵ, τ)-density-based adaptive spatiotemporal clusters can be incrementally extracted in real time.

Suppose that each geotagged tweet arrive one by one and the observed emergency topic is etp. There are two steps in the proposed analysis method.

1. A Naive Bayes classifier classifies the i-th geotagged tweet gt_i. If and only if gt_i is classified as "relevant," which means that gt_i is a relevant geotagged tweet and $rgt_j^{(etp)} = gt_i$, where $\phi_j^{(etp)} = i$ is related to the observed emergency topic, go to the next step.
2. An incremental algorithm for extracting (ϵ, τ)-density-based adaptive spatiotemporal clusters is executed, where there are two input data items; one is $rgt_j^{(etp)}$ and the other is a set of extracted (ϵ, τ)-density-based adaptive spatiotemporal clusters. The incremental algorithm outputs a set of current (ϵ, τ)-density-based adaptive spatiotemporal clusters.

4.2   Naive Bayes Classifier

To extract geotagged tweets that include the observed emergency topic related to a natural disaster as relevant geotagged tweets, we utilize the Naive Bayes classifier technique. A Naive Bayes classifier is a simple probabilistic classifier based on the application of Bayes' theorem, which is derived from Bayesian statistics with naive independence assumptions. The Naive Bayes classifier manually classifies geotagged tweets as either "relevant" or "irrelevant" where "relevant" class is related to the observed emergency topic and the "irrelevant" class is not.

In this study, the Naive Bayes classifier is based on keywords in the text data included in a geotagged tweet. Let $CLASS^{(etp)} = \{relevant, irrelevant\}$ be a set of classes for an emergency topic etp. The posterior probability that the geotagged tweet gt belongs the class $class \in CLASS^{(etp)}$ is as follows:

$$Pr(class|gt) = \frac{Pr(class)Pr(gt|class)}{Pr(gt)}$$
$$\propto Pr(class)Pr(gt|class), \qquad (5)$$

where $Pr(class)$ is the prior probability of $class$ and $Pr(gt|class)$ is the likelihood.

The Naive Bayes classifier requires a training data set including multiple georeferenced documents that are classified in one class in $CLASS^{(etp)}$. Let a training data set $TGT^{(etp)}$ be $TGT^{(etp)} = \{(tgt_1^{(etp)}, c_1), (tgt_2^{(etp)}, c_2), \cdots, (tgt_m^{(etp)}, c_m)\}$, where $c_i = \{relevant, irrelevant\} \in CLASS^{(etp)}$. the set of all

input : $nrtg_k^{(etp)}$ - a newly relevant geotagged tweet, $RGTS^{(etp)}$ - a set of
relevant geotagged tweets, $CSTC^{(etp)}$ - a set of extracted
spatiotemporal clusters, ϵ - user-specified value, τ - user-specified value,
$MinRGT$ - the minimum number of relevant geotagged tweets

output: $NSTC^{(etp)}$ - a set of updated spatiotemporal clusters

$NSTC^{(etp)} \leftarrow CSTC^{(etp)}$;
$RD \leftarrow$ GetRecentData$(nrtg_k^{(etp)},\tau,RGTS^{(etp)})$;
for $i \leftarrow 1$ **to** $|RD|$ **do**
    $nrtg^{(etp)} \leftarrow rd_i \in RD$;
    $lstd(nrtg^{(etp)}) =$ GetDens$(nrtg^{(etp)})$;
    $STN \leftarrow$ GetNeighborhood$(nrtg^{(etp)},\epsilon,\tau)$;
    if $|STN| \geq$ AT$(nrtg^{(etp)}, MinRGT)$ **then**
        if IsClustered$(nrtg^{(etp)}) == false$ **then**
            | $stc \leftarrow$ MakeNewCluster$(cid,nrtg^{(etp)})$;
        end
        else
            | $stc^{(etp)} \leftarrow$ GetCluster$(nrtg^{(etp)},NSTC^{(etp)})$;
        end
        EnQueue(Q,STN);
        while Q *is not empty* **do**
            $nrtg^{(etp)} \leftarrow$ DeQueue(Q);
            if IsClustered$(nrtg^{(etp)}) == true$ **then**
                $N \leftarrow$ GetNeighborhood$(nrtg^{(etp)},\epsilon,\tau)$;
                $lstd(nrtg^{(etp)}) =$ GetDens$(nrtg^{(etp)})$;
                if $|STN| \geq$AT$(nrtg^{(etp)}, MinRGT)$ **then**
                    $stc^{(etp)'} \leftarrow$ GetCluster$(nrtg^{(etp)},NSTC^{(etp)})$;
                    $stc^{(etp)} \leftarrow$ AppendClusters$(stc^{(etp)},stc^{(etp)'})$;
                end
            end
            else
                $stc^{(etp)} \leftarrow stc^{(etp)} \cup nrtg^{(etp)}$;
                $STN \leftarrow$ GetNeighborhood$(nrtg^{(etp)},\epsilon,\tau)$;
                $lsrd(nrtg^{(etp)}) =$ GetDens$(nrtg^{(etp)})$;
                if $|STN| \geq$AT$(nrtg^{(etp)}, MinRGT)$ **then**
                  | EnNniqueQueue(Q,STN);
                end
            end
        end
        $NSTC^{(etp)} \leftarrow NSTC^{(etp)} \cup stc^{(etp)}$;
    end
end
return $NSTC^{(etp)}$;

Algorithm 1. Incremental (ϵ,τ)-density-based adaptive spatiotemporal clustering algorithm

words in $class \in TGT^{(etp)}$ is denoted by $W_{class}^{(etp)} = \{word_1^{(etp)}, word_2^{(etp)}, \cdots, word_l^{(etp)}\}$.

The geotagged tweet gt represents a bag-of-words $\{wd_1, wd_2, \cdots, wd_{numw(gt)}\}$, where $numw(gt)$ denotes the number of words in gt.

$$Pr(gt|class) = Pr(wd_1 \wedge wd_2 \wedge \cdots \wedge wd_k|class) = \qquad (6)$$

$$\prod_i^{numw(gt)} Pr(wd_i|class)$$

where, $Pr(wd_i|class)$ is the probability that $word_i$ occurs in $class$. $Pr(wd_i|class)$ is defined as

$$Pr(wd_i|class) = \frac{OW^{(etp)}(wd_i, class) + 1}{\sum_{j=1}^{|W_{class}^{(etp)}|}(OW^{(etp)}(word_j^{(etp)}, class) + 1)}, \qquad (7)$$

where $OW^{(etp)}(word, class)$ is the number of occurrences of a word $word$ in $class \in CLASS^{(etp)}$.

The assigned class of gt is denoted by $c_{gt} \in CLASS^{(etp)}$ and is determined by finding the maximum posterior probability.

$$c_{gt} = \arg\max_{class} Pr(class|gt)$$

$$= \arg\max_{class} \left(Pr(class) \prod_i^{numw(gt)} Pr(wd_i|class) \right) \qquad (8)$$

4.3 Incremental Algorithm

In the incremental algorithm, the algorithm updates the states of the extracted spatiotemporal clusters and extracts new spatiotemporal clusters every time a relevant geotagged tweet arrives. Algorithm 1 describes the incremental (ϵ, τ)-density-based adaptive spatiotemporal clustering algorithm, which extracts (ϵ, τ)-density-based adaptive spatiotemporal clusters. There are two steps in the incremental (ϵ, τ)-density-based adaptive spatiotemporal clustering algorithm: limited re-clustering and merging.

When a new relevant geotagged tweet arrives, the existing (ϵ, τ)-density-based adaptive spatiotemporal clusters are updated; however, the arrived geotagged tweet only affects its (ϵ, τ)-density-based neighborhood directly. In this algorithm, the (ϵ, τ)-density-based neighborhood is extracted to generate seeds and these relevant geotagged tweets are re-clustered. In the incremental algorithm, during re-clustering, some (ϵ, τ)-density-based adaptive spatiotemporal clusters are needed to append to other (ϵ, τ)-density-based adaptive spatiotemporal clusters. Suppose that the algorithm a (ϵ, τ)-density-based adaptive spatiotemporal clusters $stc^{(etp)}$ is expanding. If a core-relevant geotagged tweet in $stc^{(etp)}$ includes another relevant geotagged tweet, which is clustered in $stc^{(etp)'}$, $stc^{(etp)'}$ is appended to $stc^{(etp)}$.

Table 1. Example of training data set.

"relevant" class	"irrelevant" class
It is raining hard now	It was rainy yesterday
After the heavy rain there was a big flood	We had a heavy rain yesterday
This is an amazing heavy rain	I bet it will rain tomorrow
The river overflowed because of the heavy rain.	I wonder if it will be rainy tomorrow.

Table 2. Data set and detected bursty areas ($MinRGT = 5$.)

Date	Number of tweets	Number of relevant tweets	Number of heavily rainy areas(A)	Number of detected heavily rainy areas(B)	Number of detected heavily rainy areas(C)	Detection rates (B/A 100)	Detection rates (C/A 100)
2014/6/4	325095	2249	9	2	5	22.2	55.6
2014/6/6	312145	6401	7	4	6	57.1	85.7
2014/6/7	330540	4433	2	0	1	0.00	50.0
2014/6/13	346507	2589	2	0	1	0.00	50.0
2014/6/16	340675	750	3	2	2	66.7	66.7
2014/6/22	411863	4172	1	0	1	0.00	100
2014/6/23	355384	700	4	2	3	50.0	75.0
2014/6/25	393441	2331	12	11	12	91.7	100
2014/6/29	441959	4838	6	4	5	66.7	83.3
2014/7/3	341770	4738	12	5	10	41.7	83.3
2014/7/7	376734	4173	4	3	4	75.0	100
2014/7/8	366887	1405	3	2	3	66.7	100
2014/7/9	374707	4704	3	2	2	66.7	66.7
2014/7/10	395061	4803	1	0	0	0.00	0.00
2014/7/11	383704	1763	3	0	1	0.00	33.3
2014/7/19	412403	5369	3	2	3	75.0	100

5 Experimental Result

To evaluate our new spatiotemporal analysis method, we used crawling geo-tagged tweets and analyzed situation awareness of weather topics in Japan in real time. We collected geotagged tweets from the Twitter site using its API. In experiments, we had analyzed the natural disaster of "rain" in Japan. We observed from June, 2014 to July, 2014 and evaluated whether or not the proposed method can identify bursty areas and be situation awareness according to the topic "rain." Especially, we focused on 16 days for two months in which it rained heavily. Moreover, we compared the (ϵ, τ)-density-based adaptive spatiotemporal clustering algorithm (denoted by DBASTC) with the (ϵ, τ)-density-based spatiotemporal clustering algorithm (denoted by DBSTC). The parameters in the experiments are set to as follows: ϵ is 5km, τ is 3600sec. Moreover, we compared the results of changing $MinRGT$ from 2 to 10.

The spatiotemporal space for local spatiotemporal densities is a rectangle consisted of the westernmost(24.4494,122.93361) and the northernmost tip(45.55 72,148.752) of Japan. The rectangle was equally divided into several spatiotem-

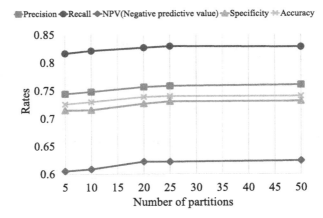

Fig. 2. Cross-validation.

poral grids of $div_{lng} = 1,000$, $div_{lat} = 1,000$ and $div_{time} = 24$. We utilized 3,301,605 geotagged tweets from December 13 to December 23, 2013 and counted in each spatiotemporal grids.

First, we evaluate the classifier. The training data set $TGT^{(etp)}$ is composed of 2,555 geotagged tweets included "rain" as a keyword, which were posted on June 4. The geotagged tweets in $TGT^{(etp)}$ were labeled manually. Table 1 shows the example of $TGT^{(etp)}$. To extract relevant geotagged tweets, geotagged tweets related to actually-occurred natural disaster of "rain" like the left side of Table 1 belong in "relevant" class. Conversely, geotagged tweets like the right side of Table 1 belong in "irrelevant" class. The numbers of geotagged tweets in "relevant" class and "irrelevant" class are 1,458 and 1,097 respectively.

Table 2 lists the number of crawled geotagged tweets and extracted relevant geotagged tweets using the Naive Bayes classifier at each day. To evaluate the Naive Bayes classifier, we performed a cross-validation. The number of partitions for the cross-validation are 5, 10, 20, 25 and 50. Figure 2 shows the precisions, recalls, NPVs(Negative predictive values), specificities and accuracies at each number of partitions. The precisions and recalls are about 75 % and 83 % respectively and, the NPVs and specificities are about 62 % and 73 % respectively. Moreover, the accuracies is about 74 %. The precisions and recalls are higher than the NPVs and specificities. In this study, the Naive Bayes classifier was able to extract the natural disaster of "rain" related to geotagged tweets in high-precision and high-recall because we utilized geotagged tweets in "relevant" class as relevant geotagged tweet.

We collected areas heavily rained that were reported from the newspaper articles in 16 days in which it rained heavily. There are 75 areas that is reported to be a heavily rainy area in newspaper from June to July in 2014. Figure 3 illustrates the detection rates of DBASTC and DBSTC in each $MinRGT$. In DBASTC, the detection rates is almost unchanged from about 80 % from 2 to 10 in $MinRGT$. On the other hand, the detection rates fell from about 70 % to about 30 % in DBSTC. Moreover, we considered details of the results in

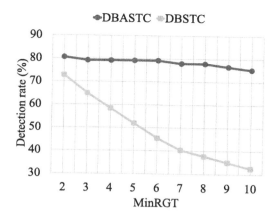

Fig. 3. Detection rate.

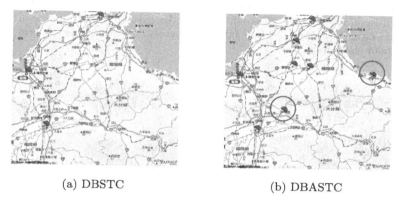

(a) DBSTC   (b) DBASTC

Fig. 4. Extracted Heavily Rained Areas. These figures show extracted heavily rained areas by DBSC and DBASTC. Circles show areas heavily rained that were reported from newspaper articles in the countryside (Asakura city, Fukuoka and Nakatsu city, Oita.)

$MinRGT = 5$. Table 2 lists number of heavily rainy areas (A), sorted by date. Moreover, number of detected heavily rainy areas (B) by using DBSTC and (C) by using DBASTC, respectively. Value of ($B/A \times 100$ and $C/A \times 100$) indicate the detection rates of the identification. In DBASTC, the detection rates are larger than 50 % for 14 out of 16 days. On the other hand, the detection rates are larger than 50 % for 9 out of 16 days in DBSTC. These results indicate that DBASTC can detect spatiotemporal clusters with high rates compared with DBSTC.

Figures 4 and 5 show screen shots of the implemented web-based application in Kyushu on July 3, 2014. The web-based application can display locations, texts and pictures of geotagged tweets in real-time. Figures 4(a) and (b) show extracted spatiotemporal clusters by DBSTC and DBASTC on the map. Spatiotemporal clusters indicated by circles show areas heavily rained that were reported from newspaper articles in the countryside. Moreover, Fig. 5 shows a

Fig. 5. Extracted Geotagged Tweet. This figure shows the screen shot of the web-based application. This screen shot shows the extracted spatiotemporal cluster in Asakura city, Fukuoka at 7:47 on July 3. The geotagged tweet says "A heavy rain!. I wish school was out."

geotagget tweet related to emergency topic heavy rain in the extracted spatiotemporal cluster by DBASTC.

6   Conclusion

In this paper, we propose a new spatiotemporal analysis method for emergency situation awareness for natural disasters awareness using the (ϵ, τ)-density-based adaptive spatiotemporal clustering algorithm. The aim of this study is identifying local high-density areas depending on their neighboring regions and, grasping the emergency situation of natural disasters. To evaluate the proposed analysis method, an actual crawling geotagged tweets posted on the Twitter site was used. The experimental results indicate that (ϵ, τ)-density-based adaptive spatiotemporal clustering algorithm can identify local high-density areas and the implemented web-based application can grasp the emergency situation of "rain" in Japan. In our future work, we intend to develop automatically parameters parameters setting method and automatically notifying danger situations in real-time.

Acknowledgments. This work was supported by JSPS KAKENHI Grant Number 26330139 and Hiroshima City University Grant for Special Academic Research (General Studies).

References

1. Aramaki, E., Maskawa, S., Morita, M.: Twitter catches the flu: Detecting influenza epidemics using twitter. In: Proceedings of the Conference on EMNLP 2011, pp. 1568–1576 (2011)

2. Avvenuti, M., Cresci, S., Marchetti, A., Meletti, C., Tesconi, M.: Ears (earthquake alert and report system): A real time decision support system for earthquake crisis management. In: Proceedings of the 20th ACM SIGKDD International Conference on Knowledge Discovery and Data Mining, pp. 1749–1758 (2014)
3. Ester, M., Kriegel, H.P., Sander, J., Xu, X.: A density-based algorithm for discovering clusters in large spatial databases with noise. In: Second International Conference on Knowledge Discovery and Data Mining, pp. 226–231 (1996)
4. Hui, C., Tyshchuk, Y., Wallace, W.A., Magdon-Ismail, M., Goldberg, M.: Information cascades in social media in response to a crisis: a preliminary model and a case study. In: Proceedings of the 21st International Conference Companion on WWW, pp. 653–656 (2012)
5. Hwang, M.H., Wang, S., Cao, G., Padmanabhan, A., Zhang, Z.: Spatiotemporal transformation of social media geostreams: a case study of twitter for flu risk analysis. In: Proceedings of the 4th ACM SIGSPATIAL IWGS, pp. 12–21 (2013)
6. Kamath, K.Y., Caverlee, J., Lee, K., Cheng, Z.: Spatio-temporal dynamics of online memes: a study of geo-tagged tweets. In: Proceedings of the 22nd International Conference on WWW, pp. 667–678 (2013)
7. Kim, K.S., Lee, R., Zettsu, K.: mTrend: discovery of topic movements on geo-microblogging messages. In: Proceedings of the 19th ACM SIGSPATIAL International Conference on Advances in GIS, pp. 529–532 (2011)
8. Kreiner, K., Immonen, A., Suominen, H.: Crisis management knowledge from social media. In: Proceedings of the 18th ADCS, pp. 105–108 (2013)
9. Kumar, A., Jiang, M., Fang, Y.: Where not to go?: detecting road hazards using twitter. In: Proceedings of the 37th International ACM SIGIR Conference on Research & Development in Information Retrieval, pp. 1223–1226 (2014)
10. Mendoza, M., Poblete, B., Castillo, C.: Twitter under crisis: can we trust what we rt? In: Proceedings of the First Workshop on SOMA, pp. 71–79 (2010)
11. Naaman, M.: Geographic information from georeferenced social media data. SIGSPATIAL Special 3(2), 54–61 (2011)
12. Sakai, T., Tamura, K.: Identifying bursty areas of emergency topics in geotagged tweets using density-based spatiotemporal clustering algorithm. Proceedings of the IWCIA 2014, 95–100 (2014)
13. Sakaki, T., Okazaki, M., Matsuo, Y.: Earthquake shakes twitter users: Real-time event detection by social sensors. In: Proceedings of the 19th International Conference on WWW, pp. 851–860 (2010)
14. Sander, J., Ester, M., Kriegel, H.P., Xu, X.: Density-based clustering in spatial databases: the algorithm gdbscan and its applications. Data Min. Knowl. Disc. 2(2), 169–194 (1998)
15. Thom, D., Bosch, H., Koch, S., Worner, M., Ertl, T.: Spatiotemporal anomaly detection through visual analysis of geolocated twitter messages. In: Pacific Visualization Symposium (PacificVis), 2012 , pp. 41–48. IEEE (2012)
16. Vieweg, S., Hughes, A.L., Starbird, K., Palen, L.: Microblogging during two natural hazards events: what twitter may contribute to situational awareness. In: Proceedings of the SIGCHI Conference on Human Factors in Computing Systems, pp. 1079–1088 (2010)
17. Yin, J., Lampert, A., Cameron, M., Robinson, B., Power, R.: Using social media to enhance emergency situation awareness. IEEE Intell. Syst. 27(6), 52–59 (2012)

Distributed Data Managing in Health Care Social Network Based on Mobile P2P

Ye Wang[1], Hong Liu[1], and Lin Wang[2(✉)]

[1] Second Military Medical University, Information Center, Shanghai, China
[2] The 85th Hospital of PLA, Shanghai, China
wanglininn@163.com

Abstract. Nowadays, more and more public health care information is being stored and transferred on Internet and mobile devices. However in developing countries, there are many rural residents could not afford the cost for commercial network. To solve the problem of being lack of a cheap and stable communication infrastructure directly between hospital servers and rural village residents' cellphones, this system managed to leverage mobile P2P and social network to build the health care information system. Based on the open-source P2P framework Alljoyn, social network engine Elgg (and Elgg Mobile), and distributed system HBase/Hadoop, we implemented HealthSocialNet, which focuses on immunization and antenatal care with high electrical efficiency and scalability. This research on health care social network system could help poor areas in developing countries implement and deploy a low-cost personalized health care data managing system fast.

Keywords: Mobile P2P · Health care · Social network

1 Introduction

In recent years, with the fast development of Internet and electrical public health care systems, quality of health care has been improved much. However, health care challenges still remain the greatest in rural areas of developing countries. Many rural areas are plagued with poverty, lower education levels, and reduced availability of health care professionals and health care centers and hospitals. For instance, more than 350,000 women in the world die annually from complications during pregnancy or childbirth; 99 % of these deaths occur in developing countries [1]. More than two million children under five die each year from preventable diseases due to inadequate access to vaccinations [2]. In China 2010, every 100,000 pregnancy would result 30.1 women die, and the mortality of children under-five-year is 2.01% [3].

Being lack of basic IT communication infrastructure, antenatal and children health care are big problems in Chinese rural areas. Residents in small villages are living far away from the city hospitals. And there are no certificated health workers living in the small villages. Thus it's very difficult for rural people to get enough public health care information or service.

© Springer International Publishing Switzerland 2015
A. Liu et al. (Eds.): DASFAA 2015 Workshops, LNCS 9052, pp. 170–180, 2015.
DOI: 10.1007/978-3-319-22324-7_14

According to the Chinese health care policy, there are 3 levels of rural health care system [4] in China which are listed below:

- County Level: Including county hospital, Center for Disease Control and prevention (CDC) and the Maternal and Child Health Hospital.
- Town Level: Including the township health center.
- Village Level: Including village clinic and health posts in remote areas.

At the village level, there are communities health workers working for village clinic and health posts. Township health center workers go to health posts (or village clinic) periodically. For example, Hanzhong is an old city located in Shaanxi, a northwest province of China. A small county named Liuba [5] is one of the 11 counties of Hanzhong. There is an estimated population of 53,000 within an approximate 25 km radius around Liuba. The Maternal and Child Health Hospital f Liuba is in charge of antenatal and childcare in a big region including Liuba itself and nine towns consist more than 200 villages. Most of these villages have their own village clinic with professional health worker living and working there. However there are still about 30 villages in remote areas (in some big mountains) do not have village clinic but only have health posts (one health post for several nearby villages) without health workers. So every month health workers from the township health centers go to the health posts conduct routine exams of pregnant women and child nutrition monitoring and immunizations. In the township health centers, health workers have computers and access to Internet helping them handle digitalized health care information. But in many remote areas, people do not have modern communication services as in the urban areas. In the health posts of remote villages, the health workers do not have any computers and even mobile phone service.

In recent years there have been a number of notable projects that have worked to improve health care in remote communities [6, 7]. Using mobile devices, these projects build distributed remote health care systems. However, what many of these projects have in common is that they leverage existing broadband or cellular networks to provide limited data connectivity and interaction with health professionals and patients. Unfortunately, although mobile phones are getting cheaper and cheaper, cellular access is still a problem in rural area: coverage can be spotty, or even completely non-existent; and voice, SMS and data costs can be out of reach for poor residents.

So our aim is building a low-cost personalized health care information managing and sharing system (focus on immunization and antenatal care) based on mobile devices for resource-constrained, rural regions such as in small villages of Liuba in western China. This system is called HealthSocialNet (HSN), which does not rely on the existing broadband or cellular networks. The following innovations are central to our system:

- **Leveraging Mobile P2P Technology to Integrate Physical Communications and Data Communications.** Commercial networks or Internet do not have adequate coverage in rural regions (or too expensive for rural people), hence we shall build some cheaper information bridges over the gap between urban areas and rural areas. In our system, people are not only users but also data sources and carriers. Their physical movements collect and spread system data using hybrid communication infrastructures including mobile P2P network and commercial networks.

- **Bottom-up Social Network Built Based on Peer-to-Peer Connections for Health Care.** In the P2P system, people are connected with some other people (including health workers and village residents) if they share health care data with each other. Using this social network system, people can get personalized information. Village residents can get the personalized immunization or antenatal care information, and the health workers can get their own working information. And the system managers could efficiently make different analysis using the social network data.
- **Fault-tolerant System on Unreliable Electricity and Communication Infrastructure.** Electricity and communication systems are not stable at rural areas in developing countries. Not only the base stations in the villages but also the servers in the city hospitals would probably encounter some electricity problems. We implemented a fault-tolerant distributed database and a health care data managing system based on open-source framework Hadoop/HBase at the server side to preserve data consistency and availability.
- **Supporting Multi-platforms Devices and using Open-source Software Architecture.** Our mobile P2P system support multi-platforms devices, including smart phones, Pads and even laptop PCs. And all the software architectures used in the system are open-source frameworks. Hence the entire system is low-cost and easy to be implemented and deployed in the poor areas of developing countries.

2 Methods

2.1 Workflow of Health Workers

Health worker's workflow in Liuba is a very common workflow for Chinese rural area. Doctors from town hospitals go to health posts periodically. They would do regular examinations for pregnant women and bring vaccination for small babies. Each pregnant woman and small baby has a health-care card with a unique ID. And each village has one or more community health workers helping them. The health workers would write down the record at health post and then take them back to the town hospital where these records could be digitalized and stored.

Not putting additional burdens on the health workers, we are not changing the basic workflow but using some new technology to help health workers simplify their work with mobile devices. The main workflow in HealthSocialNet is as below (see Fig. 1):

- The health worker download the new health care data from hospital servers to his mobile device, then goes to the health post.
- The health worker distributes the health care data to the community health worker (local village resident who help the health worker) and the residents at the health post using P2P network.
- Community health workers go back to the village and distribute the public health care data to all the mobile phones used by village residents through mobile P2P network.
- Residents go back to their own places to distribute the health care data to their friends using mobile P2P network.

Fig. 1. Workflow of HealthSocialNet

As it's shown in Fig. 1, the resident families live in the village cannot communicate with the hospital health workers directly unless they go to the city hospitals by themselves. Our system builds a bridge between the remote village and the city hospital. In this workflow, we integrate health care data communication with people's real physical communication.

2.2 Multi-Platform P2P Ad Hoc Network

In this health care system, we use peer-to-peer technology to build a proximity-based system sharing health care data. Peer-to-peer (P2P) is a kind of computing or networking distributed application architecture that partitions data, tasks or workloads among peers. With the significant advances of mobile devices and wireless communication technology, mobile P2P (or wireless P2P) has been a hot research topic [8]. Mobile P2P use different kinds of wireless communication technology including Bluetooth, WLAN, WiMAX, WifiDirect, NFC (Near Field Communication) and some other 3G or 4G communication techniques. P2P nodes in wireless systems are different mobile devices including multi-platform smart phones, laptops and Pads. In our system, we choose multi-platform mobile P2P network as a key part in HealthSocialNet mainly for 4 reasons as below.

- Communication Infrastructures are not Developed well Enough in Rural Areas of Developing Countries. Hence we can't distribute health care information through commercial network system. Mobile P2P would be a good supplementary communication method.
- Because of the Electricity Issue in Developing World, we Need a more Flexible and Decentralized System Need Less Electricity Power-supply and Less Communication Infrastructure. Mobile P2P system could preserve the data consistency better.
- Using 3G or 4G in Developing Country is Very Expensive (in China, it's 1$ for 40 MB/month and Additional 1$ for each extra 6 MB data). If we are going to share data, images, animations or videos of health care information on mobile devices, P2P would be a better choice for people no matter they have commercial mobile services or not.
- Mobile Devices are Getting Cheaper and Cheaper, however the Operating System for these Devices Varies. We don't know what kind of mobile device a doctor or a

village people would have. So we should build HealthSocialNet a cross-platform architecture.

Smart phones now have got the ability of being a portable Wi-Fi hotspot, which means a smart phone could be an access point to provide Wi-Fi service for several other devices in a small area. So it's convenient for the doctors or village residents only using their smart phones or other kinds of mobile devices to build small mobile ad hoc networks (MANETs). Combination of P2P systems and ad hoc networks allows for the creation of highly dynamic, self-organizing, mobile P2P systems. Nodes in the P2P system could use proximity-based ad hoc Wi-Fi networks to communicate with each other. And if it's not possible to use Wi-Fi, mobile devices can also use Bluetooth to build bilateral relation to connect with each other.

Even though the evolution of mobile technology is impressive, today's mobile phones still have substantial constraints that we must take into consideration when developing a mobile P2P design. A P2P system implementation could be very complicated cause it should consider multi factors such as device discovery, service discovery, security, connection pairing, and some other platform issues. There are several previous open source projects on building a general P2P protocol or software framework.

For example, JXTA (Juxtapose) is the first open source peer-to-peer protocol specification begun by Sun Microsystems in 2001 [9]. It's defined as a set of XML messages, which allow any device (including sensors, cellphones and computers) exchange messages and collaborate independently of the underlying network topology. And JXTA does not limit development to a specific language or OS environment.

However, a JXTA peer would be too complex to run on most mobile devices cause it needs to take care of many tasks and process messages at the XML-over-socket level. [IBM] Being a subproject of JXTA, JXME (JXTA for Java Micro Edition) is a developing platform mainly designed for constrained mobile devices using wireless communication. JXME was done in J2ME/MIDP 1.0 to implement a small-sized compatible JXTA on MIDP mobile device [10]. Cause it's based on J2ME/JAVA, JXME could be implemented on any mobile devices that support JAVA.

And in 2011, based on JXME Qualcomm Innovative Center (QIC) presented Alljoyn, an open-source application development framework that enables ad hoc, proximity-based device-to-device communication with an emphasis on mobility, security, and dynamic configuration [11]. Now supporting Microsoft Windows/WindowsRT, Linux, Apple IOS, and Android, it's easy to be embedded in mobile applications and has better performance.

We implement the P2P system mainly on Alljoyn framework to have more compatibility and efficiency. Hospital health workers, community health workers, and residents in villages with mobile phones can easily organize different proximity-based ad hoc P2P networks in different situations as Fig. 3 depicts. When they (Alljoyn peers) communicate with each other, they can use either WIFI or Bluetooth (Fig. 2).

In the P2P system, the hospital health workers download new health care data on an Alljoyn mobile device from hospital servers. At the health post, health workers distribute the health care data to the community health worker and the village residents. And in the rural village, community health worker and village residents can exchange the health care data using proximity-based Alljoyn P2P ad hoc network on smart phones.

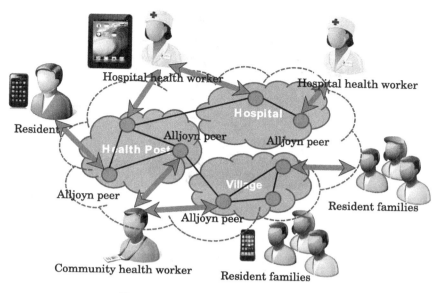

Fig. 2. Alljoyn P2P system architecture

2.3 Social Network and Distributed Data Storage

The mobile P2P connections are stored in HealthSocialNet system in which users could login and get personalized health care information. These communications automatically build a directed graph model stored in servers while the clients only store the direct connections. The data storage and managing layer of HSN is maintained in different hospital's servers. These servers (or virtual machines) are located in the hospital's private network, and different hospitals are connected via commercial networks. Health workers could get information about their work and share health care data with other health workers. Village residents could get immunization or antenatal care reminding information, and share health care information with their neighbors or friends. HSN could be reached via P2P network or Internet using computers and smart phone clients or browsers.

We use Elgg and Elgg mobile while implementing HSN system. Elgg is a well-known open-source social networking platform, which powers hundreds of web applications and social networks worldwide [12]. And Elgg Mobile could extend Elgg site into a mobile version that works faster and is easier to use on mobile devices. There are also application clients designed for Elgg mobile on IOS and Android platform. Based on open-source Alljoyn and Elgg mobile, we developed multi-platform HSN clients, which could be used for village residents to get immunization or antenatal care information from HSN system. And health workers on Internet could use the Elgg social network system to manage health care information.

And we also used a distributed database system as the data storage layer. Today distributed systems are quickly moving to the Cloud infrastructure to get more stability

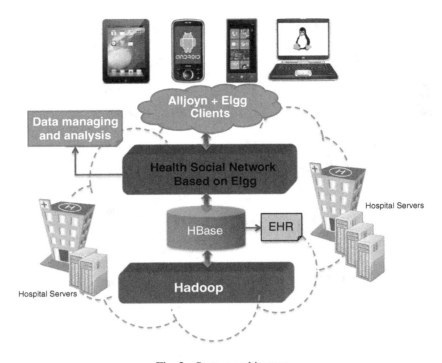

Fig. 3. System architecture

and scalability. In our system, we implemented a Private Cloud based distributed system using open-source HBase/Hadoop [13]. It has three advantages as below.

- **Data Availability and Fault-tolerance.** As we discussed above, electricity and communication systems are not stable in developing areas. Therefore data availability and fault-tolerance would be the most important factors we should take into consideration. In a SAAS (Software as a Service) cloud model, system data are stored in distributed nodes. Every part of data has several replicas saving in different nodes. When some nodes get offline due to electricity or network problems, the other data nodes can still keep working as a complete system.
- **Complex Big Data** HSN system would store and manage a big amount of health care information data. Since HL7 (health level 7) was presented, XML gradually became the most adopted health care information-exchanging standard. For example, EHR (Electrical Health Record) and CDA (Clinical Data Architecture) documents are XML based files that have complex hierarchical structures and each document has thousands of attributes [14]. And they often carries different media content such as videos and images, so each capacity of a health care document file could reach 50 MB. So we present a hybrid cloud based architecture, which integrates SQL and NOSQL databases.
- **Scalability and Efficiency.** The most distinguished advantage of cloud model is its ability to scale up with the increasing need. Health care data in HSN would be

increasing fast while system deploying. Hospitals and servers will be added into the system. Cloud model could distribute and balance the data and workload automatically according to the hardware and network infrastructure. And the distributed index and caching system would help to increase the querying and analyzing efficiency over big amount of data.

We now describe the detailed implementation of the cloud-computing infrastructure for the HSN data storage layer (HSNStore). It is a Hadoop/HBase based NOSQL system integrated with a MySQL based relational database. HSNStore system is composed of 2 major parts.

The first part of HSNStore is a JAVA Tomcat application uses MySQL as the underlying relational database. Considering the constrained hardware and software of the mobile clients in HSN system, we'll use Apache Axis to implement a SOAP ("Simple Object Access Protocol") Web Service interface [15] communicating with P2P network and commercial network. Some basic health information is stored in MySQL such as username, ID, phone number and so on for further query and analysis. It has complete replications in different hospital servers and would synchronize data periodically.

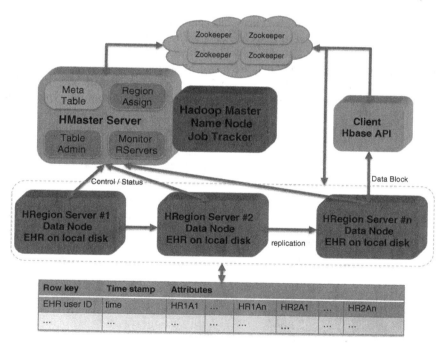

Fig. 4. HSNStore architecture

More complicated EHR data is stored in HBase. EHR document (usually XML document) has a big amount of attributes and many empty values. Built on top of Hadoop/HDFS, HBase provides a distributed, column-oriented data store modeled as Google's Bigtable, which fit for EHR data storage. And HBase has already implemented the low-level distributed storage of the HBase table data and nodes fault-tolerant processing. In HBase system, HRegion is the smallest unit for distributed storage, which is maintained in HRegionServers (Data Nodes in Hadoop system). And the HMaster server is responsible to assign the HRegions to each HRegionServer and balance the workload. In our system, HMaster server also acts as the Hadoop Name Node. Zookeeper clusters are responsible for coordinating the whole distributed system. Architecture of HSNStore system is shown in Fig. 4.

We model EHR documents into a big table, using user ID as the row key. When EHR table size reaches the threshold, HRegion split will be automatically triggered and data will be automatically distributed in HRegionServers (Data Nodes) by the load balancing of HMaster. As long as new HRegionServer (Data Node server) is added in, the storage capacity and the read-write throughput of the system will be increased as well.

3 Results

HealthSocialNet (HSN) is a public health care data sharing and management system focused on rural area in developing world. Individuals using this system include health workers (both community health workers and health workers form city hospitals) and local families. Each individual in HSN has data with various attributes reflecting their specific characteristics. From a modeling point of view, our system is very different from other electric health care systems that are based on a single unitary network. HSN is based on a hybrid network infrastructure including commercial network and mobile P2P network. Four major components (shown in Fig. 5) are designed in our prototype system.

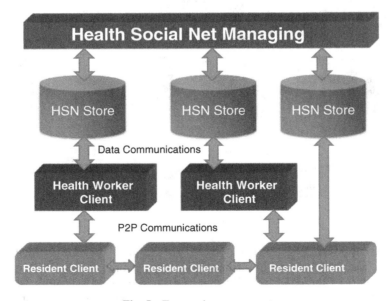

Fig. 5. Four major component

- **Health Worker Client:** We implement health worker client on multi-platforms including different smart phones, pad, and usual computer OS like Linux or Window. It's responsible for getting data from HSNStore and distributing data to resident client using P2P network based on WIFI or Bluetooth. And it could gather information from resident client then submit to the HSNStore.
- **Resident Client:** Resident client could get data from health worker client and distribute data to their friends using P2P network. All resident clients and health worker clients compose a directed social network. Data of this social network graph will be stored in the mobile phone clients and will be submitted to the health worker's client when they connect.
- **Distributed Data Storage:** Health care data are stored in distributed databases on different servers. These servers located in different hospitals have duplicated data storages, which are fault-tolerant. We implement the distributed data storage system (HSNStore) at the server side based on HBase.
- **HealthSocialNet Managing System:** We implement a distributed health care data managing system at the top of server side. It's in charge of managing and analyzing health care data, and generating health care notifications for every village people. And also it could communicate with national public health system on Internet to get information needed.

4 Conclusions

Combining mobile P2P network and commercial network to be the system's hybrid communication infrastructure, and also an open-source social network system based on distributed database to be the system's managing and data storage layer, a low-cost electrical health care information managing and sharing system which focuses on immunization and antenatal care in rural areas of China is presented. After analyzing the health workers' workflow in rural area of Liuba in China, we implement a prototype system for the health workers and village residents to efficiently manage and share immunization and antenatal information using mobile devices. This research on health care social network system based on mobile ad hoc network and distributed database could help poor areas in developing countries implement and deploy a personalized health care system fast.

References

1. Black, R.E., Morris, S.S., Bryce, J.: Where and why are 10 million children dying every year? Lancet **361**(9376), 2226–2234 (2003)
2. Jones, G., Steketee, R.W., Black, R.E., et al.: How many child deaths can we prevent this year? Lancet **362**(9377), 65–71 (2003)
3. Li, Z., Liu, J., Ye, R., et al.: Maternal prepregnancy body mass index and risk of neural tube defects: a population-based case–control study in Shanxi province, China. Birth Defects Res. A Clin. Mol. Teratol. **88**(7), 570–574 (2010)

4. Liu, Y.: Development of the rural health insurance system in China. Health Policy Plann. **19**(3), 159–165 (2004)
5. http://en.wikipedia.org/wiki/Liuba_County
6. Anderson, R., Blantz, E., Lubinski, D., O'Rourke, E., Summer, M., Yousoufian, K.: Smart connect: last mile data connectivity for rural health facilities. In: NSDR 2010, San Francisco, CA, June 2010
7. Greensfelder, L.: New wireless networking system brings eye care to thousands in India. http://www.berkeley.edu/news/media/releases/2006/06/06_telemedicine.shtml
8. Buchegger, S., Le Boudec, J.Y.: A Robust Reputation System for P2P and Mobile Ad-hoc Networks (2004)
9. Maibaum, N., Mundt, T.: JXTA: a technology facilitating mobile peer-to-peer networks. In: MobiWac 2002 International Mobility and Wireless Access Workshop, pp. 7–13. IEEE (2002)
10. Bisignano, M., di Modica, G., Tomarchio, O.: A JXTA compliant framework for mobile handheld devices in ad-hoc networks. In: Proceedings 10th IEEE Symposium on Computers and Communications, ISCC 2005. IEEE, pp. 582–587 (2005)
11. Namiot, D., Sneps-Sneppe, M.: Proximity as a service Future In: 2012 2nd Baltic Congress on Internet Communications (BCFIC), pp. 199–205. IEEE (2012)
12. Sharma, M.: Elgg Social Networking: Create and manage your own social network site using this free open-source tool. Packt Publishing, Berlin (2008)
13. Taylor, R.C.: An overview of the Hadoop/MapReduce/HBase framework and its current applications in bioinformatics. BMC Bioinform. **11**(Suppl 12), S1 (2010)
14. Dolin, R.H., Alschuler, L., Boyer, S., et al.: HL7 clinical document architecture, release 2. J. Am. Med. Inf. Assoc. **13**(1), 30–39 (2006)
15. Curbera, F., Leymann, F., Storey, T., et al.: Web services platform architecture: SOAP, WSDL, WS-policy, WS-addressing, WS-BPEL, WS-reliable messaging and more. Prentice Hall PTR, Englewood Cliffs (2005)

Survey of MOOC Related Research

Yanxia Pang[1,2]($\boxtimes$), Min Song[1], YuanYuan Jin[1], and Ying Zhang[1]

[1] East China Normal University, Shanghai, China
ladypang@163.com
[2] Shanghai Second Polytechnique University, Shanghai, China

Abstract. MOOC expands fast in recent years so that it shows both advantages and bottlenecks. MOOC platforms try to solve their massive learning dealings and make further research on their MOOC data. E-learning research organizations combine MOOC research into their area to find better learning models of MOOC. Some MOOC related research organizations follow the frontier of MOOC. Research of MOOC is on the way.

Keywords: MOOC · Research · Provider · Learning · Survey

1 Introduction

MOOC(Massive Open Online Course) develops fast in recent years. In the fall of 2011, Stanford University launched 5 courses online. The instructor of course artificial intelligence Sebastian Thrun started his MOOC company Udacity. The instructors of MOOC machine learning Andrew NG and probabilistic graph model Daphne Koller built Coursera. Shortly after that MIT and Havard University launched edX too. Now no one is sure how many MOOC providers there are. Universities open their MOOC platforms one by one, and lots of MOOC providers were built for different types of learners. MOOC platforms spread to different countries around the world.

In Apr. 2014 Coursetalk (http://www.coursetalk.com) a MOOC platform with courses from different providers) has 36 providers. In Jan.2015 there were 59 providers on Coursetalk. In about 9 months it has increased almost double.After early 2012, the number of MOOC courses keeps climbing fastly. By Jul. 2014 there have been 400 Universities/Institutions, 2500 Instructors, 1800 courses started.2 2013 and 2014 are MOOCs fast rising year.

MOOC rushes in recent 2 or 3 years and related research work on it increases too. Here we will introduce some information on MOOC research organizations and related papers.

2 Basic Research of Recent MOOC

In 2013 MOOC Research (http://www.moocresearch.com) launched an initiative project to call on researchers submit their papers for further fundamental aid. By

© Springer International Publishing Switzerland 2015
A. Liu et al. (Eds.): DASFAA 2015 Workshops, LNCS 9052, pp. 181–189, 2015.
DOI: 10.1007/978-3-319-22324-7_15

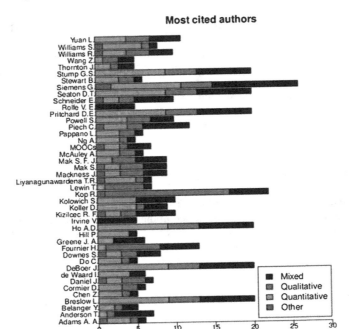

Fig. 1. Most cited authors

the submissions, MOOC Research calculates the most cited authors as Fig. 1 and papers. Authors in most cited papers are also cited more. Yuan L and Siemens G are also authors of most cited paper. Some of the authors of Fig. 1 research on learning analytics like Siemens G(George Siemens), some on computer science like Piech C (Chris Piech of Stanford University), some are not one person but a web source like MOOCs(a blog with thoughts of recent MOOC news).

In the area of MOOC, some papers show the basic reality of MOOC, and give the data of the current MOOC. 10 pieces of papers were proposed by MOOC Re-search (http://www.moocresearch.com) as a start [1–10]. Most of them talks in the role of the education researcher, meanwhile they provide big amount of data in MOOC. For an example, the paper by Lori Breslow1 about edX gives the current situation about edX including enrolling reasons, effort distribution among course learning process, learner locations and age distribution [1]. All the data shows the learning situation on edX. That can help us find the current problems of MOOC. This papers author Lori Breslowis the director of MIT Teaching and Learning Laboratory (TLL) (http://tll.mit.edu/about/lori-breslow-phd) as menas her works more on educa-tion. Another paper [3] (Mackness, J. 2010) also lists their survey result collected by questionnaire through e-mail and BBS. The paper also points out the difference in learning behavior between BBS users and blog users. The year of the MOOC(Pappano, L. 2012) is also about current

MOOC. But it focus on 4 courses from 3 platforms. It demonstrates some advantages and disadvantages of MOOC and gives advice for better interactive, review and practice. Connectivism as a new theory for online learning is always studied as a start [5]. George Siemense introduce connectivism in his paper [10] (2005) in the internet background. It proposes connectivism between learner, knowledge. Connectivism features knowledge transmissible between different objects. The theory finds a new model of learning in knowledge sea of the internet world though it is somewhat hard to be apply into the reality. Daniel analyzes MOOC history(2012) [4]. He talks about the pradox of MOOC, analyzes the problems and gives his answer. The paper focuses on the most important problems on learning mode, granted certification and completion. In the next year Kizilcec published a paper special on completion problem(2013). [2] It gives a framework according to the learning trajectory and suggest users with further advice.Among them there is one piece of paper helpful for MOOC constructor (Belanger, Y., and Thornton, J. 2013). It talks about the story of Duke Universitys first MOOC. The whole process is introduced step by step. At the end the paper evaluates the results and gives corresponding advice [7].

3   MOOC Research Organizations

With the development of MOOC, more and more research labs of MOOC emerge. They have different points on MOOC according to the percentage of submission in different areas in the MOOC Research Initiative. As expected education is the domain field. Next is computer science. Most MOOC research organizations prefer to research the education rules including assessment, learning model, etc. some come from the earlier research organizations on e-learning such as Society for Learning Analytics Research(SoLAR). A few focus on the computer science to apply in MOOC like lytics(learning analytics) lab at Stanford. Among all those some were built to serve their open course platforms like MITx research center and HarvardX serving their MOOC development. Lytics lab serves MOOC platform from Stanford Coursera. International Educational Data MiningSociety(EDMS) is an organization especially on educational data mining including Papers of this lab cover field both on education and computer science. Next we will introduce these labs and societies about their basic information, and give some of their research papers to introduce their research points.

3.1   MOOC Research Organizations in Educational Area

Society for Learning Analytics Research (SoLAR). Society for Learning Analytics Research (SoLAR) is a sister organization of EDMS. The difference lies on the area. SoLAR pays more attention on the education. It has ever held many lectures for teachers to discuss and learn teaching techniques. Also it is an inter-disciplinary network of leading international researchers who are exploring the role and impact of analytics on teaching, learning, training and development.

SoLAR has been active in organizing the International Conference on Learning Analytics & Knowledge (LAK) and the Learning Analytics Summer Institute (LASI), launching multiple initiatives to support collaborative and open research around learning analytics, promoting the publication and dissemination of learning analytics research, and advising and consulting with state, provincial, and national governments. Its journal is named as Journal of Learning Analytics (http://solaresearch.org/about/). Compared with EDMs, SoLAR tends to work more on education than techniques. The Journal of SoLAR also collects papers of LAK.Papers in this journal are mainly educational research especially learning analytics. The research area is more than MOOC [21,22]. Some research on psychometric data to modelling achievement [20].

HarvardX. There are many organizations on online learning and more in the area of education. Teaching and Learning Laboratory of MIT (http://tll.mit. edu) is led by Lori Breslow who wrote the well-known paper Studying Learning in the Worldwide Classroom Research into edXs First MOOC. Open education-consortium (http://www.oeconsortium.org/) is a big circle including MIT and many other open education departments in countries. Harvard and MIT built edX together. HarvardX (http://harvardx.harvard.edu) supports faculty innovation in the use of technology in teaching & research on campus, online, and beyond. HarvardX researchs mainly on their open course platform from MOOC learning to course information. Sometimes papers here are like reports more. Anne Lamb from Harvard University compared big data from their platform and triedmethods to improve the forum actions [13].

MOOC Research. HarvardX and MITx both keep researching projects on MOOC foundation MRI. In 2013 a project named MOOC Research Initiative (MRI, http://www.moocresearch.com/) was funded by the Bill & Melinda Gates Foundation as part of a set of investments that intended to explore the potential of MOOCs to extend access to postsecondary credentials through more personalized, more affordable pathways. The conference was held in Dec. 2013 at University of Texas, Arlington. The following 3 project candidates submitted to MRI show more research is supposed to solve detailed MOOC application problems for better learner experience. One accepted project analyses MOOC corpus with CDA(Critical Discourse Analysis) and classifies 15 themes to observe some problems of MOOC [14]. Another project(John Whitmer, Ed.D.) [15] majors in a Remedial English Writing MOOC to find the reason of student engagement in the MOOC. In a particular MOOC course it visualizes the whole activity data of certain students and gives the relation between student engagement and course design. We can find a projection psychology [16]. It analyses the 3 weeks psychology intervention in the MOOC and finds the close relationship between psychology,interest and motivation. The projects of MRI aim to improve MOOC learning with solution in different fields.

3.2   MOOC Research Organizations in Computer Science Area

IAIED (International Artificial Intelligence in Education Society).
IAIED (http://iaied.org/) is an affiliate of EDMS. Some papers from Lytics
Lab are published on IAIEDS journal IJIAIED (International Journal of Artificial Intelligence in Education) [31]. IAIED is an interdisciplinary community
at the frontier of the fields of computer science, education and psychology. It
promotes rigorous research and development of interactive and adaptive learning environments for learners of all ages, across all domains. The society brings
together a community of members in the field through the organization of Conferences, Journal, and other activities of interest. IAIED Society is governed by
an Executive Committee according to the IAIED Constitution, which seeks to
support AI in Education developments throughout the international community.

IJIAIED publishes papers on the application of artificial intelligence techniques and concepts to the design of learning support systems. Much work on
practical learning experience is analysed and experimented for better learning
result [23,24]. An experimental lecture for pilot study was conducted and online
tools and learning mode were put into learning as a way to promote learning [25].

Lytics Lab of Stanford University. Lytics lab (http://lytics.stanford.edu/)
is not about conference or journal. It came from a discussion based on the data
from the new MOOC offered by Stanford faculty by 5 Stanford doctoral students
from Education, Communication, Electrical Engineering and Computer Science
in 2012. Professor John Mitchell was appointed Vice Provost for Online Learning
(VPOL) at Stanford. It engages in use-driven research and data-driven design to
develop collective knowledge around improving online learning. Papers to solve
MOOC application problems were soon published. The big data from the MOOC
platform Coursera was researched for online learning improvement.

In 2013 it published 14 papers by the list of their website. By Jan.2015,
they have published more than 30 pieces of publications on top international
conferences. Many of them focus on detailed problems of MOOC platform like
homework dealing, code submission, etc. The platform requirement and MOOC
data make lytics lab a frontier research lab. Their papers are also published on
EDMS and IAIED.

Papers of this lab can be divided into different areas about online learning
including computer science, psychology and education. It may focus on a detailed
problem like the code submission problem of course "machine learning" [31].
Some try to find better education modes for online education [32]. Some do
research work of data science on the big educational data from Stanford MOOC
platforms [26]. Some try new computer research in MOOC environment and
most use MOOC data [26,27] for better MOOC dealings of the platform like
instructors' face show problem [28].

The area of their research includes mainly education and computer science.
John Michell from computer science and Roy Pea from education are the two
main leader of the lab. The lab keeps providing seminars open to faculty, stu-

dents, researchers, and others from academia or industry. And they keep showing their project on the website to discuss on the seminar.

Rakesh Agrawal from Microsoft Research. Another research from computer science area is learning material recommendation on textbook by Rakesh Agrawal from Microsoft Research (http://www.educationaldatamining.org/). Professor Rakesh Agrawal does some research work on learning material recommendation including images, videoes, etc. All this work was done on the usage of concepts extracted by algorithms and NLP tools. In one of the recent papers Rakesh talks about similarity search with concept graph [11]. In this paper the complex query object was transmitted into concept graph. Series of key words from concept graph are used for query. The combined selection results of key words forms the final result. This paper gives a better way to find related material in learning without comparing all the internet content. And the group tries to attach videos into textbook at appropriate locations. Both location in textbook and granularity of video assignment are considered [12].

3.3   MOOC Research Organizations Combining both Education and Computer Science

International Educational Data Mining Society. International Educational Data MiningSociety (EDMS, http://www.educationaldatamining.org/) was first proposed as a workshop in 2005 (at AAAI). In 2008 the first conference on EDM was held in Montreal of Canada.The society was formally founded in July 2011. By 2013 it has held 7 conferences on educational data mining. The Society is incorporated as a non-profit corporation in the State of Massachusetts. It has a non-profit tax-exempt status under U.S. . The director is Ryan Baker from Columbia University, USA. He holds a course education in big data on Coursera about educational data mining methods and tools. Most members of EDMS are young men from USA, Canada and some European and Asian countries.

EDMS focuses more on the application of data mining theory and tools to solve educational problem. Though the research is not all about MOOC, in recent years with the MOOC research developing some papers about MOOC can be found on JEDM [33]. And EDMS adapts their main goal from educational support research to increasing opportunities for participation and input from the broader EDM com-munity, and maintaining scientific quality and our community's core values and focus, while setting up an organization that can function smoothly, indefinitely. EDMs aims to research on educational data to serve learning for better assessment, real-time support, etc. The research mainly on the area of educational technology by the rules of educational theory.

EDMS also posts data resources and mining tools free on their website. Maybe thats one of the reason it becomes popular so fast. Coursebuffet (a MOOC platforms) invited one member of EDMS to support their instruction design. In 2009 EDMS built their journal JEDM(Journal of Educational Mining) free to both authors and readers. Professor Rakeshs group from Microsoft Research published 2 pieces of papers on vol 6 2014 of JEDM [11,12].

4 Summary

MOOC related research covers many fields. We introduce only part of ralted research organizations about MOOC. For every organization we write about the main research points and refer some recent papers. MOOC was proposed in 2008 aiming at e-learning and got popular in 2011 by the work of MOOC platforms. So the related MOOC research started in 2012. But much work has been done on e-learning. Here we talk about a few e-learning related organizations which may not on MOOC. Though we classify organizations by education and computer science, one organization may research on both areas. MOOC area calls for more concern on its research to solve problems on MOOC platforms and better MOOC learning modes and learning design.

References

1. Breslow, L., Pritchard, D.E., DeBoer, J., Stump, G.S., Ho, A.D., Seaton, D.T.: Studying learning in the worldwide classroom: research into edXs first MOOC. Res. Pract. Assess. J. **8**, 13–25 (2013). Summer
2. Kizilcec, R.F., Piech, C., Schneider, E.: Deconstructing disengagement: analyzing learner subpopulations in massive open online courses. In: Third International Conference on Learning Analytics and Knowledge, LAK 2013 Leuven, Belgium (2013)
3. Mackness, J., Mak, S., Williams, R.: The ideals and reality of participating in a MOOC. In: Proceedings of the 7th International Conference on Networked Learning 2010, University of Lancaster, Lancaster, pp. 266–275 (2010)
4. Daniel, J.: Making sense of MOOCs: Musings in a maze of myth, paradox and pos-sibility. Korean National Open University, Séoul (2012)
5. Pappano, L.: The year of the MOOC. New York Times, 2 November 2012. http://www.nytimes.com/2012/11/04/education/edlife/massive-open-online-courses-aremultiplying-at-a-rapid-pace.html
6. McAuley, A., Stewart, B., Siemens, G., Cormier, D.: The MOOC Model for Digital Practice (2010). http://oerknowledgecloud.org/sites/oerknowledgecloud.org/files/MOOC_Final.pdf
7. Belanger, Y., Thornton, J.: Bioelectricity: A Quantitative Approach. dukespace (2013). http://dukespace.lib.duke.edu/dspace/bitstream/handle/10161/6216/Duke_Bioelectricity_MOOC_Fall2012.pdf
8. Kop, R., Fournier, H., Mak, J.S.F.: A pedagogy of abundance or a pedagogy to support human beings? participant support on massive open online courses. Int. Rev. Res. Open Distance Learn. **12**(7), 74–93 (2011). Special Issue
9. Siemens, G.: MOOCs are really a platform. eLearnspace (2012). http://www.elearnspace.org/blog/2012/07/25/moocs-are-really-a-platform/. Accessed 21 Sept 2012

10. Siemens, G.: Connectivism: a learning theory for the digital age. Int. J. Instr. Technol. Distance Learn. **2**(1), 2005 (2005)
11. Agrawal, R., Gollapudi, S., Kannan, A., Kenthapadi, K.: Similarity search using concept graphs. In: International Conference on Information and Knowledge Management (CIKM), ACM – Association for Computing Machinery (2014)
12. Kokkodis, M., Kannan, A., Kenthapadi, K.: Assigning educational videos at appropriate locations in textbooks. In: International Conference on Educational Data Mining (EDM) (2014)
13. Lamb, A., Smilack, J., Ho, A., Reich, J.: Addressing common analytic challenges to randomized experiments in moocs: attrition and zero-Inflation. In: ACM Conference onLearning @ Scale (2015)
14. Selwyn, N., Bulfin, S.: The discursive construction of MOOCs as educational opportunity and educational threat. MOOC Research Initiative (2013)
15. Whitmer, J.: Patterns of Persistence: What Engages Students in a Remedial English Writing MOOC? MOOC Research Initiative (2013)
16. Greene, D.: Learning Analytics for Smarter Psychological Interventions. MOOC Research Initiative (2013)
17. Jackson Schultz, W.: Security Outlook: Look Out! What to watch for in 2015. Security Consultant – Information Security Practice. GraVoc Associates, Inc., Peabody (2015)
18. EVERGAGE: The Perceptions of RealTime Marketing and How It's Achieved (2014). http://www.evergage.com
19. EVERGAGE: RPI Versus RFM The Difference Between Real-Time and Near-Time Marketing March (2014). http://www.evergage.com
20. Gray, G., McGuinness, C., Owende, P., Carthy, A.: A review of psycho-metric data analysis and applications in modelling of academic achievement in tertiary education. J. Learn. Analytics **1**(1), 75–106 (2014)
21. Halatchliyski, I., Hecking, T., Goehnert, T., Ulrich Hoppe, H.: Analyzing the main paths of knowledge evolution and contributor roles in an open learning community. J. Learn. Analytics **1**(2), 72–93 (2014)
22. Jayaprakash, S.M., Moody, E.W., Laurí, E.J.M., Regan, J.R., Baron, J.D.: Early alert of academically at-risk students: an open source analytics initiative. J. Learn. Analytics **1**(1), 6–47 (2014)
23. Yoo, J., Kim, J.: Can online discussion participation predict group project performance? investigating the roles of linguistic features and participation patterns. J. Artif. Intell. Educ. **24**(1), 8–32 (2014)
24. Walker, E., Rummel, N., Ruhr-Universität Bochum, Koedinger, K.: Adaptive intelligent support to improve peer tutoring in algebra. Int. J. Artif. Intell. Educ. **24**(1), 33–61 (2014)
25. Tegos, S., Demetriadis, S., Tsiatsos, T.: A configurable conversational agent to trigger students productive dialogue: a pilot study in the CALL domain. Int. J. Artif. Intell. Educ. **24**(1), 62–91 (2014)
26. Kizilcec, R.F., Schneider, E.: Motivation as a lens to understand online learners: data-driven design with the OLEI scale. ACM Trans. Comput. Human Interact. (ACM TOCHI), 22(2). (2015, in press)
27. Thille, C., Schneider, E., Kizilcec, R. F., Piech, C., Halawa, S., Greene, D.: The Future of Data-Enriched Assessment. Research and Practice in Assessment, Special Issue on Big Data and Learning Analytics. (2014, in press)
28. Kizilcec, R.F., Bailenson, J.N., Gomez, C.J.: The instructors face in video instruction: evidence from two large-scale field studies. J. Educ. Psychol. (2015, in press)

29. Yuan, L., Powell, S.: MOOCs and open education: implications for higher Education. In: CETIS, International Conference on Advanced Learning Technologies (2013)
30. Allen, E., Seama, J.: Changing course: ten years of tracking online education in the United States. BABSO Survey Research (2013)
31. Huang, J., Piech, C., Nguyen, A., Guibas, L.: Syntactic and functional variability of a million code submissions in a machine learning MOOC. In: The MOOCshop Workshop, International Conference on Artificial Intelligence in Education, Memphis, TN (2013)
32. Kizilcec, R.: Collaborative learning in geographically distributed and in-person groups. In: The MOOCshop Workshop, International Conference on Artificial Intelligence in Education, Memphis, TN (2013)
33. Piech, C., Huang, J., Chen, Z., Do, C., Ng, A., Koller, D.: Tuned models of peer assessment in MOOCs. In: the International Conference on Educational Data Mining, Memphis, TN (2013)
34. Siemens, G.: An Overview of the MOOC Research Initiative: The project, literature, and landscape. Society for Learning Analytics Research (2013)

Modeling Large Time Series for Efficient Approximate Query Processing

Kasun S. Perera[1(✉)], Martin Hahmann[1], Wolfgang Lehner[1],
Torben Bach Pedersen[2], and Christian Thomsen[2]

[1] Database Technology Group, Technische Universität Dresden, Dresden, Germany
{kasun.perera,martin.hahmann,wolfgang.lehner}@tu-dresden.de
[2] Department of Computer Science, Aalborg University, Aalborg, Denmark
{tbp,chr}@cs.aau.dk

Abstract. Evolving customer requirements and increasing competition
force business organizations to store increasing amounts of data and
query them for information at any given time. Due to the current growth
of data volumes, timely extraction of relevant information becomes more
and more difficult with traditional methods. In addition, contemporary
Decision Support Systems (DSS) favor faster approximations over slower
exact results. Generally speaking, processes that require exchange of data
become inefficient when connection bandwidth does not increase as fast
as the volume of data. In order to tackle these issues, compression techniques have been introduced in many areas of data processing. In this
paper, we outline a new system that does not query complete datasets
but instead utilizes models to extract the requested information. For
time series data we use Fourier and Cosine transformations and piece-
wise aggregation to derive the models. These models are initially created
from the original data and are kept in the database along with it. Subsequent queries are answered using the stored models rather than scanning
and processing the original datasets. In order to support model query
processing, we maintain query statistics derived from experiments and
when running the system. Our approach can also reduce communication
load by exchanging models instead of data. To allow seamless integration
of model-based querying into traditional data warehouses, we introduce
a SQL compatible query terminology. Our experiments show that querying models is up to 80 % faster than querying over the raw data while
retaining a high accuracy.

1 Introduction

In the current age of information, data is generated at an unprecedented state
and stored in increasing volume. As today's economy and society run on knowledge, organizations not only need to keep massive amounts of data but also must
be able to extract information from them. This concerns every imaginable area
of modern life from stock market fluctuations and news feeds in social networks
to renewable energy supply and vehicle travel patterns in traffic control systems.
As the volume of gathered data grows bigger in all these application scenarios

© Springer International Publishing Switzerland 2015
A. Liu et al. (Eds.): DASFAA 2015 Workshops, LNCS 9052, pp. 190–204, 2015.
DOI: 10.1007/978-3-319-22324-7_16

two major problems emerge. First, data must be kept in order to utilize it, which requires sophisticated storage systems. Second, querying the data becomes more demanding and poses new challenges for the information extraction process.

While the type of generated data varies with each application domain, time series are one of the most common formats of data as they capture the change/behaviour of certain measures/objects over time. This is why analyzing and querying time series data is an important part in many decision making processes. For example, a company that sells certain products is interested in sales patterns occurring in the past in order to make decisions about the future. Such queries often include average values, counts and sums for items in certain periods of time or geographic regions. In order to obtain the most up-to-date information, these queries are issued repeatedly which means large amounts of data must be loaded, copied and processed frequently. As data volume increases while computational resources are often constrained, query processing of large data becomes inefficient with regards to time. Although the answers that are generated are exact, decision making often favours an approximate answer at the right time over an exact one that is available too late. Before introducing our model-based database system we would like to define what a model is in our context. *A model is a representation, generally a simplified description, especially a mathematical one, of a system or process to assist in calculations and predictions* [5]. As an example systematic sample from a Fast Fourier Transformation (FFT) of a complex signal can be seen as a model of the original signal.

Contemporary literature suggests different methods for different approximate queries and generally associate one model with one or more queries. This approach limits the utilization of models as a model designed for one query cannot be used for a different query. In this work we propose a system that maintains a pool of models within database itself such that queries can either be answered with individual models or with a combination of 'general-purpose' models. By utilizing a model-based representation of the underlying data we are able to generate approximate answers much faster than exact answers from the original data. In addition our representative models have a small memory footprint and still provide query results with high accuracy. In the remainder of this paper we first outline the concept of our system, then describe early results for model creation and querying on time series data and finally give an outlook regarding the future potential of our approach.

2   Model-Based Database System Concept

It is obvious that large volumes of data can significantly reduce the runtime efficiency of complex queries in traditional database systems. One approach to tackle this issue is to materialize query results. While being a useful approach, query materialization does not provide full flexibility and scalability as materialized queries have to be updated whenever there are changes to the query or the data. To overcome this limitation while still increasing query performance we propose a model-based database system which runs queries against representative models to produce approximate results. This offers multiple benefits:

(1) models have a smaller memory footprint than the raw data, (2) maintaining a pool of different models for the same data allows to select an optimal model for each issued query, and (3) models can be used to re-generate the approximated original data with high accuracy. In this section we outline the concept of a system that enhances query efficiency by representing data with models and executing queries over these models.

When a query is issued against our model-based database system, the database engine decides whether to use a model or the conventional database approach to answer the query. The conventional approach is to scan and fetch the required raw data from the storage (usually disk storage) and execute the query over the dataset. If the decision is to use models, an appropriate stored model is accessed and approximate results are produced. Figure 1 depicts the overall architecture of the query processing system. After the query engine decides that a query will be answered with a model, it selects those models that produce the best possible results for the given query. In order to make this decision, the system maintains statistics of past queries that were executed on both models as well as raw data. This means, our proposed system undergoes an initial learning phase before it can fully apply the model based approach. During this phase, query statistics for queries running over raw data as well as models are recorded.

Fig. 1. Overall system architecture with both model and data query engines

The major part of the result-generation process takes place in the query processor. It is responsible for parsing the user query, extracting query components, optimizing the operations and finally executing the optimized query. Similar to the data query processor, the first step of the model query processor is to parse the user query into identifiable query objects. During parsing it is necessary to identify parts that relate to model query processing (optimization and execution) and other query operations. The proposed query parser extracts query objects such as Projections (*SELECT avg(productsales)*), Source Relations (*FROM sales*), Selections (*WHERE product='xxx'* or *WHERE date BETWEEN start to end*), Model Selections (*USE MODEL modelCategory*) as well as error- and time-bound parameters. An example of such a query is given below:

SELECT avg(sales)
FROM sales
WHERE date **BETWEEN** 01-01-2010 to 31-12-2013
USE MODEL modelCategory
ERROR WITHIN 10 %
RUNTIME WITHIN 5 SECONDS;

One of our main goals is the seamless integration of models into the database system in order to enable users to perform the same SQL query that they would run on raw data. In addition we add specific parameters to the query syntax namely TIMEOUT, ACCURACY, and MODEL. A sample query syntax that employs these new parameters is shown above. Our taxonomy is similar to [1] and [6]. The given example shows that users can explicitly define which model should be used to answer the query. In addition the user can define an error bound for the results. This directly affects query performance as approximate results can be generated from high-level models with less accuracy in a very short time, or from low-level models with high accuracy at the cost of a longer running time. In our model hierarchy, high-level models refer to coarse grained models with low accuracy and low-level model with high accuracy that represent the original data with fine granularity.

An important part of our proposed system is the query optimization which uses stored query statistics to decide the optimal set of models to answer a query within given time and accuracy constraints.

3   Model Querying for Time Series Data

While we focus on the modeling of time series data in this paper, our system can be extended to other types of data with different sets of models. In this section we describe an early stage of our system for time series data, where we focus on query efficiency and accuracy on large time series.

Our system employs a model pool where different models, built on the original data, are stored to answer user queries. Our goal is to store multiple models of the same data, so that queries with different parameters can be answered by a stored model or combination of them. For simplicity we define our time series as $TS = (t_1, v_1), (t_2, v_2), ..., (t_n, v_n)$ where TS defines the whole time series and (t_x, v_x) defines an individual (time, value) pair of the time series. A final model built on the original time series TS is referred to as $\Sigma(TS)$ and is a collection of sub-models of $\sigma(ts)$ such that each represents a smaller part of the larger time series. The transformation of TS into the model $\Sigma(TS)$ is described in the following sub section.

3.1   Model Construction

In the current version of our system, we use the described approaches to represent and query large time series data. In this section we describe how we construct a model pool to represent time series data. While we focus on individual time series in this paper, we plan to extend our model pool for multiple time series so that one model can represent different similar time series. We focus on large time series with millions of points. Generally, different patterns emerge in different periods of long time series. This is why it is necessary to consider a large time series as a collection of continuous smaller time series, which in general show more manageable patterns within a given range. Models derived from these sub-time-series have a higher accuracy as they only need to represent a compact set of local changes. The model construction process that turns the original time series into a model is depicted in Fig. 2. It starts by decomposing the long original time series into smaller parts and then models each part using the modeling approaches we describe in the next section. Each part of the original time series is considered as a single model and stored in a matrix along with a partition ID that indicates the part of the original time series the model represents.

The models created that way are then combined to generate query results. When a given time series TS is partitioned into smaller parts $(ts_1, ts_2, ...)$, each part is going through a model construction phase. In this phase, an initial model is constructed on each partition ts_i and passed on to the model evaluation. There, it is tested for accuracy against the original data and added to the model pool if it reaches a user-defined accuracy level. If the model fails the accuracy test, an update is performed in order to adjust model parameters and increase accuracy. If a model fails to maintain adequate accuracy bounds then raw data for that

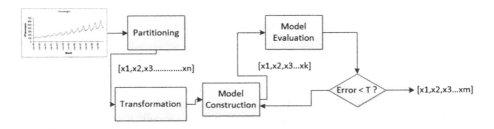

Fig. 2. Step-by-step construction of models from time series.

particular part of the time series is stored. The sequence of model evaluation and update is repeated until the given accuracy criteria are fulfilled. In the following, we describe the set of representative models that we use to transform TS into $\Sigma(TS)$.

Aggregation over Time Granularity. Segmenting a time series into several meaningful smaller parts and using these individual parts for queries over the original time series is a widely used approach in many applications. An example of a segmentation-based system for similarity pattern querying is Landmark [8] by Perng et al. It proposes a method to choose landmark points—points of the time series which best describe its pattern—and uses them in similarity checks. For our implementation we use time granularity aggregation, i.e. the mean, to represent the underlying time series. Piece-wise polynomial integration and regression has been used as a data reduction technique where it is used to reduce the frequency by aggregating nearby points. One such example is Best-Time [10] where the authors use piece-wise aggregation to reduce the frequency of original time series in to manageable frequency. For example if the frequency in original time series is 5 Hz and required frequency is 1 Hz, the deduction process aggregates 5 points using interpolation or regression method and present it with 1 point. In our aggregated mean time series each original segment is represented by its mean value. Thus an approximated version of the original time series is created. The number of points used for aggregation can be defined by the user. Using more points for the aggregation leads to a more compressed but also more coarse-grained model that can compromise accuracy. In our implementation we used 5, 10 and 20 points as aggregation levels which are similar to compressed levels in other types of models.

Discrete Fourier Transformation. Discrete Fourier Transformation (DFT) is a widely known technique to represent complex signals, using a collection of symmetric signals, where each signal is represented with two components. A mathematical representation of DFT decomposition can be described as:

$$X_k = \sum_{n=0}^{N-1} x_n e^{-i2\pi k \frac{n}{N}} \tag{1}$$

where $x_0, ..., x_{N-1}$ are complex numbers. Evaluation of this equation has $O(N^2)$ time complexity that can be reduced to $O(N\log N)$ with Fast Fourier Transformation algorithms such as CooleyTukey or Prime-factor FFT.

A time series can be viewed as a signal when we consider it as a continuous signal rather than discrete values in discrete time points. Thus, we use FFT to analyze and represent our time series. When using DFT/FFT it returns a collection of components with the same number of points in the input time series. Not all these components are important as the latter part is a repetition of the first part in a mirror form. In order to obtain an approximation, a fewer number of FFT components is adequate.

Discrete Cosine Transformation. Similar to DFT, Discrete Cosine Transformation also decomposes a given signal to simpler components and uses these components to get an approximated signal. DCT is widely used in image and signal compression as it has a strong energy compression behavior. DCT-II is the most common compression technique and can be formalized as:

$$X_k = \sum_{n=0}^{N-1} x_n \cos\left[\frac{\pi}{N}\left(n + \frac{1}{2}\right)k\right] \tag{2}$$

where $k = 0, ..., N - 1$ and $x_0, ..., x_{N-1}$ are real numbers in contrast to the complex numbers in DFT/FFT.

The selection of DCT or DFT is based on user requirements. Model construction for DCT is more expensive than for DFT. In contrast DCT models have a smaller memory footprint compared to DFT models with the same number of components. As we suggest offline and online model construction, we evaluate both approaches in our experiments.

3.2   Query Computation over Models

In its current state, our implementation supports standard SQL aggregate queries like AVG, SUM and Histogram analysis. Because our query results are approximations, they significantly increase the run-time efficiency of query processing. As SUM queries are one of the most widely used aggregation query types, we first demonstrate how they can be answered with FFT and DCT models. Assume a query such as;

SELECT avg(sales)
FROM sales
WHERE date BETWEEN 01-01-2010 to 31-12-2013
USE MODEL DCT
ERROR WITHIN 10 %

After parsing the query, the DCT model category is selected to run the query and decides the model parameters using the given error boundaries. The number of sub-models which are used in the query depends on the *WHERE* predicate.

We denote that each model has its own way of calculating aggregation parameters. We begin by defining our models as $\Sigma(TS) = [\sigma(ts_1), \sigma(ts_1), ..., \sigma(ts_n)]$ where $\sigma(ts_i) = [c_1, c_2, c_3, ..., c_k]$ and k is the number of FFT/DCT components. Each $\sigma(ts_i)$ represents a single partition (size C) of the main time series and its index corresponds to the partition ID. The inherent feature of FFT and DCT transformation is the first component which is also known as the base component. This component is a summation of all points from the original signal. Thus calculating SUM—and also AVERAGE—queries can be performed as follows: We define Start Point (SP) and End Point(EP) of the queried time series, '%' as the modulo operator and '/' as the integer division. The reason for integer division is, the resulting value can directly used as the index to access the model. For simplicity we consider an FFT model but the same applies for DCT models too.

$$Q_{SUM} = \sum_{n=SP\%C}^{C} \frac{\sum_{i=0}^{C} \sigma(ts_{\lfloor SP/C \rfloor})[i]e^{-i2\pi k\frac{n}{C}}}{C} + \sum_{p=\lceil SP/C \rceil}^{\lfloor EP/C \rfloor} \sigma(ts_p)[1]+$$
$$\sum_{n=1}^{EP\%C} \frac{\sum_{i=0}^{C} \sigma(ts_{\lceil EP/C \rceil})[i]e^{-i2\pi k\frac{n}{C}}}{C} \tag{3}$$

We consider $\sigma(ts_{\lfloor SP/C \rfloor})$ as a vector taken from the matrix representation of the larger model $\Sigma(TS)$ with its partition ID matching the raw ID $\lfloor SP/C \rfloor$. As shown here it is not necessary to approximate the full length of the queried time series but only the head and tail parts. This is due to the inherent feature of DCT and DFT transformation. In both DCT and DFT the inverse transformation from only the first component defines the base components which is the mean value of all the points being considered for transformation.

4   Evaluation

We are currently in the process of implementing the model query processor, thus our initial results are based on finding sufficient query statistics to support the model query processing. For example we want to evaluate when we should use the model query processor based on queried data length etc. We also store statistics such that given the query parameters, we can select different models. In order to show the efficiency and effectiveness of our approach we conduct extensive experiments with different time series, under different settings. We have implemented two models from the signal processing domain known as DFT (also known as FFT) and DCT to use with our time series. We also implemented an approximated aggregate model to be used with time series data.

Data: We have conducted our experiments with different datasets. We obtained NREL wind data set[1] where we extract data from 50 different wind turbines. We also use the UK Energy Demand data set which has a pretty nice seasonal pattern. Most of our results are based on the NREL wind dataset. As individual time series from the NREL wind dataset contains only around 157000 data points each with 10mins resolution. In order to derive sufficiently large time series, we extend the original time series by permutation with other time series in the dataset. The resulted time series have around 3140000 data points.

Experimental Setup: We used a standard personal computer with an Intel Core i5 processor, 8GB RAM with Windows 8.1 Operating System. We used R to implement our system. In order to implement standard query processing, which reads raw data and produce results, we use standard R functions SUM and AVERAGE. Thus, base results of our experiments are from standard R functions.

---

[1] http://www.nrel.gov/gis/data_wind.html

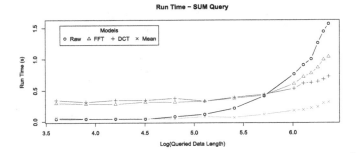

Fig. 3. Running time for different query lengths on different models and running time of the original raw data for Wind Data time series.

4.1 AVG and SUM Queries

One of our query parameters is the query run time. When a query is being issued with time bounds it is necessary to select suitable models. Thus we conduct experiment on query run time for SUM query against different models and the results are depicted in Fig. 3. As depicted in Fig. 3, we observe a significant improvement in performance wrt. query run time for SUM query when using the model based approach. One reason for this improvement is that IO which is significantly lower for models than for raw data. We can justify from our results that models with low memory footprint have better run-time performance. In this experiment models are one order of magnitude smaller than the raw data. We maintain these statistics in our query statistics such that future queries can be answered using the optimal approach (Raw Data or Models). We notice a break-even point at around 5.7 on the x-axis such that if the queried data length is larger than $10^{5.7}$ then the model query processor should be used to answer the query. The selection of models depends on the individual model characteristics and specified query parameters such as error boundaries.

Our models are adjustable such that a more compressed representation decreases query run time but also increase the error. Thus we evaluate query run time against compression factor. The results are depicted in Fig. 4 where it is visible that run time decreases significantly when we increase the compression. Having these results in our statistics helps to decide which model(s) to use for a particular query depending on the query parameters. Comparing Figs. 4(a) and (b) we can see that the model with compression factor 20 is roughly 3–4 times faster than the models with compression factor of 5. From that we can devise, Run Time and the Compression Factor are linearly proportional.

Use of models for information querying has an inherent problem. The models are abstract representations of the underlying data. The results obtained from the model do not exactly match with the result obtained from the raw data. Thus, we perform experiments to evaluate the accuracy of the results with respect to the same query issued on the raw data and experiment's results are depicted in Fig. 5. It can be seen that the larger the queried data, the better the

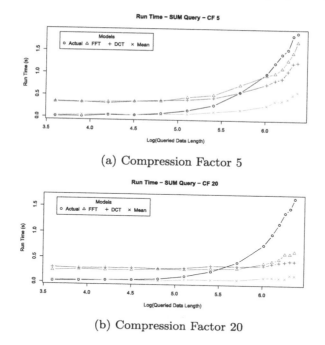

(a) Compression Factor 5

(b) Compression Factor 20

Fig. 4. Comparison of query run time changes with respect to compression

accuracy. Using our models, we approximate the points in head and tail parts of the time series being queried. The body part of the time series is calculated directly from the models associated with that part. There is no approximation needed for that part.

We also noted statistics on accuracy changes when the compression factor differs. In the previous experiment we show that higher compression leads to models with a lower memory footprint that have a lower execution time during query processing. It is also necessary to store the accuracy of different compression level to better answer user queries. Figure 6 depicts accuracy changes for compression factors 5 and 20 where we see high error in more compressed models. With these statistics the user always has the ability to trade off between performance and accuracy.

In our proposed method, we provide the functionality to the user where he/she can specify the query parameters like time boundary and accuracy boundary. Thus, in our model pool we store different models to support such specific parameters. In some cases we create simpler models from original stored models. Due to this flexibility it is necessary to evaluate the result accuracy with respect to the model specification. We use the compression factor as the model complexity where less compression means the model is more detailed and high compression means model is less detailed, thus, provides a more abstract representation. The results are shown in Table 1.

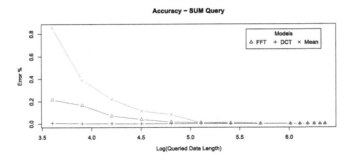

Fig. 5. Accuracy of SUM query on NREL wind dataset

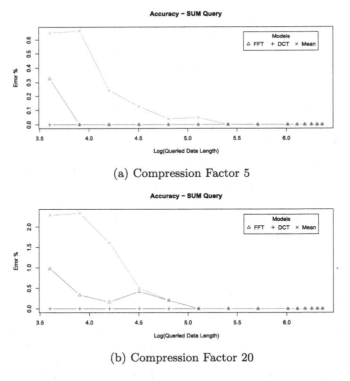

(a) Compression Factor 5

(b) Compression Factor 20

Fig. 6. Comparison of query accuracy changes with respect to compression

Table 1. De-compression error of different models in different compression levels

Compression	Measure	Model		
		FFT	DCT	Aggregate
10	RMSE	4.269471	7.42458	15.00886
	MAPE	4.27367	8.048909	17.77352
5	RMSE	2.236411	3.552057	9.980734
	MAPE	1.851273	4.050521	10.51729
4	RMSE	1.879704	2.84323	6.990188
	MAPE	1.399919	3.025598	6.728125

4.2 Histogram Query

Histograms are a widely used analysis tool in the business domains and used as a metadata statistic in database and data warehouse domains for query optimization. A histogram shows the data distribution and we generate histogram from our models and analyze the accuracy of the generated histograms compare to histograms created using the original data. An example is shown in Fig. 7 where top left histogram represents original data, top right histogram represents reconstructed data using FFT model. Bottom left histogram and bottom right histogram represent DCT model and Mean aggregated model reconstructed histograms respectively. We want to analyze our decompression accuracy with respect to data distribution and similar to [9] we conduct Histogram Error Rate (HER) analysis where $HER(x) = \frac{\sum_{B}^{i=1} |H_i(x) - H_i(x)|}{\sum_{B}^{i=1} H_i(x)}$, defines error with respect to total bin differences in the histogram. The HER is 0.044, 0.063 and 0.152 for

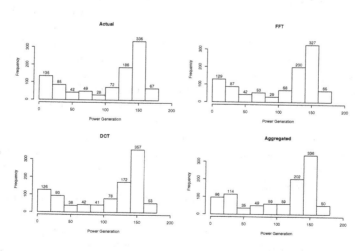

Fig. 7. Histogram representation of raw data and decompressed data from models in NREL Wind Dataset

FFT, DCT and Mean Aggregated models respectively. Mean aggregated model has three times higher error compared to FFT model. The compression factor in this experiment is 10 for all 3 models.

From our findings we can devise that different models have different characteristics. When we maintain all these models in our model pool together with their characteristics on different queries, it is necessary to select the optimal model to answer given query. Thus, our next step is to implement the model adviser which is responsible selecting the model to be used.

5   Related Work

Approximate query processing has been used in many application domains and new methods have been introduced to deal with general and application specific problems. Among these approaches, sampling based methods, histogram based methods and wavelet based methods are to name few.

BlinkDB [1] is a sampling based database system which provides approximate answers with time and error bounds. BlinkDB proposes stratified sampling which supports offline (samples are pre-calculated for the given query) as well as online sample creation when a new query is issued and cannot be answered by existing samples in the database. The system handles WHERE and GROUP BY queries and create samples accordingly. BlinkDB also maintains Error Latency Profile which are created with respect to sample size and their response time and relative error. BlinkDB cannot be used with time series data, which doesn't have multiple dimensions in contrast our models supports time series data. Interpreting a time series as a signal is a common technique in time series analysis. This approach gives the flexibility to use signal analysis methods to use on time series. One such approach is presented in [9] where they used FFT is used to compress original time series and represent it then through sampled Fourier components with residuals of original and approximated signal. They propose only histogram query analysis where we define SUM and AVERAGE queries over transformed models. We also introduce DCT models. Wavelet analysis (both continuous and discrete) is widely used in data analysis [2], in particular time series analysis. Discrete Wavelet transform-based time series and mining has been proposed in [3]. The proposed system can only handles multidimensional data and doesn't work well with time series data in comparison to our system. More recent work on approximate query over temporal data is presented in [7] which focuses on Frequency Domain Transformation which includes DFT, DCT and DWT. MauveDB [5] and its early implementation of model based data access [4] is another attempt to represent the underlying data through the models to provide a complete view of the data. MauveDB uses regression based models to represent the underlying data and does not provide compression. MauveDB presents a new abstraction level over the real data with errors and missing values to present a complete view to the users. This abstraction is built by using models represent to underlying data. As this layer provides transparency to the user, the user can still use SQL queries to derive information, and this information is provided by using models in the abstraction layer.

6 Conclusion and Future Work

Due to the large amount of data being generated by today's applications, it is necessary to introduce new information extraction methods as opposed to traditional database systems. In this paper we suggests a model based database system which underlying data is represented by models and queries are answered using models instead of using data. As a start we use models on time series data and show that it is possible to improve the performance by using models. Querying over proposed DFT (FFT), DCT and aggregation models have run-time efficiency gain upto 80 % as compared to querying over the original raw data and query results have high accuracy with less than 5 % of Mean Absolute Percentage Error (MAPE) with reference to exact answers.

As directions for future work, we suggest maintaining query statistics to determine when to use model based query processor and when to use data query processor. Our model-based database system implementation starts with time series representation as time series is the simplest form of data. Even though our model-based database system can apply only to few scenarios in time series domain, our system will have higher impact when representing data with many dimensions. In such cases models used for simple time series cannot be used as implemented now. Thus an extension of the current models to capture multiple dimensions as well as completely new set of models will be introduced in our future research.

Acknowledgment. This research has been funded by the European Commission through the Erasmus Mundus Joint Doctorate, Information Technologies for Business Intelligence - Doctoral College (IT4BI-DC).

References

1. Agarwal, S., Mozafari, B., Panda, A., Milner, H., Madden, S., Stoica, I.: Blinkdb: queries with bounded errors and bounded response times on very large data. In: Proceedings of the 8th ACM European Conference on Computer Systems, EuroSys 2013, pp. 29–42. ACM (2013)
2. Chakrabarti, K., Garofalakis, M.N., Rastogi, R., Shim, K.: Approximate query processing using wavelets. In: Proceedings of the 26th International Conference on Very Large Data Bases, VLDB 2000, pp. 111–122. Morgan Kaufmann Publishers Inc. (2000)
3. Chaovalit, P., Gangopadhyay, A., Karabatis, G., Chen, Z.: Discrete wavelet transform-based time series analysis and mining. ACM Comput. Surv. **43**(2), 6:1–6:37 (2011)
4. Deshpande, A., Guestrin, C., Madden, S.R., Hellerstein, J.M., Hong, W.: Model-driven data acquisition in sensor networks. In: Proceedings of the 30th International Conference on Very Large Data Bases - Volume 30, VLDB 2004, pp. 588–599. VLDB Endowment (2004)
5. Deshpande, A., Madden, S.: Mauvedb: supporting model-based user views in database systems. In: Proceedings of the 2006 ACM SIGMOD International Conference on Management of Data, SIGMOD 2006, pp. 73–84. ACM (2006)

6. Khalefa, M.E., Fischer, U., Pedersen, T.B., Lehner, W.: Model-based integration of past and future in timetravel. Proc. VLDB Endow. **5**(12), 1974–1977 (2012)
7. Khurana, U., Parthasarathy, S., Turaga, D.S.: FAQ: a framework for fast approximate query processing on temporal data. In: Proceedings of the 3rd International Workshop on Big Data, Streams and Heterogeneous Source Mining: Algorithms, Systems, Programming Models and Applications, BigMine 2014, 24 August 2014, pp. 29–45 (2014)
8. Perng, C.-S., Wang, H., Zhang, S., Parker, D.: Landmarks: a new model for similarity-based pattern querying in time series databases. In: 2000 Proceedings of 16th International Conference on Data Engineering, pp. 33–42 (2000)
9. Reeves, G., Liu, J., Nath, S., Zhao, F.: Managing massive time series streams with multi-scale compressed trickles. Proc. VLDB Endow. **2**(1), 97–108 (2009)
10. Spiegel, S., Schultz, D., Albayrak, S.: BestTime: finding representatives in time series datasets. In: Calders, T., Esposito, F., Hüllermeier, E., Meo, R. (eds.) ECML PKDD 2014, Part III. LNCS, vol. 8726, pp. 477–480. Springer, Heidelberg (2014)

Personalized User Value
Model and Its Application

Gang Yu[1], Zhiyan Wang[1], and Yi Cai[2(✉)]

[1] School of Computer Science and Engineering,
South China University of Technology, Guangzhou, China
[2] School of Software Engineering,
South China University of Technology, Guangzhou, China
ycai@scut.edu.cn

Abstract. With the increase of telecom services, it is becoming increasingly difficult for either an individual customer or a business to find his/her suitable one. To facilitate telecom service providers to recommend suitable telecom services to customers, we propose a five-dimension user value model in this paper. Our model is evaluated by a very large real-life data set. The experimental results show that the precision of our model outperform the baseline model by 2 %.

Keywords: User value model · Telecom service

1 Introduction

Nowadays, with the fierce competition among telecom service providers (i.e. telecom operators, such as Verizon, AT&T and China Mobile), telecom service providers have offered many telecom services to meet the needs of different customers. Facing such a number of options, it is usually difficult for either an individual customer or a business to find his/her suitable telecom services [2]. Therefore, in order to recommend suitable telecom services to current customers and attract potential customers, telecom service providers are exploring models to analyze customers' values. These models are referred to *User Value Model* in the literature [6,9]. Traditionally, user value models analyze a user's value based on two parameters: average revenue per user (ARPU) and minutes of user value (MOU). However, given that many customers may gain the same ARPU value and MOU value, existing user value models are coarse-grained for telecom service providers to recommend telecom services to customers accurately. As a result, in this paper, we propose a five-dimension personalized user value model to analyze a user's value. To evaluate our model, we conduct experiments on a very large real-life data set. The experimental results show that the precision of our model outperforms the baseline model by 2 %.

The rest of our paper is organized as follows. We discuss some related works in Sect. 2, present our model in Sect. 3 and show our experimental results in Sect. 4.

A. Liu et al. (Eds.): DASFAA 2015 Workshops, LNCS 9052, pp. 205–214, 2015.
DOI: 10.1007/978-3-319-22324-7_17

2   Related Works

In the face of continuous growing business competition environment, operators need to develop a comprehensive cognition for users, which can be executed in the marketing activities in measuring targeted marketing and the priority implementation process of it. This help complete the user value enhancement with lower cost and enhance the stability of the user, realize the maximization of the efficiency of marketing activities.

In this article, the comprehensive cognition of the user is based on the theory of value chain, from the user's value degree, user credit and user loyalty all three aspects. The Guangzhou company, explore actively and increase the elastic characteristics and user value marketing, user preferences on the basis of it, and it combined the "five-dimensional user evaluation model based on unified user information view" project with life cycle management, providing clear insight into user information.

Value chain concept proposer, Porter [1] believes that the competitiveness of the enterprises, is decided by influencing the user value chain, which is formed in the process of creating value for users. In Porter's [1] opinion, agencies for business or personal user is the value chain. The difference of enterprise advantage is derived from how to connect its own value chain to the user's value chain. Therefore, for gaining competitive advantages, it is necessary to build their own value chain, and the construction of enterprise value chain must conform to the enterprises strategy, around the user create value for users.

Liu et al. [3] believe that at present the study of user value model mainly has two directions. The first direction is to research the user perceived value from the user's perspective. The value is a evaluation of enterprises that provides products and services, namely, the benefits enterprise bring to the customer. Zeithaml et al. [4] give the definition to the customer perceived value:

1. The value can be easy to fetched
2. The value is what the customer needs
3. The value is the service quality the customer gets
4. The value is something that can be measured

The second direction is the value customer provides in the angle of the enterprise, according to customer's consumer behavior and consumer features, to measure benefits for the enterprise. The customer value scales the relative importance of the enterprise. On the basis of previous study, researches on interaction relationship of customers and companies are carried recently. Blattberg and Deighton [5] point out that enterprises should keep users of high value and high returns, and should consider customer as assets to conduct asset management. Customer assets is defined as the sum of all the customer lifetime discount value of the enterprise [5].

From the perspective of customers, customer perceived value for promotion, customer loyalty will increase, causing users to repeat purchase behavior, which benefits to the enterprise. In addition there are two main types of customer perceived value promotion measures, one is the increase of the client's return value, the other is a user pay cost is reduced, and the cost of users pay decreases, the user's credibility will ascend, bring the actual payment behavior, enterprise profit. From the perspective of enterprise, the customer can create value for the enterprise, and can also improve their product/service as well as customer perceived value, where acts as the motor for continuous value creation. Therefore it can form a good circulation of the value chain.

In this article, the research is carried on the perspective of mobile operators, evaluation of user value, subsequently it mainly aims at the second customer value. A mobile communication operator, by providing service that meets the needs of information products and services, influence users to improve user perceived value chain, where enterprise create value for customers, thus the service operator offers improves customer loyalty, and encourage users to purchase products and services repeatedly in this enterprise. This is a win-win situation of their own value chain and customer value chain.

User loyalty and credit management, is what can sustain the user life cycle, implementation and guarantee of the user perceived value to the enterprise.

Based on the concept of user value chain, our model strengthen the elasticity of users marketing preferences and user value dimension.The Five-dimensional Personalized User Value Model based on Unified Information View Model is introduced from following dimensions:

1. User contribution value which helps enterprises bundled user value more clearly, rather than the user itself,
2. User value elasticity: to help enterprises invest marketing resources more reasonably and precisely,
3. User loyalty: to help enterprise handle users prioritization,
4. Users credit: to help user quality control,
5. User preference: marketing to guide enterprises in users selection.

Based on the theory of value chain built above, the five dimensions of user personalized value will be articulated with following specification.

There are many users value model have been studied. A variety of users lifetime value model is put forward in cite1. In Chen et al. [3]'s research, customer value is decided by users current value and potential value. Dong et al. [4] believe that user value is the comparison of benefits and costs in the whole process of products purchase and consumption. Prior to this study, there are several definitions for the user value in mobile operator so differences in algorithm and application methods exists:

1. User input-output ratio model: Targeted on input-output ratio production of the user group , it computes the history total output of customer value, and the proportion of enterprise marketing for user input resources, combined with the customer value elastic model for application. In following sections, we refer it as MA model for short.
2. User behavior value model: From the aspect of user behavior, the model calculates customer value based on its current income in the composition of the current behavior, and calculates the user's future value, including customer loyalty, credit model. We calls it MB model.

3. Net value model: It proposes user-customer net present value concept, focusing on the customer's current profit contribution, and result of the customers revenues minus inputs (discount, cost) , to get the net value of the customer. It is applied to the group in the assessment of customers, channels, marketing solutions.

In following sections, it is called as MC model for short.

3   Five-Dimension Personalized User Value Model

3.1   User Input-Output Ratio Model

This model mainly measures the relationship between the outputs from users to business and the investment costs from business to customers, namely, $Users\ value = output\ from\ userCinput\ from\ business.$

In practice, this model defines the users value as the total value of user, which is decomposed into two parts: the users historical value and the user's current value. Users historical value mainly is based on stability and loyalty of user. User's current value is mainly represented by the current net output of user. The equation is represented as follows:

$$THV = w1 * HV + w2 * CV \tag{1}$$

where w is the value impact factor. Although for a simple model, w can directly be equal to 1, for complex models, w is determined by business experts evaluate or linear regression. A large amount of experimental data shows that when $w2 > w1$, the equation can more accurately describe the total historical value of user. THV means the total historical value of the user. CV is the current user value, which can be represented by the current net output value. In this paper, CV is equal to the income that the user contributed to business minus the marketing investment that business put into user. HUV stands for Users historical value.

The following describes the user historical value model and its implementation.

User Historical Value. The historical value refers to the amount of the cash flow (also called profit) that the user brings to the business among the days from the new user opening a new account to remarking. It manifests the stability of the customer and the ability to make contribution. Historical value is based upon the total time the customer using the network and is cumulated by the value coefficient which decreases each year. General speaking, the historical value is the total value that the customer contributed to the mobile network from the first day using the network to last year. It is calculated as follows:

$$HUV = \sum_{i=1}^{t} A_i Q_i = A_0 * [\frac{Q(1 - Q^n)}{(1 - Q)} * 12 - (m - 1) * Q^n] \tag{2}$$

where A means the monthly average profit value in that year, which can be approximated to the average monthly profit value in this year, t is the number of months from using network to January of this year, Q is the value of the attenuation coefficient. Due to the fact that the value decreases annually, we will break down t as m/n. The default value of Q is 0.5.

However, in Some applications, another method is used to define the user value in the form of production ratio, namely, the user life-cycle value.

Customer Life-Cycle Value. The Customer Life-cycle Value, for short CLV, means the total benefit that the business can gain from the user in the future under the current circumstances, which is put forward by Gupta et al. [6]. But they define customer life-cycle value as the following equation:

$$HUV = \sum_{t=0}^{T} \frac{(p_t - c_t)r_t}{(1 + i)^t} - AC \tag{3}$$

where t means the times the user bought the product, p_t is the price of product, c_t is the cost the business serves the customer, r_t is the probability user keep purchasing after t times, AC is the Acquisition costs, and T is the number of purchases estimate in the CLV.

From the quations, we can see, this model takes into consideration the income user brings to the enterprise from past to now and the investment the business puts into the user from past to now. Nevertheless, this model has its limitations because it did not take into account the prediction about the development of the future value for the user.

3.2   User Behavior Value Model

The customers current value is estimated by the weighted costs of the 7 items, which will be showed in the next, using the Analytic Hierarchy Process. The weight value of those items is based on the factors that have impact on the customers value, such as the income of the customers and the current profit contribution and so on. The estimate of the current value is based on the main sources of the inputs that the customer contributed to the business. According to the percentage of the sources, we select only 7 items, namely, the fixed voice fee (FVF), the voice communications fee (VCF), the fixed data services fee $(FDSF)$, the new data services revenue $(NDSR)$, the traditional data services fee $(TDSF)$, the average monthly cost (AMC), the package minimum guarantee fee $(PMGF)$. The user's potential value is calculated as follows: $UsersPotentialValue = (FutureIncomeContributionCFutureCostPrediction) * credibility * loyalty$ $(5 - 4)$.

The computing process of the current value is as follows:

1. we select the 7 items as the estimate indexes according to the experts judgments,
2. we get the final estimate matrix (as follows) according to the importance of each item.

Table 1.

Items	FVF	VCF	FDSF	NDSR	TDSF	AMC	PMGF	WEight Value
FVF	1	1	0.5	1	0.5	0.5	2	0.11115936
VCF	1	1	2	2	1	0.5	2	0.15809814
FDSF	2	0.5	1	2	0.5	0.5	0.333	0.11881339
NDSR	1	0.5	0.5	1	0.5	0.5	2	0.10017035
TDSF	2	1	2	2	1	0.5	2	0.17313573
AMC	2	2	2	2	2	1	3	0.24898539
PMGF	0.5	0.5	2	0.5	0.5	0.3333	1	0.08963765

In Table 1, if the value of $u, v = 1$, it means that they have the same importance. When the value of $u, v > 1$, the higher the value of u, v is, the more important u can be than v. On contrary, when the value <1, the lower the value is , the more important v can be than u. We use the eigenvalues computing method to normalized eigenvectors of the largest eigenvalues w. At the same time, we use the CI as the consistency test indicators, where

$$AW = \lambda W \tag{4}$$

$$W = w_1, w_2, ..., w_n \tag{5}$$

$$\sum_{i-1}^{n} w_i = 1 \tag{6}$$

$$CI = \frac{\lambda - n}{n - 1} \tag{7}$$

At last, compute the current value CV of the customer:

$$CV = \sum_{i=1}^{N} x_i * wt \tag{8}$$

where N is the number of the estimate items (indexes), in this model, $N = 7$, X is the standardized score of the indexes, for example, x_1 is the standardized score of $FVF(fixed voice fee)$, w is the weight value of the corresponding indexes, for example, w_1 is the weight value of FVF.

After computing the current values of all customers, we sort them based on the scores and standardize them into the numeric between 0 and 100.

3.3   User Net Present Value Model

User NPV model is pure output of customers, respectively, to achieve individual customers, customer value assessment of channels and corporate customers. Customer value is equal to the net income minus customer billing customer input.

According to the results of research, customer input is mainly composed of gift discounts, network costs, marketing costs and management costs four categories. The NPV is defined as follows: $Users\ NPV = Revenue\ of\ bill - Discount - Network\ costs - Marketing\ costs - Administrative\ costs.$ where

– *Revenue of bill* is the revenue of bills in recent months in the billing system,
– *Discount* is the amount of consumption of a single user accounts of complimentary charge in that month,
– *Network Costs* include the circuit and NE rental fees, business technology support costs, interconnection expenses, net settlement expenses,
– *Marketing costs* include customer service costs, gratuities of channels, communication costs, Terminal subsidies
– *Management costs* are staff salaries, travel expenses, office expenses, depreciation expense.

This model is too simple; just consider the current period net output of users, ignoring the history and future trends for comprehensive consideration.

3.4   Five-Dimension Personalized User Value Contribution

Based on the value model introduced before, in this paper re-defined one of the five dimensions of user personalized value contribution: User Value Contribution. We define customer value contribution (UV) from the historical value (HV), the current value (CV), the future value (FV), affect the value (IV) construct four values, namely:

$$UV = w_1 * HV + w_2 * CV + w_3 * FV + w_4 * IV \qquad (9)$$

where

– HV refers to the amount of cumulative cash flow which the user historical cycle brought for the business,
– CV refers to the amount of cumulative cash flow which the user current cycle brought for the business,
– FV means if you keep user stickiness, the increased amount of cash flow which the user will purchase in the future for the business,
– IV refers to that when user satisfaction increases; they will not only buy our products, but also through their guidelines influence other customers to bring the value for the business.

The weight of all types of user value contribution re-calculated with reference to the methods of calculating all kinds of the weights of income of customer, are calculated in the following steps:

1. business experts confirm the weights of historical value, present value, future value and the affect value of customer value contribution. According to the business experts scoring weights gained weight matrix A,
2. figure out the weight matrix eigenvalue and eigenvector of matrix A,

3. take the largest eigenvalues λ_{max} corresponding eigenvector as a weight vector and normalized,
4. for calculation purposes to retain two decimal places, and finalize its constituent weights are 0.25, 0.55, 0.1 and 0.1.

Figure 1 shows the framework of the customer value model.

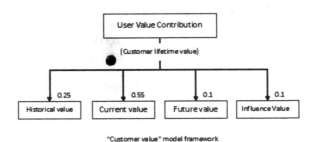

Fig. 1. "Customer value" model framework

Historical Value. The total historical value of $user = user\ \ current\ \ value + user\ \ historical\ \ value$. Meanwhile, we make the following adjustments:

- The adjustment of the split point of value and the present value. The time range of historical value changes from accessing network to one year before the statistics period to accessing network to three months ago of statistical period,
- The removal of the invalid operation. Meanwhile, according to data regression analysis, we found that the network time of 80 % of users is less than two years; on the other hand, because of the historical value of the formula (5-2), Q is the attenuation factor. For more than 2 times after the first n iterations the historical value of the rapid decay of factors, only reached 0.125 (initially 1, according to an annual 0.5 attenuation), to accelerate the speed of data processing paper attempts to remove user history billing revenue calculations which more than two years,
- Improvement of the data accuracy. Because the system exist for high-speed access in the first six months of user billing data, therefore, the income data six months before the user is directly replaced by accurate data.

4 Experiments

In this paper, we use 18 million user data(2009–2013) to make regression calibration. We discovery that relative to the computing historical value in the user value of Production ratio model, the average time of computing historical value in the five-dimensional model reduced form 58 milliseconds to 25 ms. The efficiency increased by more than 100 %, and with the increasing amount of data to

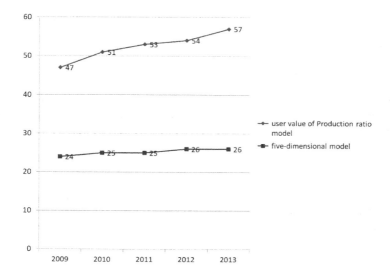

Fig. 2. The comparison of processing time of two models

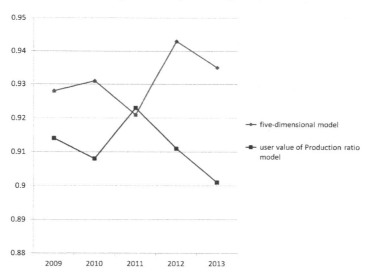

Fig. 3. The comparison of accuracy of two models

calculate the speed of a five-dimensional model tends to be more stable. Figure 2 which compares the processing time of two models is shown as follows.

Because of the adjustment of the current and historical value of the division of time window, while the introduction of the first 6 months of data accurate calculation, the average precision rate shown in Fig. 3 is increased by two percentage points.

Experiments show that using a five-dimensional user model to fit the user historical value is more referential.

Acknowledgement. This work is supported by National Natural Science Foundation of China (Grant NO. 61300137), the Guangdong Natural Science Foundation, China (NO. S2013010013836), Science and Technology Planning Project of Guangdong Province China NO. 2013B010406004 the Fundamental Research Funds for the Central Universities, SCUT(NO. 2014ZZ0035).

References

1. Porter, M.E.: Comptetitive Advantage: Creating and Sustaining Superior Performance. Simon and Schuster, New york (2008)
2. Zhang, Z., et al.: A hybrid fuzzy-based personalized recommender system for telecom products/services. Inf. Sci. **235**, 117–129 (2013)
3. Liu, Y., et al.: Analysis of Customer Value Based on Value Chain. J. Ind. Eng. Eng. Manage., 99–1011 (2004)
4. Zeithaml, V.A.: Consumer perceptions of price, quality, and value: a means-end model and synthesis of evidence. J. Mark., 2–22 (1988)
5. Blattberg, R.C., Deighton, J.: Manage marketing by the customer equity test. Harvard Bus. Rev. **74**(4), 136 (1996)
6. Gupta, S., et al.: Modeling customer lifetime value. J. Serv. Res. **9**(2), 139–155 (2006)
7. Chen and Li: Study on Value Segmentation and Retention Strategies of Customer. Group Technol. Prod. Modernization, 23–27 (2001)
8. Dong, Q., et al.: Theory of customer value and its formation. J. Dalian Univ. Technol., 18–20 (1999)
9. Gupta, S., Lehmann, D.R.: Customers as assets. J. Interact. Mark. **17**(1), 9–24 (2003)

Posters

Flexible Aggregation on Heterogeneous Information Networks

Dan Yin[✉], Hong Gao, Zhaonian Zou, Xianmin Liu, and Jianzhong Li

Massive Data Computing Research Lab, Harbin Institute of Technology,
Harbin, China
yindan630@163.com, {honggao,znzou,liuxianmin,lijzh}@hit.edu.cn

Abstract. With the advent of heterogeneous information networks that consist of multi-type, interconnected nodes, such as bibliographic networks and knowledge graphs, it is important to study flexible aggregation in such networks. In this paper, we investigate the flexible aggregation problem on heterogeneous information networks, which is defined on multi-type of nodes and relations. We develop an efficient heuristic algorithm for aggregation in two phases: informational aggregation and structural aggregation. Extensive experiments on real world data sets demonstrate the effectiveness and efficiency of the proposed algorithm.

1 Introduction

Heterogeneous information networks are newly emerged graph data, which involve multi-type nodes and relations. Figure 1(a) shows a bibliographic network which contains three types of nodes: *Paper*, *Author* and *Venue*, and four types of relations exist among these nodes. They are *Cooperate* relation between *Author*, *Write* relation between *Author* and *Paper*, *Cite* relation between *Paper*, *Publish* relation between *Paper* and *Venue*. In addition, each type of nodes has a set of attributes, *i.e*, *Paper(ID, Title, Topic, Keywords)*; *Author(ID, Institute, Field, Location)*; *Venue(Name, Year)*. Values of part attributes are given in Fig. 1(b).

Aggregations allow users to observe and model data in different dimensions, to perform drill-down, roll-up and other OLAP operations further. We investigate the problem of aggregation on multi-type nodes and relations on heterogeneous information networks. Next we give two aggregation queries over the bibliographic network. The aggregate functions are COUNT.

Query 1. Aggregate on *Paper* node and *Cite* relation, the selected attribute of *Paper* is *Topic*.

Query 2. Aggregate on *Paper*, *Author* nodes and *Write* relation, the selected attribute of *Paper* is *Topic*, and the selected attribute of *Author* is *Location*.

Figure 2(a) displays the aggregate result of query 1. *Paper* nodes in the same aggregate nodes have the same values of *Topic*. Meanwhile, they also have the same *Cite* relations with other aggregate nodes. Node 9 and 10 are not in the same aggregate node, because node 9 cites 'DM' paper, while node 10 does not.

A. Liu et al. (Eds.): DASFAA 2015 Workshops, LNCS 9052, pp. 217–222, 2015.
DOI: 10.1007/978-3-319-22324-7_18

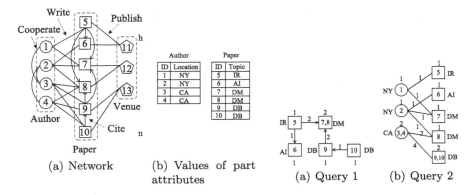

Fig. 1. Bibliographic network   Fig. 2. Aggregation results

Figure 2(b) gives the aggregate result of query 2. *Author* nodes in the same aggregate nodes have the same values of *Location* and the same *Write* relation with aggregate *Paper* nodes.

From the two examples, we can see flexible aggregation on multi-type nodes and relations is meaningful. The main contributions of this paper are:

1. Flexible aggregation problem for heterogeneous information networks is proposed, which can aggregate multi-type nodes and relations;
2. A novel function based on graph entropy is proposed, which is effective to measure the structural similarities with regard to different types of relations;
3. An efficient aggregation algorithm from two phases is proposed: informational aggregation and structural aggregation.
4. Experiments demonstrate the effectiveness and efficiency of algorithm.

2   Preliminaries

Definition 1 Heterogeneous Information Network. *A heterogeneous information network is defined as a directed graph* $G = (V, E, T, R, \phi_V, \phi_E, A, D, \phi_A)$, *where* V *is node set,* $E \subseteq V \times V$ *is edge set.* T *is set of node types, and* R *is set of edge types.* $\phi_V : V \longrightarrow T$ *is node type mapping function and* $\phi_E : E \longrightarrow R$ *is edge type mapping function.* A *is attribute set of nodes and* D *is domain of* A. $\phi_A : T \longrightarrow A$ *is mapping function from node types to attributes.*

Definition 2 Graph Projection. *Given selected node types* $Q = \{T_1, T_2, \ldots, T_l\}$, $Q \subseteq T$, *and selected edge types* $L = \{R_1, R_2, \ldots, R_k\}$, $L \subseteq R$. *The graph projection of* G *on* Q *and* L *is a graph* $G_{pj} = (V_{pj}, E_{pj})$, *where* V_{pj} *is node set* $V_{pj} = \{v | v \in V, \phi_V(v) \in Q\}$, E_{pj} *is edge set,* $\forall u, v \in V_{pj}$, $(u, v) \in E_{pj}$ *iff* $(u, v) \in E$ *and* $\phi_E(u, v) \in L$.

Given an attribute set $S = \{A_1, A_2, \ldots, A_k\}$, $S \subseteq A$, for $\forall u, v \in V$, if $A_i(u) = A_i(v)$ $(1 \leq i \leq k)$, then we say $S(u) = S(v)$.

Definition 3 Graph Partition. *Given selected attributes $S = \{S_1, S_2, \ldots, S_l\}$ of Q, where $S_i \subseteq \phi_A(T_i)$. The partition of G_{pj} w.r.t Q, S and L is a set of graphs $G_p = \{G_1, G_2, \cdots, G_m\}$, satisfying:*

1. *For $\forall G_i \in G_p$, $G_i = (V_i, E_i)$, G_i is a subgraph of G_{pj};*
2. *$\bigcup_{i=1}^m V_i = V_{pj}$;*
3. *For $\forall G_i, G_j \in G_p$, $i \neq j$, $V_i \cap V_j = \emptyset$;*
4. *For $\forall u, w \in G_i$, $\exists T_j$, $\phi_V(u) = \phi_V(w) = T_j$, $S_j(u) = S_j(w)$;*
5. *For $\forall u, w \in G_i$, for $\forall G_j$, if $\exists u' \in G_j$, $(u, u') \in E$, $\phi_E((u, u')) \in L$, then $\exists w' \in G_j$, $(w, w') \in E$, $\phi_E((w, w')) \in L$;*
6. *For $\forall u, w \in G_i$, $(u, v) \in E_i$ iff $(u, v) \in E$ and $\phi_E(u, v) \in L$;*

Definition 4 Aggregate Graph. *The aggregate graph of G on Q, S and L is a graph $G_c = (V_c, E_c, f_1, f_2)$, where V_c is node set, E_c is edge set, f_1 is aggregate function on V_c and f_2 is aggregate function on E_c, satisfying:*

1. *$|V_c| = |G_p|$;*
2. *$\forall a \in V_c$, a corresponds to a subgraph $G_a \in G_p$;*
3. *$\forall a, b \in V_c$, a corresponds to a subgraph G_a and b corresponds to a subgraph G_b, if $a \neq b$, then $G_a \neq G_b$;*
4. *$\forall a \in V_c$, a corresponds to a subgraph G_a, $f_1(a) = f_1(V_a)$;*
5. *$\forall a, b \in V_c$, a corresponds to a subgraph G_a and b corresponds to a subgraph G_b. $(a, b) \in E_c$ iff $\exists u \in V_a$, $w \in V_b$, $(u, w) \in E$ and $\phi_E(u, w) \in L$;*
6. *$\forall (a, b) \in E_c$, a corresponds to a subgraph G_a and b corresponds to a subgraph G_b. $f_2((a, b)) = f_2(\{(u, w) | u \in V_a, w \in V_b, (u, w) \in E \text{ and } \phi_E(u, w) \in L\})$.*

For $\forall a \in V_c$, we call it aggregate node, and for $\forall e \in E_c$, we call it aggregate edge. Aggregate functions can be selected freely, *e.g.*, COUNT, AVERAGE.

Graph entropy [1] has been widely used in graph mining. So we employ graph entropy to measure the structural consistency of nodes.

Definition 5 Graph Entropy. *$G_c = (V_c, E_c, f_1, f_2)$ is an aggregate graph of G on Q, S and L, for $\forall a = (V_a, E_a), b = (V_b, E_b) \in V_c$, the entropy from a to b is*

$$H_b(a) = \begin{cases} -\frac{|V_b(a)|}{|V_a|} \cdot log_2 \frac{|V_b(a)|}{|V_a|} & |V_b(a)| \neq 0 \\ 0 & |V_b(a)| = 0 \end{cases}. \tag{1}$$

where $V_b(a) = \{v | v \in V_a, \exists w \in V_b, (v, w) \in E, \phi_E((v, w)) \in L\}$.
   The entropy of a is

$$H(a) = \sum_{b \in V_c} \lambda_{a,b} H_b(a). \tag{2}$$

$\lambda_{a,b}$ represents the weight of relation from nodes in a to nodes in b.

Definition 6 C-function. *The C-function of aggregate graph G_c is*

$$F(G_c) = \sum_{i=1}^{l} \alpha_i \sqrt{NUM_{T_i}} + \sum_{a \in V_c} H(a). \tag{3}$$

where $NUM_{T_i} = |\{a | a \in V_c, \forall u \in V_a, \phi_V(u) = T_i\}|$.

α_i distinguishes the importance of different types of nodes. Fewer aggregate nodes are easier for users to understand, while Graph entropy will increase.

The definition of flexible aggregation problem is given as follow.

Input: Given a heterogeneous information network $G = (V, E, T, R, \phi_V, \phi_E, A, D, \phi_A)$, selected node types $Q = \{T_1, T_2, \cdots, T_l\}$, $Q \subseteq T$, with selected attributes $S = \{S_1, S_2, \ldots, S_l\}$, $S_i \subseteq \phi_A(T_i)$, and selected edge types $L = \{R_1, R_2, \ldots, R_k\}$, $L \subseteq R$, f_1, f_2;

Output: Aggregate graph $G_c = (V_c, E_c, f_1, f_2)$.

Object: Minimize $F(G_c)$.

3   Aggregation Algorithm

To distinguish nodes in semantics of attributes and structures, we design a two-phase aggregation: Informational aggregation and Structural aggregation.

Informational aggregation. This process guarantees that nodes aggregated together have the same types and attribute values. It partitions nodes of selected types according to the their selected attributes.

Structural aggregation. Based on the informational aggregation, we should make nodes in the same aggregate nodes have similar structures. We can reduce C-function by decreasing graph entropy. This process has three challenges: how to choose the partitioned aggregate nodes; what strategy should be used for partition; and when does iteration stop. We discuss how to tackle the challenges.

Challenge 1. In order to decrease C-function, we may choose the aggregate node with the largest graph entropy. Because nodes in it have diverse structures. In order to improve the readability of aggregate graphs, for the aggregate nodes with the same graph entropy, we are prior to choose the one with larger size. Each iteration, we choose the aggregate graph with the largest partition level, where the partition level of a is $P(a) = \sqrt{|a|} \cdot H(a)$.

Challenge 2. In order to respond the quickly, in each iteration, we divide the aggregate node into two aggregate nodes according to the nodes' neighbors with t, where $t = \arg \max_b \{\lambda_{a,b} H_b(a)\}$.

Challenge 3. In view of C-function minimization, the sizes of aggregate graphs should be moderate. The iteration doesn't terminate until C-function reaches the first maximal value or the size of aggregate graph exceeds a specific threshold.

4   Experiments Evaluation

Amazon[1] network contains three types of nodes: *Customer, Product, Category*. Four types of relations exist, which are: *CoPurchase* relation between *Product*, *Purchase* relation between *Customer* and *Product*, *Classify* relation between *Product* and *Category* and *Like* relation between *Customer*. Each node type has a set of attributes, *Customer*(*ID, Purchase times*), *Product*(*ID, Rank, Reviews*),

---

[1] SNAP: http://snap.stanford.edu/data/.

Category(*Name*). The data set has 53,182 customers, 5,000 products and 4 categories. There are 147 edges of *CoPurchase*, 77,997 edges of *Purchase*, 7,231 edges of *Like* and 5,000 edges of *Classify*. We set $\alpha_{customer}$, $\alpha_{product}$ and $\alpha_{category}$ to 1, $\lambda_{product \leftrightarrow product} = 15$, $\lambda_{customer \leftrightarrow product} = 2$, $\lambda_{product \leftrightarrow category} = 1$ and $k = 20$. Experiments are done on a Microsoft Windows 7 machine with an Intel Core i5-2400 CPU 3.1 GHz and 4 GB main memory by Microsoft Visual Studio 2010.

We compare our algorithm with the reference [2]. We apply the compared algorithm on each type of nodes, respectively, without taking the structures into consideration.

Query 1. A set of node types $Q = \{Product, Category\}$ with attributes $S = \{S_{Product}, S_{Category}\}$ and relations $L = \{CoPurchase, Classify\}$, where $S_{Product} = \emptyset$ and $S_{Category} = \{Name\}$.

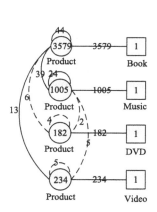
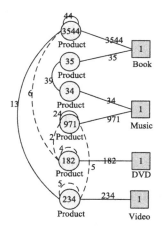

Fig. 3. Aggregate graph of query 1 by compared algorithm

Fig. 4. Aggregate graph of query 1 by our algorithm

Figure 3 shows the aggregate graph of query 1 of the compared algorithm. The values of nodes and edges represent their aggregate values. We use dotted lines to represent the edges whose function values are below 10, and other edges are represented by solid lines. The most products are books, music stands the second, DVD and videos are the least. The co-purchased DVD products are not co-purchased with music and DVD. Figure 4 presents the aggregate result of query 1 in this paper. Our algorithm presents a deeper result than the compared algorithm. Books that co-purchased with music products are not likely co-purchased with DVD and videos. Meanwhile, the co-purchased music products are not co-purchased with books, which may be co-purchased with DVD and videos. Aggregate results are interesting after considering structural information.

Query 2. A set of node types $Q = \{Customer, Product, Category\}$ with attributes $S = \{S_{Customer}, S_{Product}, S_{Category}\}$ and relations $L=\{CoPurchase, Classify, Purchase\}$, where $S_{Customer} = \{Purchasepower\}$, $S_{Product} = \emptyset$ and $S_{Category} = \{Name\}$.

Figure 5 presents runtime comparisons of different queries by different algorithms. The x-axis represents the queries 1 and 2 and the y-axis stands for the average runtime of queries. Each query is run 10 times and we compute the average running time. Previous work only focuses on attribute of nodes, while our work also considers the structures of networks. In query 2, our algorithm costs 13.2 s more than the compared algorithm.

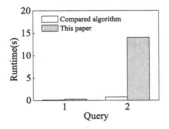
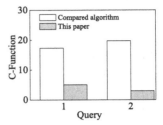

Fig. 5. Runtime   **Fig. 6.** C-function

Figure 6 displays the comparison of C-function values of different algorithms. The x-axis represents the queries 1 and 2 and the y-axis shows the values of C-function. Although aggregation including structural aggregation costs more time, the C-function values are much smaller than the compared algorithm. The values of the compared algorithm increases, while our algorithm decreases. Because graph entropy is decreased by taking structures into consideration.

5   Conclusions

We introduced the flexible aggregation problem on heterogeneous information networks. In order to aggregate efficiently, we propose a two-phase aggregation algorithm to aggregate nodes with similar attributes and structures. Experiment results demonstrate our algorithm can provide more accurate and implicit knowledge with a wealth of information.

Acknowledgement. This work is supported by the National Grand Fundamental Research 973 Program of China under grant 2012CB316200, the Key Program of National Natural Science Foundation of China under grant 60933001, the Major Program of National Natural Science Foundation of China under grant 61190115, the General Program of National Natural Science Foundation of China under grant 61173023.

References

1. Shetty, J., Adibi, J.: Discovering important nodes through graph entropy the case of enron email database. In: Proceedings of the 3rd International Workshop on Link Discovery, pp. 74–81. ACM (2005)
2. Zhao, P., Li, X., Xin, D., Han, J.: Graph cube: on warehousing and olap multidimensional networks. In: Proceedings of the 2011 ACM SIGMOD International Conference on Management of data, pp. 853–864. ACM (2011)

Discovering Organized POI Groups in a City

Yanxia Xu[1], Guanfeng Liu[1], Hongzhi Yin[2],
Jiajie Xu[1], Kai Zheng[2], and Lei Zhao[1]([✉])

[1] School of Computer Science and Technology, Soochow University, Suzhou, China
`xyx.edu@gmail.com`, {`gfliu,xujj,zhaol`}`@suda.edu.cn`
[2] School of ITEE, The University of Queensland, Brisbane, QLD, Australia
`h.yin1@uq.edu.au`, `kevinz@itee.uq.edu.au`

Abstract. With the development of urban modernization, a great number of hot spots, such as buildings, business streets and shopping malls, scatter over the city which have a great influence on people's lives and modern civilization. All of these hot spots consist of a set of point of interests (POIs). In this paper, we propose a new concept, i.e., Organized POI Group (OPG) and present a method to find them out. In addition, we classify the OPGs as three categories: building, street and village, according to their features.

Keywords: Urban computing · Organized POI groups · Spatio-textual similarity · Density-based clustering

1 Introduction

The development of urban modernization fosters a great number of large hot spots, such as shopping malls and business buildings, which are larger than a single POI and smaller than functional regions [1]. The distribution of large hot spots not only gives a general understanding of a city, but also is of great importance for urban planners. Intuitively, a large hot spot usually have a set of POIs which locate close to each other. Hence, we call a large hot spot as an Organized POI Group (OPG) and try to find them out. The challenge of this work is two POIs locate close to each other doesn't mean that they are actually in the same OPG. For example, two POIs with similar geo tags most likely locate in two adjacent buildings, respectively. To avoid putting these two POIs in the same cluster, we also take some textual information (e.g. street address) into consideration while clustering the POIs. A relevant work [2] makes significant efforts on dealing with both spatial and textual information of POIs. However, different from it, we focus on unstructured textual information which is bringing a big challenge to our work. After discovering OPGs, we classify them as three categories, street, building and village, by a decision tree.

In this paper, we firstly define the hybrid similarity which takes both spatial and textual information into consideration. Secondly, we describe the algorithms for discovering OPGs. Thirdly, we show the classification of OPGs. Finally, the result of our work is visualized.

A. Liu et al. (Eds.): DASFAA 2015 Workshops, LNCS 9052, pp. 223–226, 2015.
DOI: 10.1007/978-3-319-22324-7_19

2   Problem and Approach

Given a set of POIs $\mathcal{O} = \{o_1, o_2, o_3, ..., o_{|\mathcal{O}|}\}$, each $o_i \epsilon \mathcal{O}$ is a triple in the form of $\langle o_i.x, o_i.y, o_i.t \rangle$ where $o_i.x$ and $o_i.y$ respectively denote longitude and latitude and $o_i.t$ represents textual information of street address. The result of our work is a set of clusters $\mathcal{C} = \{c_1, c_2, c_3, ..., c_{|\mathcal{C}|}\}$ where c_i is a set of POIs which are in the same OPG.

2.1   Definition of Hybrid Similarity

We have to introduce spatial similarity and textual similarity before presenting hybrid similarity.

Definition 1 *Spatial Similarity.* *Given a set of points* $\mathcal{O} = \{o_1, o_2, o_3, ..., o_{|\mathcal{O}|}\}$, *for any two objects* o_i *and* o_j, *the spatial distance is* $d_s(o_i, o_j) = |o_i.x - o_j.x| + |o_i.y - o_j.y|$ *and then spatial similarity between* o_i *and* o_j *is defined as:*

$$w_s(o_i, o_j) = \frac{MaxD - d_s(o_i, o_j)}{MaxD} \tag{1}$$

where $MaxD = \max_{o_m \in \mathcal{O}, o_n \in \mathcal{O}} d_s(o_m, o_n)$.

Definition 2 *Longest Common Subsequence.* *Given two sequences* $\mathcal{S}_i, \mathcal{S}_j$, *we use* $\mathcal{T}_{ij}$ *to denote a common subsequence [3] of* $\mathcal{S}_i$ *and* $\mathcal{S}_j$. *If add any character to* $\mathcal{T}_{ij}$, $\mathcal{T}_{ij}$ *is not a common subsequence of* $\mathcal{S}_i$ *and* $\mathcal{S}_j$, *then* $\mathcal{T}_{ij}$ *is the longest common subsequence of* $\mathcal{S}_i$ *and* $\mathcal{S}_j$.

For example, consider two sequences "abcbcf" and "abfcab", the longest common subsequence of these two strings is "abcb". Computation of longest common subsequence can be seen in [3].

Definition 3 *Textual Similarity.* *Given two objects* o_i *and* o_j, *suppose* $\mathcal{T}_{ij}$ *denotes the longest common subsequence between* $o_i.t$ *and* $o_j.t$, *then textual similarity is define as follows:*

$$w_t(o_i, o_j) = \frac{|\mathcal{T}_{ij}|}{|o_i.t| + |o_j.t| - |\mathcal{T}_{ij}|} \tag{2}$$

Definition 4 *Hybrid Similarity.* *The hybrid similarity of* o_i *and* o_j *is represented by* $\tau(o_i, o_j)$, *which is defined as:*

$$\tau(o_i, o_j) = (1 - \sigma) \times w_s(o_i, o_j) + \sigma \times w_t(o_i, o_j) \quad (0 < \sigma < 1) \tag{3}$$

2.2   Algorithm for Discovering OPGs

Before describing the solution in details, we firstly formulate the problem and define the OPGs.

Definition 5 *Organized POI Group(OPG)*. *Let $\mathcal{O}$ denote the whole POI set and $\mathcal{P} \subseteq \mathcal{O}$ where $|\mathcal{P}| > \widehat{\varphi}$. $\mathcal{P}$ is an OPG only if $\forall o_i \in \mathcal{P}$, $\nexists o_j \in \mathcal{O} - \mathcal{P}$ satisfies*

$$\begin{cases} w_s(o_i, o_j) \geqslant \widehat{w_s} \\ w_t(o_i, o_j) \geqslant \widehat{w_t} \\ \tau(o_i, o_j) \quad \geqslant \widehat{\tau} \end{cases} \tag{4}$$

where $\widehat{\varphi}$ is the threshold of size for each cluster, $\widehat{w_s}$ is the threshold of spatial similarity, $\widehat{w_t}$ is the threshold of textual similarity and $\widehat{\tau}$ is the threshold of hybrid similarity.

The solution to the problem is a variant of density-based clustering algorithm, i.e., DBSCAN [4], using hybrid similarity. Different from traditional DBSCAN which traverses the whole data space to find neighbors of a point, we take advantage of grid index (the details can be seen in [5]) to obtain a set of candidate points which may fall in the $\widehat{\tau}$-neighborhood of a given point o_i where $\widehat{\tau}$-neighborhood denotes a set of points whose hybrid similarities with o_i are larger than $\widehat{\tau}$.

Our algorithm can be divided into three parts. Firstly, it randomly selects a unclustered POI o_j as potential core point and obtains the $\widehat{\tau}$-neighborhood $\mathcal{N}_j$ of the potential core point o_j. Secondly, if the size of $\mathcal{N}_j$ is larger than $\widehat{\varphi}$, o_j becomes a core point and all points in $\mathcal{N}_j$ become potential core points. Otherwise, o_j becomes a boundary point. Finally, we obtain a complete cluster until there doesn't exist any potential core point. The union of core points and boundary points makes up an OPG while its size exceeds $\widehat{\varphi}$.

2.3 Algorithm for Classification

Intuitively, the difference between village and building is POIs of a village scatter over a large region while those of a building gather in a small area. Besides, streets are narrower and longer than villages and buildings. Hence, we define length-width ratio and density of cluster to construct a decision tree shown in Fig. 1.

Definition 6 *Length-Width Ratio*. *If all POIs in a cluster $\mathcal{C}$ are enclosed by a rectangle r, then the length-width ratio of $\mathcal{C}$ is defined as:*

$$LWR(\mathcal{C}) = \frac{length \ of \ r}{width \ of \ r} \tag{5}$$

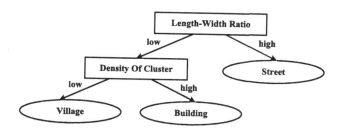

Fig. 1. Decision tree of OPG classification

Definition 7 Density Of Cluster. *Given a cluster $\mathcal{C}$ containing $\mathcal{N}$ POIs, if a rectangle r encloses all POIs of $\mathcal{C}$, then the density of $\mathcal{C}$ can be figured out by:*

$$DOC(\mathcal{C}) = \frac{N}{width\ of\ r \times length\ of\ r} \tag{6}$$

We select maximum latitude, maximum longitude, minimum latitude and minimum longitude among an OPG to construct the enclosing rectangle r.

3   Visualization

In this section, we visualize the distribution of OPGs in two large cities. Figure 2(a) and (b) show the OPGs of Beijing and Shanghai in 2014, respectively(blue indicates street, red stands for building and green represents village). These pictures reflect the urban layouts and levels of modernization of the cities to some extent.

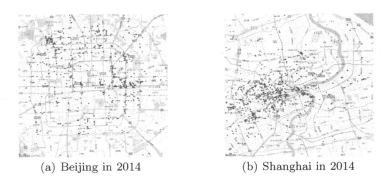

(a) Beijing in 2014   (b) Shanghai in 2014

Fig. 2. OPGs in Beijing and Shanghai

4   Conclusions

This paper studies a novel problem of discovering OPGs in a city using spatial and textual information of POIs. We propose a clustering-and-classification approach to address the problem.

References

1. Yuan, J., Zheng, Y., Xie, X.: Discovering regions of different functions in a city using human mobility and pois. In: SIGKDD, pp. 186–194. ACM (2012)
2. Fan, J., Li, G., Zhou, L., Chen, S., Hu, J.: Seal: spatio-textual similarity search. Proc. VLDB Endow. **5**(9), 824–835 (2012)
3. Bergroth, L., Hakonen, H., Raita, T.: A survey of longest common subsequence algorithms. In: String Processing and Information Retrieval, pp. 39–48. IEEE (2000)
4. Ester, M., Kriegel, H., Sander, J., Xu, X.: A density-based algorithm for discovering clusters in large spatial databases with noise. In: KDD, pp. 226–231 (1996)
5. Shi, J., Mamoulis, N., Wu, D., Cheung, D.W.: Density-based place clustering in geo-social networks. In: SIGMOD, pp. 99–110. ACM (2014)

Multi-roles Affiliation Model
for General User Profiling

Lizi Liao[1,2(✉)], Heyan Huang[1], and Yashen Wang[1]

[1] Beijing Engineering Research Center of High Volume Language Information
Processing and Cloud Computing Applications,
Beijing Institute of Technology, Beijing, China
liaolizi.llz@gmail.com
[2] Living Analytics Research Center,
Singapore Management University, Singapore City, Singapore
{hhy63,yswang}@bit.edu.cn

Abstract. Online social networks release user attributes, which is important for many applications. Due to the sparsity of such user attributes online, many works focus on profiling user attributes automatically. However, in order to profile a specific user attribute, an unique model is built and such model usually does not fit other profiling tasks. In our work, we design a novel, flexible general user profiling model which naturally models users' friendships with user attributes. Experiments show that our method simultaneously profile multiple attributes with better performance.

Keywords: General user profiling · Multi-roles affiliation model · Social networks

1 Introduction

The rapid growth of social network websites such as Facebook, LinkedIn and Twitter attracts a large number of Internet users. However, only a small proportion of these users intentionally or unintentionally disclose their attributes like occupation, education and interests, which are important to many online applications, such as recommendation, personalized search, and targeted advertisement. Research on user profiling has focused on various kinds of user attributes, ranging from demographic information like gender [2,8], age [5,10] and location [1–3], to user preference information like political orientation or interests.

In most of these works, in order to profile a specific user attribute, a unique model is built and such model usually does not fit other profiling tasks. In our paper, we propose a general user profiling model to profile multiple user attributes simultaneously. Since social network users connect to other users regularly, many works [1,4,11] leverage the principle of homophily [7] to profile attributes via social connections. The basic assumption is that users are more likely to connect with those sharing same attribute values. Based on the observed

© Springer International Publishing Switzerland 2015
A. Liu et al. (Eds.): DASFAA 2015 Workshops, LNCS 9052, pp. 227–233, 2015.
DOI: 10.1007/978-3-319-22324-7_20

features from the connected friends, user's attributes can be obtained by directly applying a majority vote or its variations [6].

However, this assumption oversimplifies the complexity of online social networks. In real life scenarios, users become friends only because of certain attributes and those attributes make different degrees of contribution. Thus, as our second contribution, we quantify the different linking factors for each attribute entry. For instance, compared to both being *Democrats*, both working at *Google* might be more likely to link two users together. That is to say, the linking factor of attribute entry *Google* is larger than that of attribute entry *Democrat*. Note that which attributes are more likely to link users are automatically inferred from data rather than pre-defined.

2   Model

In this section, we proceed to introduce our multi-roles affiliation model(MRA). Figure 1 illustrates the essence of our model. We start with a bipartite graph $\mathcal{A}$ where the nodes in the top represent users, the nodes in the bottom represent attribute values(or roles), and the edges indicate attribute affiliations. We also observe the social network $\mathcal{G}$ of those users. In our model, there are two important intuitions.

First, each attribute entry of the users corresponds to a specific role. MRA models users' preference towards each role with a bipartite attributes affiliation network as Fig. 1(a). Formally, we assume that there is a set of N users. Each user $u = 1, 2, \cdots, N$ has a latent group membership indicator $z_{uk} \in \{0, 1\}$ for each attribute entry (or role) $k = 1, 2, \cdots, K$. In Fig. 1(a), roles are indicated as a, b, c. Note that each indicator z_{uk} of user u is independent. Each user can belong to multiple roles simultaneously.

Second, we use a set of link factors π to capture the probability that users sharing a certain attribute value are linked together. For example, π_k is the probability of users taking the same role k to be linked together. Note that for different roles $k1$ and $k2$, their contribution to the link formation is different,

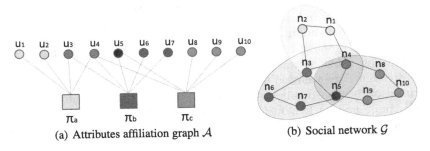

(a) Attributes affiliation graph $\mathcal{A}$ (b) Social network $\mathcal{G}$

Fig. 1. (a) Bipartite attributes affiliation graph. Squares: attribute values, Circles: users. (b) Social network of users.

which is quantitatively measured by π_{k1} and π_{k2}. As in [14], we define the probability of creating an edge (u, v) between a pair of users u, v as:

$$\delta_{uv} = 1 - \prod_{k \in \{C_{uv}\}} (1 - \pi_k) \tag{1}$$

where C_{uv} is the set of attribute entries u and v share (or roles they both take). We can see that the equation above already ensures that pairs of users that share more attributes are more likely to link together. To allow for edges between users who do not share any attribute, we also introduce an additional role, called the ϵ-role, which connects any pair of users with a very small probability ϵ. We simply set it to be the random link probability.

3   Inference

Given partially observed binary user attribute entries $F = \{f_{uk} : u \in \{1, \cdots, N\}; k \in \{1, \cdots, K\}\}$ and the user social network $\mathcal{G}$, we aim to find the full attributes affiliation graph $\mathcal{A}$ and link factors $\pi = \pi_k : k = 1, \cdots, K\}$. We apply the maximum likelihood estimation, which finds the optimal values of π and graph $\mathcal{A}$ so that they maximize the likelihood $L(\mathcal{A}, \pi) = P(\mathcal{G}, F|\mathcal{A}, \pi)$:

$$\arg\max_{\mathcal{A}, \pi} L(\mathcal{A}, \pi) = \prod_{(u,v) \in E} p(u, v) \prod_{(u,v) \notin E} (1 - p(u, v)) \tag{2}$$

We employ the coordinate ascent algorithm to solve the above optimization problem. The algorithm iterate the following two steps. First, we update π by keeping $\mathcal{A}$ fixed. Then we update $\mathcal{A}$ while keeping π fixed. To start the process, we need to initialize $\mathcal{A}$. Note that $\mathcal{A}$ is indeed a set of latent group membership indicator $z_{uk} \in \{0, 1\}$. For those partially observed binary user attribute entries F, we keep those z_{uk} to be the same as f_{uk}. For others, we randomly generate z_{uk} by using the ratio calculated from F and $\mathcal{G}$.

3.1   Update of Link Factors π

By keeping the attributes affiliation graph $\mathcal{A}$ fixed, we aim to find π by solving the following optimization problem:

$$\arg\max_{\pi} \prod_{(u,v) \in E} (1 - \prod_{k \in \{C_{uv}\}} (1 - \pi_k)) \prod_{(u,v) \notin E} (\prod_{k \in \{C_{uv}\}} (1 - \pi_k)) \tag{3}$$

where the constraints are $0 \leq \pi_k \leq 1$. We transform this non-convex problem into a convex optimization problem. We maximiza the logarithm of the likelihood and change the variables $e^{-x_k} = 1 - \pi_k$:

$$\arg\max_{x} \sum_{(u,v) \in E} log(1 - e^{- \sum_{k \in C_{uv}} x_k}) - \sum_{(u,v) \notin E} \sum_{k \in C_{uv}} x_k \tag{4}$$

where the constraints $0 \leq \pi_k \leq 1$ become $x_k \geq 0$. This problem is a convex optimization of x. We can solve it by gradient descent.

3.2   Update of Attributes Affiliation Graph $\mathcal{A}$

Given the link factors π, we aim to find appropriate attributes affiliation graph $\mathcal{A}$ for all the users. We use the Metropolis-Hastings algorithm [9] where we stochastically update $\mathcal{A}$ using a set of 'transitions'. Given the current attributes affiliation graph $\mathcal{A}$, we consider two kinds of transitions to generate a new attributes affiliation graph $\mathcal{A}'$. One is that a latent group membership indicator z_{uk} change from 1 to 0. The other is that a latent group membership indicator z_{uk} change from 0 to 1. Note that we fix z_{uk} for those already observed binary user attribute entries F. Once we have generated new attributes affiliation graph $\mathcal{A}'$, we accept $\mathcal{A}'$ with probability :

$$Min(1, L(\mathcal{A}', \pi)/L(\mathcal{A}, \pi)). \tag{5}$$

In other words, we initialize $\mathcal{A}_1 = \mathcal{A}$. We start the process with some $\mathcal{A}$ and then perform a large number of steps, where each step i we take $\mathcal{A}_i$ and apply a random 'transition' generating a new attributes affiliation graph $\mathcal{A}'_i$. At each step, we accept the transition probabilistically based on the ratio of log-likelihoods. If the transition is not accepted, we do not update $\mathcal{A}_i$.

4   Experiments

4.1   Experimental Setup

We use the Facebook networks of 4 colleges and universities: Georgetown, Oklahoma and Princeton and UNC Chapel Hill from a date in Sept. 2005 [12]. The edges here are only intra-school links between users. In these Facebook datasets, there are 8 user attributes which are ID, student/faculty status, gender, major, dorm or house and high school. Since ID information is very user specific as well as high school, we ignore these two attributes. We run experiments on these data sets.

Next, we introduce natural baseline as well as the state-of-the-art method. Taking the homophily phenomenon into consideration, we use random guess within direct neighbors (RA) as our natural baselines. Thus, we predict the k-th missing attribute entry value by randomly selecting a value of the k-th attribute entry from the users neighbors. Another baseline is CESNA [13]. This method detects overlapping communities in networks with node attributes. It statistically models the interaction between the network structure and the node attributes , which makes it capable of determining community membership as well as recover missing attributes. Using the community membership result and the association weights between attributes and each community, we obtain the final probability of each missing attribute for each user. Since attributes assignment probability obtained from CESNA are continuous values varying from 0 to 1. We choose the threshold as the one which gives us the largest F1 value. Then we treat this F1 value as the CESNA results.

4.2 Results and Discussion

We first focus on comparing our model MRA with natural baseline RA here in Table 1. In order to gain a comprehensive view of the performance, we randomly hide 10 % of user attributes to 50 % of user attributes in a gradual way. Then we take an average result for comparison. Note that most of the user attribute entries are 0, which is determined by the flattening procedure. Also, people care more about the entries with value 1 in real-world. Thus we provide the precision, recall and F1 value for value 1 respectively. From the Table 1, we observe that MRA can achieve higher performance.

Table 1. Results comparing with natural baseline

	Georgetown			Oklahoma			Princeton			UNC		
	P	R	F1	P	R	F1	P	R	F1	P	R	F1
RA	0.34	0.42	0.38	0.25	0.41	0.31	0.30	0.41	0.35	0.30	0.41	0.35
MRA	0.52	0.36	**0.43**	0.36	0.37	**0.36**	0.43	0.36	**0.39**	0.47	0.37	**0.41**

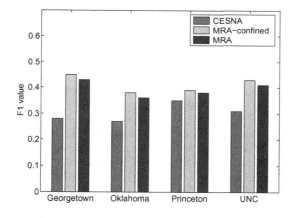

Fig. 2. Experiment results on 4 universities datasets.

Figure 2 shows the comparison between MRA and CESNA in terms of the F1 value. The x-axis refers to different datasets as mentioned above. Note that CESNA ignore those users who have no link with others. In the prediction results, the performance of CESNA is only based on those users who eventually get community assignment, which is a subset of the whole input users. To compare performance with CESNA on the same set of users who obtain community assignment successfully, the result of our model is named as MRA-confined. Since our model is able to handle both loosely connected users as well as unconnected users, we also give the performance detail named as MRA for the whole set of input users. We can see that the performance starts to deteriorate in some amount due to those uninformative users, which conforms the reality. However, those results are still better than the baseline results.

5 Conclusion

In this paper, we developed a general user profiling framework to simultaneously profile multiple attributes. Our multi-roles affiliation model (MRA) naturally captures the relationship between users friendship links and user attributes. It effectively profiles missing attributes for social network users. Moreover, the way we treat each attribute entry enables our model easily adapting to various kinds of attributes profiling.

Acknowledgments. This research is supported in part by Chinese National Program on Key Basic Research Project (Grant No. 2013CB329605). This research is also supported by the Singapore National Research Foundation under its International Research Centre@Singapore Funding Initiative and administered by the IDM Programme Office, Media Development Authority (MDA).

References

1. Backstrom, L., Sun, E., Marlow, C.: Find me if you can: improving geographical prediction with social and spatial proximity. In: Proceedings of the 19th International Conference on World Wide Web, pp. 61–70. ACM (2010)
2. Burger, J.D., Henderson, J., Kim, G., Zarrella, G.: Discriminating gender on twitter. In: Proceedings of the Conference on Empirical Methods in Natural Language Processing, pp. 1301–1309. Association for Computational Linguistics (2011)
3. Eisenstein, J., O'Connor, B., Smith, N.A., Xing, E.P.: A latent variable model for geographic lexical variation. In: Proceedings of the 2010 Conference on Empirical Methods in Natural Language Processing, pp. 1277–1287. Association for Computational Linguistics (2010)
4. Li, R., Wang, S., Deng, H., Wang, R., Chang, K.C.C.: Towards social user profiling: unified and discriminative influence model for inferring home locations. In: Proceedings of the 18th ACM SIGKDD International Conference on Knowledge Discovery and Data Mining, pp. 1023–1031. ACM (2012)
5. Liao, L., Jiang, J., Ding, Y., Huang, H., Lim, E.P.: Lifetime lexical variation in social media. In: Twenty-Eighth AAAI Conference on Artificial Intelligence (2014)
6. Liao, L., Jiang, J., Lim, E.P., Huang, H.: A study of age gaps between online friends. In: Proceedings of the 25th ACM Conference on Hypertext and Social Media, pp. 98–106. ACM (2014)
7. McPherson, M., Smith-Lovin, L., Cook, J.M.: Birds of a feather: homophily in social networks. Ann. Rev. Sociol. **27**, 415–444 (2001)
8. Mukherjee, A., Liu, B.: Improving gender classification of blog authors. In: Proceedings of the 2010 Conference on Empirical Methods in Natural Language Processing, pp. 207–217. Association for Computational Linguistics (2010)
9. Newman, M.E., Barkema, G.T., Newman, M.: Monte Carlo Methods in Statistical Physics, vol. 13. Clarendon Press, Oxford, Wotton-under-Edge (1999)
10. Nguyen, D., Gravel, R., Trieschnigg, D., Meder, T.: "How old do you think i am?" a study of language and age in twitter. In: ICWSM (2013)
11. Pennacchiotti, M., Popescu, A.M.: Democrats, republicans and starbucks afficionados: user classification in twitter. In: Proceedings of the 17th ACM SIGKDD International Conference on Knowledge Discovery and Data Mining, pp. 430–438. ACM (2011)

12. Traud, A.L., Kelsic, E.D., Mucha, P.J., Porter, M.A.: Comparing community structure to characteristics in online collegiate social networks. SIAM Rev. **53**(3), 526–543 (2011)

13. Yang, J., Leskovec, J.: Community-affiliation graph model for overlapping network community detection. In: 2012 IEEE 12th International Conference on Data Mining (ICDM), pp. 1170–1175. IEEE (2012)

14. Yang, J., McAuley, J., Leskovec, J.: Community detection in networks with node attributes. In: 2013 IEEE 13th International Conference on Data Mining (ICDM), pp. 1151–1156. IEEE (2013)

Needle in a Haystack: Max/Min Online Aggregation in the Cloud

Xiang Ci, Fengming Wang, and Xiaofeng Meng[✉]

School of Information, Renmin University of China, Beijing, China
{cixiang,rucwfm1991,xfmeng}@ruc.edu.cn

Abstract. As the development of social network, mobile Internet, etc., an increasing amount of data are being generated, which beyond the processing ability of traditional data management tools. In many real-life applications, users can accept approximate answers accompanied by accuracy guarantees. One of the most commonly used approaches of approximate query processing is online aggregation. Most existing work of online aggregation in the cloud focuses on the aggregation functions such as Count, Sum and Avg, while there is little work on the Max/Min online aggregation in the cloud now. In this paper, we measure the accuracy of Max/Min online aggregation by using quantile which is deduced by Chebyshev's inequality and central limit theorem. We implement our methods in a cloud online aggregation system called COLA and the experimental results demonstrate our method can deliver reasonable online Max/Min estimates within an acceptable time period.

Keywords: Online aggregation · Cloud computing · Chebyshev's inequality · Central limit theorem

1 Introduction

Data-driven activities are rapidly growing in various applications, including Web access logs, sensor data, scientific data, etc. All these big data have brought great challenges to traditional data management in terms of both data size and significance. Fortunately, in many real-life applications, people just want to obtain a bird's eye view of the whole dataset. This situation has brought more attention to the already-active area of Approximate Query Processing (AQP).

OLA is one of the most widely used AQP techniques and our work also focuses on OLA. OLA is first proposed in the area of relational database management system (RDBMS). The basic idea behind OLA is to estimate the result

This research was partially supported by the grants from the Natural Science Foundation of China (No. 61379050, 91224008); the National 863 High-tech Program (No. 2013AA013204); Specialized Research Fund for the Doctoral Program of Higher Education(No. 20130004130001), and the Fundamental Research Funds for the Central Universities, and the Research Funds of Renmin University(No. 11XNL010). This work was supported in part by Noah's Ark Lab and DNSLAB, China Internet Network Information Center.

© Springer International Publishing Switzerland 2015
A. Liu et al. (Eds.): DASFAA 2015 Workshops, LNCS 9052, pp. 234–239, 2015.
DOI: 10.1007/978-3-319-22324-7_21

by sampling data and the approximate answer should be given some accuracy guarantees. One important problem is that current work of OLA lacks support for all the commonly used aggregate functions. A representative trace of 69438 Hive queries from Facebook [5] shows that Min, Count, Avg, Sum and Max are the most popular aggregate functions at Facebook constituting 33.35 %, 24.67 %, 12.20 %, 10.11 % and 2.87 % of the total queries respectively. Most work of OLA now focuses on the functions that reflect the overall characteristics of the dataset, e.g., Count, Sum. In contrast, there is little work of Max/Min OLA although they are very important queries from above trace. Motivated by above requirements and challenges, we assert that Max/Min online aggregation in the cloud is very useful but it is almost impossible to use confidence interval to cope with the problem. We use Chebyshev's inequality to give a lower (upper) bound of the quantile of Max (Min) online aggregation. Then we correct the estimation error of the quantile by central limit theorem.

2   Related Work

OLA was first introduced in RDBMS [1], which focuses on single-table queries involving "group by" aggregations. Hadoop Online Prototye (HOP) [2] is a modified version of the original MapReduce framework, which is proposed to construct a pipeline between Map and Reduce. COLA [3] realizes the estimation of confidence interval based on HOP.

To the best of our knowledge, there is no work of Max/Min online aggregation in the cloud yet. The most relative work is [4]. In the paper, authors want to "guess" the extreme values in a dataset using sample. But the method it proposed is not designed for online aggregation.

3   Overview

In this section, we formalize our problem that we study in the paper. This paper mainly focuses on online aggregation for single table, so we consider a relation R and queries of the form like

SELECT $op(exp(t_{ij}))$, col FROM R WHERE $predicate$ GROUP BY col

In the above query, op is the operation of Max or Min, exp is an arithmetic expression of the attributes in R, $predicate$ is an arbitrary predicate involving the attributes, and col is one or more columns in R. t_{ij} represents the j-th tuple in block i.

4   Randomization of Blocks

The performance of online aggregation highly depends on the sampling, and the accuracy of sampling depends on the data distribution. If the data are fully random, the result will be good. Unfortunately, this is not always true for real world data. For tuple-level sampling, there are several ways to achieve random sampling from disk. For block-level sampling, the case can be worse. We still can not get random sampling by random disk access because the data layout inside the block may be not random. We must randomize the blocks firstly.

5   Query Processing

Unlike query processing in RDBMS, OLA must return the estimated result continuously with accuracy guarantees. As we know, confidence interval is a very useful statistic to measure the query accuracy because it can intuitively show the value range of result. But it is inefficient for Max/Min online aggregation. Assume probability $0 \leq P \leq 1$, the quantile z_σ of random variable X is the real number where $P(X > z_\sigma) = \delta$. In many applications, very high (low) quantile can also greatly reflect the tail characteristics of the dataset. This inspires us to use quantile to measure the accuracy of Max/Min online aggregation.

5.1   Estimation of Max/Min Value

We take random samples from the dataset, and simply use current maximum (minimum) value of sampling data as the estimated maximum (minimum) value of the dataset. At the same time, the quantile of the estimated value in the dataset is given. This method is definitively simple, but the key problem is how to obtain the quantile of the estimated value in the dataset. We solve the problem by using Chebyshev's inequality. Since we need to compute Max and Min separately, we use the following one-tailed version of Chebyshev's inequality. Let X be a random variable with finite expected value μ and finite non-zero variance σ^2. Then for any real number $\varepsilon > 0$, there is

$$\begin{cases} P[X - \mu \geq t] \leq \frac{\sigma^2}{\sigma^2 + t^2} & t \geq 0 \\ P[X - \mu \leq t] \leq \frac{\sigma^2}{\sigma^2 + t^2} & t < 0 \end{cases} \tag{1}$$

Suppose that the maximum value of sampling data is M, we can get that M is **at least** the $1 - \dfrac{\sigma^2}{\sigma^2 + (M - \mu)^2}$ quantile of the dataset. Suppose that the minimum value of sampled data is N, we can get that N is **at most** the $\dfrac{\sigma^2}{\sigma^2 + (N - \mu)^2}$ quantile of the dataset.

5.2   Error Correction

We know the error ε is generated just because average μ and standard deviation σ of the random variable are estimated. In order to meet the requirement of accuracy, we must guarantee the error between sample expectation and real expectation is no more than ε_μ under the default confidence level δ. The error between sample deviation and real deviation is no more than ε_{σ^2} under the default confidence level σ. Let z_σ be the $(\sigma + 1)/2$ quantile of the standard normal distribution, then

$$\frac{\sqrt{n}\varepsilon_u}{T_{n,2}^{1/2}} = z_\delta \tag{2}$$

Where $T_{n,2} = \frac{\sum_{i=1}^{n}\left(X_i - \overline{X}\right)^2}{n-1}$. There is $\varepsilon_\mu = (\frac{z_\delta^2 T_{n,2}}{n})^{1/2}$.

Similarly, using multivariate central limit theorem, we can get

$$\varepsilon_{\sigma^2} = \left(\frac{z_\delta^2 (T_{n,4} - T_{n,2}^2)}{n}\right)^{1/2} \tag{3}$$

where $T_{n,4} = \frac{\sum_{i=1}^{n}\left(X_i - \overline{X}\right)^4}{n-1}$.

So M is **at least** the φ_M quantile of the dataset when the confidence level is δ, where

$$\varphi_M = 1 - \frac{\sigma^2 + \varepsilon_{\sigma^2}}{\sigma^2 + \varepsilon_{\sigma^2} + (M - \mu - \varepsilon_\mu)^2} \tag{4}$$

N is **at most** the φ_N quantile of the dataset when the confidence level is δ, where

$$\varphi_N = \frac{\sigma^2 + \varepsilon_{\sigma^2}}{\sigma^2 + \varepsilon_{\sigma^2} + (N - \mu - \varepsilon_\mu)^2} \tag{5}$$

Due to limited space, all the details of above derivations are omitted.

6   Max/Min Online Aggregation in the Cloud

Cloud is different from RDBMS, and the major problem of online aggregation in the cloud is that naive MapReduce does not support pipeline operations. We use COLA to implement our method. The basic steps of Max/Min online aggregation with our method is described as follows:

- **Step1**: Read data from HDFS and get the initial samples;
- **Step2**: Use current Max/Min value as the estimated Max/Min;
- **Step3**: Compute the average and deviation of the samples;
- **Step4**: Correct the error of average and deviation;
- **Step5**: Calculate the quantile of Max/Min value by Chebyshev's inequality;
- **Step6**: Output the result and continue sampling;
- **Step7**: Repeat step 1 to 6 until users terminate the process actively. If users do not terminate, all the data will be processed.

Our method primarily focuses on the query on a single table, so the implementation only involves one MapReduce job. In the Map function, tuples of every block are filtered out according to the predicate and transformed into key-value pairs. The Reduce function is executed each time the estimate is invoked, so it is important to make the computing process incremental and make use of the results of the previous Reduce function.

7   Performance Evaluation

7.1   Experiment Overview

The testbed is established on a cluster of 11 nodes connected by a 1 Gbit Ethernet switch. One node serves as master, and the remaining 10 nodes act as slave nodes. Each node has a 2.33 GHz quad-core CPU and 7 GB of RAM, and the disk size of each node is 1.8 TB. We set the block size of HDFS to 64 MB.

Table 1 summarizes settings used in the experiments:

Table 1. Settings used in the experiments

Parameter	Values
Number of computer node	11
Data type	page view statistics of Wikipedia
Data size	100G
Platform	COLA
Map task num. per computer node	4
Reduce task num. per computer node	2

All the data are stored in HDFS, and we test online aggregation queries of Max and Min with example queries Q1 and Q2.

> **Q1= SELECT Max(pageviews), language FROM visit_log GROUP BY language**

> **Q2= SELECT Min(pageviews), language FROM visit_log GROUP BY language**

In the experiment, we set the confidence level to 95 %. There are ten kinds of languages in the dataset, and here we only show the results of English and French due to the limited space.

7.2   Performance over Real Data

In this experiment, we run above two queries on the dataset. The accuracy of estimated aggregation result is measured by *relative_error*, which is computed by following equation:

$$relative\_error = \frac{|estimateValue - actualValue|}{actualValue} \qquad (6)$$

The convergence speed of quantile is reflected by *avgtime_max* and *avgtime_min*. *avgtime_max* is the average time to get the 0.99 quantile when the confidence level is 95 %, while *avgtime_min* is the average time to get the 0.01 quantile when the confidence level is also 95 %.

7.3 Query Error

Figure 1 illustrates the update of relative_error as the online aggregation of Q1 proceeds. As Fig. 1 shows, relative_error of Max online aggregation is very high at the beginning, which is different from the Count (Sum) online aggregation. But the relative_error degrades fast as the online aggregation continues. The trend of French is similar, except that the update rate of French is a little slower, and the time to get real value is a little longer than English.

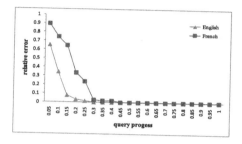 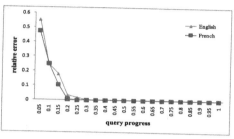

Fig. 1. Query error of Q1   **Fig. 2.** Query error of Q2

Figure 2 shows the update of relative_error as the online aggregation of Q2 proceeds. Comparing to Max online aggregation (Q1), the relative_error of Min online aggregation at the beginning is relatively small and declines a little faster.

Due to space limitations, the experiments of running time, scalability and data randomization are all omitted.

8 Conclusion

Most work of online aggregation now focuses on the functions that reflect the overall characteristics of the dataset, e.g., Count, Sum. To the best of our knowledge, this is the first work on studying Max/Min online aggregation in the cloud. The experimental results demonstrate the efficiency and scalability of our proposal.

References

1. Hellerstein, J.M., Haas, P.J., Wang, H.J.: Online aggregation. In: SIGMOD Conference, pp. 171–182 (1997)
2. Condie, T., Conway, N., Alvaro, P., Hellerstein, J.M., Gerth, J., Talbot, J., Elmelegy, K., Sears, R.: Online aggregation and continuous query support in MapReduce. In: SIGMOD Conference, pp. 1115–1118 (2010)
3. Shi, Y., Meng, X., Wang, F., Gan, Y.: You can stop early with COLA: online processing of aggregate queries in the cloud. In: CIKM, pp. 1223–1232 (2012)
4. Wu, M., Jermaine, C.: Guessing the extreme values in a data set: a bayesian method and its applications. VLDB J. **18**(2), 571–597 (2009)
5. Agarwal, S., Milner, H., Kleiner, A., Talwalkar, A., Jordan, M., Madden, S., Mozafari, B., Stoica, I.: Knowing when you're wrong: building fast and reliable approximate query processing systems. In: SIGMOD Conference, pp. 481–492 (2014)

FFD-Index: An Efficient Indexing Scheme for Star Subgraph Matching on Large RDF Graphs

Xuedong Lyu[1], Xin Wang[1](✉), Yuan-Fang Li[2], and Zhiyong Feng[1]

[1] School of Computer Science and Technology, Tianjin University, Tianjin, China
{lyuxd,wangx,zhiyongfeng}@tju.edu.cn
[2] Faculty of Information Technology,
Monash University, Clayton, VIC 3800, Australia
yuanfang.li@monash.edu

Abstract. Subgraph matching, a basic SPARQL operation, is known to be NP-complete. Coupled with the rapidly increasing volumes of RDF data, it makes efficient graph query processing a very challenging problem. In this paper, we tackle the important problem of efficient processing of star-shaped subgraph matching queries, which are a core SPARQL query pattern and usually lead to a number of costly join operations. We present a novel method to encode a star-shaped subgraph into a bit string and an indexing mechanism to improve the query answering performance, called FFD-index. Our extensive evaluation shows that FFD-index and the corresponding algorithms are effective in solving star-shaped graph queries and they significantly outperform the state-of-the-art SPARQL query engine RDF-3X.

Keywords: RDF · Subgraph isomorphism · Graph-based index

1 Introduction

Subgraph matching, also known as subgraph isomorphism, is a widely known NP-Complete problem [2]. The rapidly increasing volumes of RDF data and the high complexity of SPARQL together make efficient RDF graph query processing a significant challenge.

A SPARQL query often consists of multiple star-shaped structures as subqueries [3]. Hence, efficient evaluation of SPARQL queries, especially over large RDF graphs, critically depends on the efficient processing of star-shaped subqueries. Classical solutions generate a set of costly join operations to evaluate star-shaped graph queries, which are inefficient. In this paper, we propose a novel indexing scheme, called FFD (Filtering by Fingerprints and Degrees), for efficient processing of star-shaped queries, which form a core SPARQL query pattern. FFD-index maintains all star-shaped subgraphs of a large RDF graph and denotes each subgraph with a *fingerprint*. A fingerprint preserves structural information of a subgraph, and is used to prune the search space of subgraph

© Springer International Publishing Switzerland 2015
A. Liu et al. (Eds.): DASFAA 2015 Workshops, LNCS 9052, pp. 240–245, 2015.
DOI: 10.1007/978-3-319-22324-7_22

matching. Our work makes two contributions: (1) we propose a method to encode a star-shaped subgraph into a fingerprint, and (2) we design an efficient indexing scheme, FFD-index that can answer a query in a *one-graph-at-a-time* manner.

Moreover, since a path longer than two edges can always be split into subpaths of two edges, which can be viewed as stars with two branches, our approach can be easily adapted to support path queries.

2   FFD-Index and Query Processing

The fingerprint of a star subgraph is determined by all its edges and vertices except the central vertex. Our method is similar to the way gStore calculates the signature of a vertex [4]. However, we calculate the fingerprint of each star subgraph while gStore only does it for each vertex.

An RDF dataset $T = \{t \mid t \in S \times P \times O\}$ is a set of triples that can be modeled as a directed labeled graph $G = (V, E, \Sigma, l)$, where V is a finite set of vertices, $E \subseteq V \times V$ is a finite set of edges and $\Sigma = U \cup L$ is a set of labels. The labeling function $l : V \cup E \rightarrow \Sigma$ maps each vertex or edge to a label in Σ. Let $S = \{s \mid s = l(v), v \in V, \exists \langle v, u \rangle \in E\}$, $P = \{p \mid p = l(\langle v_i, v_j \rangle), \langle v_i, v_j \rangle \in E\}$, and $O = \{o \mid o = l(v), v \in V, \exists \langle u, v \rangle \in E\}$ represent the sets of labels of subjects, predicates and objects of RDF triples in T, respectively. For a star subgraph $G^* = (V^*, E^*, \Sigma^*, l^*)$ with $v \in V^*$ as its central vertex, all v's adjacent edges are encoded into a bit string. Let us assume that $e = \langle v, u \rangle$ is an edge. A set of hash functions H maps $l^*(e)$ and $l^*(u)$ into a fixed-length bit string represented by $Encode_{edge}(\langle e, u \rangle)$. Suppose that the length is $m + n$. The first m bits are used to represent $l^*(e)$ and the rest n bits to represent $l^*(u)$. In the case of the first m bits, $\overline{m}$ different hash functions $h_i(1 \leq i \leq \overline{m})$ are used to generate $\overline{m}$ integers $h_i(l^*(e))$. Then for each $h_i(l^*(e))$, the $(h_i(l^*(e)) \bmod m)$-th bit of first m bits is set to be "1", thus there are at most $\overline{m}$ bits set to be "1" leaving others "0". Similarly, $\overline{n}$ bits of the rest n are set to be "1".

E^* in G^* is partitioned into two parts: (1) E^*_{out}, the set of outgoing edges of v, and (2) E^*_{in}, the set of incoming edges of v. The fingerprint of G^* is composed of two parts, $fp_{out}(G^*)$ generated from E^*_{out} and $fp_{in}(G^*)$ from E^*_{in}. $fp_{out}(G^*)$ is used to denote the bit string of outgoing edges. Formally, $fp_{out}(G^*) = Encode_{edge}(\langle e_1^{out}, u_1^{out} \rangle) \mid \ldots \mid Encode_{edge}(\langle e_k^{out}, u_k^{out} \rangle)$, where $k = |E^*_{out}|$, $e_i^{out} \in E^*_{out}$ ($1 \leq i \leq k$), and $u_i^{out} \in V^*$. u_i^{out} is the corresponding vertex of e_i^{out} and "$\mid$" denotes the bitwise-or operator.

Similarly, the bit string of incoming edges, denoted by $fp_{in}(G^*)$ is defined as $fp_{in}(G^*) = Encode_{edge}(\langle e_1^{in}, u_1^{in} \rangle) \mid \ldots \mid Encode_{edge}(\langle e_l^{in}, u_l^{in} \rangle)$, where $l = |E^*_{in}|$, $e_j^{in} \in E^*_{in}$ ($1 \leq j \leq l$), and $u_j^{in} \in V^*$.

For the last step, these two bit strings are concatenated together to form the fingerprint of G^*, denoted as $fp(G^*) = fp_{out}(G^*) \circ fp_{in}(G^*)$, where "$\circ$" is the concatenation operator. Figure 1 shows the procedure of the fingerprint generation.

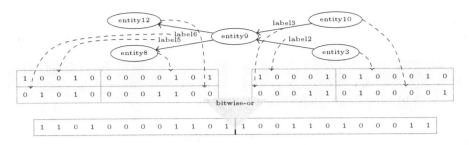

Fig. 1. The procedure of the fingerprint generation of a star-shaped graph($m = 5$, $n = 7$, $\overline{m} = 2$, $\overline{n} = 2$. The dashed edges denote $Encode_{edge}$ functions.)

Each star-shaped subgraph of an RDF graph are extracted and stored into a 3-column table T, named *neighborhood table*, whose first column is designated as the key composed of the labels of central vertices. The second column preserving adjacent edges of central vertices is denoted as $AdjSet$, of which each element is represented as a triple $(direction, l^*(e), l^*(u))$, where $direction$ denotes whether this edge is outgoing or incoming. In the third column, fingerprints are precalculated for star-shaped subgraphs, named $FgPrt$.

Given a query graph $Q^* = (V', E', \Sigma', l')$, V' is a finite set of vertices, $E' \subseteq V' \times V'$ a finite set of edges, and $\Sigma' \subseteq U \cup L \cup Var$ a label set where Var is a set of variables. The labeling function $l' : V' \cup E' \rightarrow \Sigma'$ maps each vertex or edge to a label in Σ'.

When the central vertex v of Q^* is fixed, the candidate subgraph can be fetched by looking up T with the key $l'(v)$. It is obvious that there is at most one star-shaped subgraph in T matched with Q^*. We build the candidate set by checking whether there exists a record $\langle l'(v), AdjSet_{l'(v)}, FgPrt_{l'(v)} \rangle$ in T. The detailed checking procedure is shown in Algorithm 1. If the number of edges of Q^* is larger than that of G^*, then G^* cannot be matched by Q^*. As line 3–4 of Algorithm 1 show, the candidate should be dropped.

Algorithm 1. $GenerateCandidateSet_{central\ vertex\ fixed}$

Input: T, $Q^* = \{V', E', \Sigma', l'\}$, $l'(v)$
Output: C^*: The Candidate Set for Q^*
1: $C^* \leftarrow \emptyset$;
2: $t_{l'(v)} \leftarrow \langle l'(v), AdjSet_{l'(v)}, FgPrt_{l'(v)} \rangle$;   ▷ Get $t_{l'(v)}$ from T according to $l'(v)$.
3: **if** $\left| AdjSet_{l'(v)} \right| \geqslant |E'|$ **then**
4:     $C^* \leftarrow C^* \cup \{t_{l'(v)}\}$;   ▷ Add $t_{l'(v)}$ into candidate set C^*.
5: **end if**

Otherwise, if v is variable, candidates have to be found according to Q^*'s labels and structural information. Therefore, based on how many elements are fixed out of two (except the central vertex) triple elements, we devise five indexes:

SP_star, OP_star, S_star, P_star, and O_star. We are unable to obtain candidates from T directly, so we generate the query fingerprint $fp(Q^*)$ of Q^*, using almost the same method designed for generating the fingerprint of star subgraph demonstrated in Fig. 1. The only difference is that when the label of an edge or a vertex is a variable, all the bits corresponding to this label will be set 0. If G^* is a candidate of Q^*, $fp(Q^*)$ and $fp(G^*)$ should satisfy that if i-th bit of $fp(Q^*)$ is 1, then the i-th bit of $fp(G^*)$ must be 1. We call this condition "$fp(G^*)$ *covers* $fp(Q^*)$", denoted as $fp(G^*) \succ fp(Q^*)$. Fingerprint covering is necessary but not sufficient for declaring that Q^* matches G^*. The query algorithm is shown

---

Algorithm 2. $GenerateCandidateSet_{central\ vertex\ variable}$

Input: SP_index, $Q^* = \{V', E', \Sigma', l'\}$, $\langle l'(u), l'(e) \rangle$
Output: C^*: The Candidate Set for Q^*
1: $C^* \leftarrow \emptyset$;
2: Get $\langle \langle l'(u), l'(e) \rangle, DegMap\langle centralvertex, degree \rangle \rangle$ from SP_index;
3: **for** each $\langle cv, deg \rangle \in DegMap\langle centralvertex, degree \rangle$ **do**
4:     **if** $deg \geqslant |E'|$ **then**
5:         $t_{cv} \leftarrow \langle cv, FgPrt_{cv}, AdjSet_{cv} \rangle$;   ▷ Get t_{cv} from T according to cv;
6:         **if** $FgPrt_{cv} \succ fp(Q^*)$ **then**
7:             $C^* \leftarrow C^* \cup \{t_{cv}\}$;
8:     **end if**
9:     **end if**
10: **end for**

---

in Algorithm 2 with a time complexity of $\mathcal{O}(|DegMap|)$, where $|DegMap|$ is the number of star subgraphs with an incoming $l(e)$ edge staring from u. To confirm that a subgraph is a candidate, two conditions are checked. Firstly, the degree of the central vertex of a candidate must be larger than that of the query graph (line 4). Secondly, the fingerprint covering condition is examined by checking whether $fp(G^*)$ covers $fp(Q^*)$ (line 6).

3 Verification

FFD-index allows the search space to be significantly reduced. However, whether each candidate subgraph can match a given query graph still requires verification. Suppose that $G^* = \{V^*, E^*, \Sigma^*, l^*\}$ is an element in the candidate set for star query $Q^* = \{V', E', \Sigma', l'\}$, where $|E'| = k$. For each edge $e'_i \in E'$, a candidate edge set C'_i is associated with it. We select edges from E that can match e'_i and then put them into C'_i. P is used to denote the Cartesian product of C'_i $(1 \leqslant i \leqslant k)$, i.e., $P = C'_1 \times \cdots \times C'_i \times \cdots \times C'_k$.

As an undesirable situation, one edge $e \in E$ can be matched by more than one edge in E', e.g., both e'_i and e'_j $(i \neq j)$ match e, in other words, $C'_i \cap C'_j \neq \emptyset$. So we have to remove some elements from P to guarantee that no edge is matched by different triple patterns simultaneously. We refer to this type of elements as *overlapping elements*. The verification algorithm is shown in Algorithm 3. For each edge in the query graph, we first find all edges that match it, as shown by lines 2–3. Then, the Cartesian product is calculated and examined to guarantee there is no overlapping element.

Algorithm 3. *Verification*

Input: $Q^* = \{V', E', \Sigma', l'\}$, $G^* = \{V^*, E^*, \Sigma^*, l^*\}$
Output: M: The Set of Matches From Q^* To G^*
1: $k \leftarrow |E'|$, $M \leftarrow \emptyset$;
2: **for** each $e'_i \in E'$ **do**
3:     Extract edge(s) from E^* that can match e'_i: $C'_i \subseteq E^*$;
4: **end for**
5: $P \leftarrow \prod_{i=1}^{k} C'_i$;   ▷ P is the Cartesian product of C'_i
6: **for** each $t \in P$ **do**
7:     **if** t is an overlapping element **then**
8:         $M \leftarrow M \cup \{t\}$;
9:     **end if**
10: **end for**

---

4 Experiments

Experiments were carried out on an Intel Quad-Core CPU 2.66GHZ machine with 3.8GB memory running Ubuntu 12.04 64-bit. We used both a synthetic dataset, LUBM,[1] and a real dataset, YAGO,[2] in our experiments and compared FFD-index with RDF-3X [1].

Figures 2(a) and 3(a) show the average number of candidates before and after fingerprint covering validation. Q_i is a query set in which each query has i edges. The average number of false positive candidates filtered out by fingerprint is quite large. Our approach exhibits better performance than RDF-3X with the increase in the size of query graphs.

Generally speaking, The number of costly join operations carried out by RDF-3X increases with the number of edges of a star-shaped graph query. In contrast, in our approach, more edges mean more chances, thus more bits of the query graph's fingerprint will be set "1" that makes the fingerprint more efficient in pruning search space. Average query response times are shown in Figs. 2(b) and 3(b). With the query size growing, FFD-index outperforms RDF-3X by more than 70 % in query response time on average.

---

[1] http://swat.cse.lehigh.edu/projects/lubm/.
[2] http://www.mpi-inf.mpg.de/departments/databases-and-information-systems/research/yago-naga/yago/archive/.

	# BEF. VLD.	# AFT. VLD.
Q_3	134.7	25.6
Q_5	124.2	15.5
Q_7	78.4	4.9
Q_9	42.7	3.4
Q_{11}	27.6	1.3
Q_{13}	28.2	1.4
Q_{15}	14.1	1.3

(a) Filtering Ability

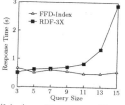

(b) Avg Response Time

Fig. 2. Performance comparison on LUBM.

	# BEF. VLD.	# AFT. VLD.
Q_3	542.6	67.4
Q_5	477.3	45.8
Q_7	237.0	21.1
Q_9	97.3	12.7
Q_{11}	104.4	4.2
Q_{13}	85.5	1.6
Q_{15}	57.2	1.8

(a) Filtering Ability

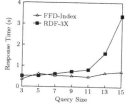

(b) Avg Response Time

Fig. 3. Performance comparison on YAGO.

5   Conclusion

In this paper, we propose FFD-index, a novel subgraph encoding and indexing scheme that reduce search space to improve SPARQL query answering performance. Extensive experiments have been conducted on widely-used synthetic as well as real-world benchmark datasets show that FFD-index is significantly more efficient over the state-of-the-art SPARQL query engine RDF-3X on star-shaped graph retrieval, especially over complex queries.

Acknowledgement. This work is supported by the National Natural Science Foundation of China (61100049), the National High-tech R&D Program of China (863 Program) (2013AA013204), CCF Opening Project of Chinese Information Processing (CCF2013-02-02), and the 3rd Baidu Subject Research Project.

References

1. Neumann, T., Weikum, G.: RDF-3X: a RISC-style engine for RDF. Proc. VLDB Endow. **1**(1), 647–659 (2008)
2. Michael, R.G., David, S.J.: Computers and Intractability: A Guide to the Theory of NP-Completeness. WH Freeman Co., San Francisco (1979)
3. Kim, H., Ravindra, P., Anyanwu, K.: From SPARQL to mapreduce: the journey using a nested triplegroup algebra. Proc. VLDB Endow. **4**(12), 1426–1429 (2011)
4. Zou, L., Mo, J., Chen, L., Özsu, M.T., Zhao, D.: gStore: answering SPARQL queries via subgraph matching. Proc. VLDB Endow. **4**(8), 482–493 (2011)

Leveraging Interactive Knowledge and Unlabeled Data in Gender Classification with Co-training

Jingjing Wang[1,2], Yunxia Xue[1,2], Shoushan Li[1,2(✉)], and Guodong Zhou[1,2]

[1] Natural Language Processing Lab, School of Computer Science and Technology,
Soochow University, Suzhou, China
{djingwang,yunxia.xue,shoushan.li}@gmail.com,
gdzhou@suda.edu.cn
[2] Collaborative Innovation Center of Novel Software Technology and Industrialization,
Nanjing, China

Abstract. Conventional approaches to gender classification much rely on a large scale of labeled data, which is normally hard and expensive to obtain. In this paper, we propose a co-training approach to address this problem in gender classification. Specifically, we employ both non-interactive and interactive texts, i.e., the *message* and *comment* texts, as two different views in our co-training approach to well incorporate unlabeled data. Experimental results on a large data set from micro-blog demonstrate the appropriateness of leveraging interactive knowledge in gender classification and the effectiveness of the proposed co-training approach in gender classification.

Keywords: Interactive knowledge · Gender classification · Co-training

1 Introduction

Gender classification, a fundamental task in social media analysis, aims to predict the user gender with the user-generated data. With the rapid growth of social media in recent years, gender classification has been drawing more and more attention for a wide range of real-life applications, such as intelligent marketing, personalization prediction, automatic advertising, and sentiment analysis (Mukherjee and Liu 2010; Burger et al. 2001; Volkova et al. 2013).

On one hand, conventional approaches conceptualize gender classification as a supervised learning problem and rely on human-annotated data for model learning. Such supervised approaches have delivered reasonable performance. However, the reliance on labeled data, which is normally hard and expensive to produce, presents a major obstacle to the widespread application of gender classification.

On the other hand, we notice that most of previous studies in gender classification focus on exploring text knowledge to infer user's gender with a statistical text classifier (Nowson and Oberlander 2006; Ciot et al. 2013). Although these studies have achieved certain success in gender classification, the utilized knowledge is always limited to non-interactive text, i.e., generated from the given user. In fact, as a well-established platform for user interaction, social media not only provides a public platform for a user to publish

© Springer International Publishing Switzerland 2015
A. Liu et al. (Eds.): DASFAA 2015 Workshops, LNCS 9052, pp. 246–251, 2015.
DOI: 10.1007/978-3-319-22324-7_23

his own experiences and opinions, but also offers an effective channel for other users to feedback through some interactive mechanisms, e.g., allowing a user to write a *comment* as a feedback to certain *message*. Such interactive process normally generates some kinds of interactive text (e.g., the *comment*) which provides a complementary resource to infer the gender of the message-publishing user.

Figure 1 illustrates an example from Twitter with the *message* text and the *comment* text. From the *comment* text, a kind of interactive knowledge, it is easy to infer the message-publishing user to be *female* from these descriptions, such as *pretty girl, your earrings* and *your boyfriend*, while it may be not obvious from the *message* text.

Message:
> *My sweet bought me a Tiffany necklace for my birthday. I am in love with it, I have always wanted one!!*

Comment:
> 1) *Happy birthday to this pretty girl! God, isn't she cute?*
> 2) *WOW, the necklace matches your earrings so perfect!!*
> 3) *Oh, your boyfriend is so sweet, you are so lucky to have him.*

Fig. 1. An example from Twitter with the *message* text and the corresponding *comment* text

In this paper, we focus on exploiting interactive knowledge, i.e., the *comment* text in social media, in gender classification. With the help of such knowledge, we propose a co-training approach to well incorporate unlabeled data and alleviate the high reliance of gender classification on labeled data by casting non-interactive and interactive texts as two views.

2   Related Work

Previous studies in gender classification mainly adopt supervised learning approaches on different text styles, such as Blog (Peersman et al. 2010; Gianfortoni et al. 2011), E-mail (Corney et al. 2002), YouTube (Filippova 2012) and Micro-blog (Rao et al. 2010; Liu et al. 2013). Specifically, besides those standard features such as character, word, and POS features, these studies focus on exploring more effective features for gender classification. For example, Nowson and Oberlander (2006) validate that using context-based n-grams tends to be more accurate than using dictionary-based features in weblog. Mukherjee and Liu (2010) propose some POS pattern features to improve the classification performance.

In comparison, there are only a few previous studies in semi-supervised gender classification. Ikeda et al. (2008) propose a semi-supervised approach to gender classification in blog. Their main idea is to utilize a sub-classifier to measure the relative similarity between two blogs so as to capture the classification knowledge in the unlabeled data. More recently, Burger et al. (2011) mention the importance of using unlabeled data and directly apply a self-training approach to perform semi-supervised learning for gender classification.

Different from above studies, our study focuses on interactive knowledge for gender classification, which has not been explored before. Furthermore, our co-training approach to semi-supervised gender classification is based on two views. Experimental results show that our two-view co-training approach is much more effective than the single-view self-training approach.

3   Gender Classification with Co-training

In supervised learning, a predictor f is trained to map an input vector x into a class label y. In this paper, the input vector x is the feature representation, generated from either the *message* text or the *comment* text.

Formally, the objective of gender classification is illustrated as follows:

$$f(x) \rightarrow y$$
$$\text{Where } y \in Y \text{ and } Y = \{male, female\} \tag{1}$$

In the literature, a variety of effective features have been proposed for gender classification. In this paper, we adopt following features due to their good performance in previous studies, e.g., Mukherjee and Liu (2010).

(1) **Bag-of-words Features:** These are basic word features. This kind of basic features is popularly utilized in not only gender classification but also many other NLP tasks where text knowledge is leveraged.

(2) **F-measure Feature:** It is defined to capture the POS usage. As a unitary measure of text's relative contextuality, this feature explores the notion of implicitness of text (Heylighen and Dewaele 2002).

(3) **POS Sequence Patterns:** These features are extracted from POS sequence of the text (Mukherjee and Liu 2010) to capture the writing styles of different genders.

In semi-supervised learning, we employ the co-training algorithm (Blum and Mitchell 1998) to leverage the two views from two kinds of text: one contains all the messages written by the user himself and the other contains the comments written by other users. Figure 2 illustrates the co-training algorithm for gender classification by using the *message* text and the *comment* text as two different views. In this figure, two different classifiers trained with the *message* text and the *comment* text are denoted as the *message* classifier C_M and the *comment* classifier C_C, respectively.

4   Experimentation

Data Setting. The data is collected from Sina Weibo,[1] the most famous Micro-blog platform in China. In a total, 12,651 users are crawled. Since many of them are not active, we remove those users who have posted less than 10 messages or received less than 10 comments since their registration or who have less than 50 followers or 50 followings.

---

[1] http://weibo.com/.

Input: L_{Mes} : labeled *message* samples; L_{Com} : labeled *comment* samples

$\quad\quad$ U_{Mes} : unlabeled *message* samples; U_{Com} : unlabeled *comment* samples

Output: L_{Mes} : New labeled *message* samples; L_{Com} : New labeled *comment* samples

Procedure: Loop for N iterations until $U_{Mes} = \varnothing$ or $U_{Com} = \varnothing$

(1). Learn classifier C_M with L_{Mes}

(2). Use C_M to label the samples from U_{Mes}

(3). Choose n_1 *positive* and n_1 *negative message* samples M_1 most confidently predicted by C_M

(4). Choose corresponding *comment* samples S_1 (the *comment* samples from the same users in M_1)

(5). Learn classifier C_C with L_{Com}

(6). Use C_C to label the samples from U_{Com}

(7). Choose n_2 *positive* and n_2 *negative comment* samples S_2 most confidently predicted by C_C

(8). Choose corresponding *message* samples M_2 (the *message* samples from the same users in S_2)

(9). $L_{Mes} = L_{Mes} + M_1 + M_2$; $L_{Com} = L_{Com} + S_1 + S_2$

(10). $U_{Mes} = U_{Mes} - M_1 - M_2$; $U_{Com} = U_{Com} - S_1 - S_2$

Fig. 2. Co-training algorithm for gender classification

Furthermore, verified organizational users are removed. As a result, we get 7658 users, from which we randomly select 2000 *male* and 2000 *female* users in our experimentation. In supervised learning, we use different sizes of labeled data as training data and 400 samples as test data. In semi-supervised learning, we select 400 samples as the initial labeled data, 400 samples as test data and the remaining as unlabeled data.

Features. The basic features are bag-of-words features. Apart from these basic features, we also consider F-measure and POS sequence pattern features, as described above. These features yield the state-of-the-art performance in gender classification (Mukherjee and Liu 2010). To get the word and POS features in Chinese text, we use ICTCLAS (http://www.ictclas.org/ictclas_download.aspx) to perform word segmentation and POS tagging on the Chinese text.

Classification Algorithm. The maximum entropy (ME) classifier implemented with the public tool, Mallet Toolkits.[2]

Experimental Results. Table 1 shows the accuracy of supervised learning with *message*, *comment* and both texts when different sizes of training data are used. From the table, we can see that although merely using *comment* text performs a bit worse than using *message* text, the performance is greatly improved when both texts are employed.

---

[2] http://mallet.cs.umass.edu/.

Table 1. Accuracy of supervised learning with different sizes of training data

	1600	2000	2400	2800	3200
Using *Message* Text	0.788	0.800	0.800	0.805	0.813
Using *Comment* Text	0.760	0.768	0.773	0.780	0.780
Using both *Message* and *Comment* Texts	**0.825**	**0.830**	**0.840**	**0.845**	**0.855**

Figures 3 and 4 show the performance of the *message* and *comment* classifiers when different numbers of unlabeled samples are added into the classifier with both self-training and co-training. Here, in each interaction, we pick 5 most confident *male* samples and *female* samples, i.e., $n_1 = n_2 = 5$. The baseline approach refers to the supervised classifier trained with only the initial labeled data (no unlabeled data is used) and the self-training approach refers to the single-view semi-supervised learning approach as applied in Burger et al. (2011).

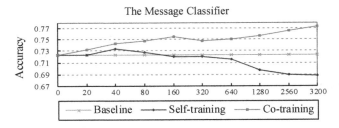

Fig. 3. Performance of the *message* classifier in self-training and co-training when different sizes of unlabeled data are added

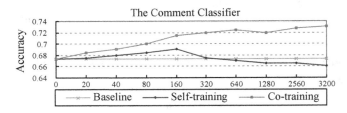

Fig. 4. Performance of the *comment* classifier in self-training and co-training when different sizes of unlabeled data are added

From these figures, we can see that the performance of both the *message* and *comment* classifiers improves gradually when more and more unlabeled data are added. Finally, the *message* classifier obtains 5 % improvements and the *comment* classifier obtains 5.7 % improvements. From these figures, we can also see that self-training fails to effectively improve the performance and even performs worse than the baseline classifier when more and more unlabeled data are incorporated.

5   Conclusion

In this paper, we perform gender classification by exploiting the interactive knowledge from the *comment* text as a complement to the non-interactive knowledge from the *message* text. On the basis, a co-training approach is proposed to integrate both kinds of knowledge and unlabeled data. Evaluation demonstrates the complementarity of the interactive knowledge and the non-interactive knowledge in gender classification. It also demonstrates that the proposed co-training approach can well incorporate unlabeled data and considerably improve the performance, while the self-training approach fails.

Acknowledgments. This research work has been partially supported by three NSFC grants, No. 61273320, No.61375073, No.61331011, and Collaborative Innovation Center of Novel Software Technology and Industrialization.

References

Blum, A., Mitchell, T.: Combing labeled and unlabeled data with co-training. In: Proceedings of the 11th Annual Conference on Computational Learning Theory, pp. 92–100 (1998)

Corney, M., Vel, O., Anderson, A., Mohay, G.: Gender-preferential text mining of E-mail discourse. In: Proceedings of the 18th Annual Computer Security Applications Conference, pp. 282–289 (2002)

Ciot, M., Sonderegger, M., Ruths, D.: Gender inference of twitter users in non-english contexts. In: Proceedings of EMNLP-13, pp. 1136–1145 (2013)

Gianfortoni, P., Adamson, D., Rosé, C.: Modeling of stylistic variation in social media with stretchy patterns. In: Proceedings of EMNLP-11, pp. 49–59 (2011)

Ikeda, D., Takamura, H., Okumura, M.: Semi-supervised learning for blog classification. In: Proceedings of AAAI-08, pp. 1156–1161 (2008)

Filippova, K.: User demographics and language in an implicit social network. In: Proceedings of EMNLP-12, pp. 1478–1488 (2012)

Heylighen, F., Dewaele, J.: Variation in the contextuality of language: an empirical measure. Proc. Found. Sci. **7**, 293–340 (2002)

Liu, N., He, Y., Chen, Q., Peng, M., Tian, Y.: A new method for micro-blog platform users classification based on infinitesimal-time. J. Inf. Computantional Sci. **10**(9), 2569–2579 (2013)

Mukherjee, A., Liu, B.: Improving gender classification of blog authors. In: Proceedings of EMNLP-11, pp. 207–217 (2010)

Nowson, S., Oberlander, J.: The identity of bloggers: openness and gender in personal weblogs. In: Proceedings of AAAI-06, pp. 163–167 (2006)

Peersman, C., Daelemans, W., Vaerenbergh, L.: Predicting age and gender in online social networks. In: SMUC 2010 Proceedings of the 2nd International Workshop on Search and Mining User-generated Contents, pp. 37–44 (2010)

Rao, D., Yarowsky, D., Shreevats, A., Gupta, M.: Classifying latent user attributes in twitter. In: Proceeding SMUC 2010 Proceedings of the 2nd International Workshop on Search and Mining User-generated Contents, pp. 37–44 (2010)

Volkova, S., Wilson, T., Yarowsky, D.: Exploring demographic language variations to improve multilingual sentiment analysis in social media. In: Proceedings of EMNLP-13, pp. 1815–1827 (2013)

Interactive Gender Inference in Social Media

Zhu Zhu[1,2], Jingjing Wang[1,2], Shoushan Li[1,2(✉)], and Guodong Zhou[1,2]

[1] Natural Language Processing Lab, School of Computer Science and Technology,
Soochow University, Suzhou, China
[2] Collaborative Innovation Center of Novel Software Technology and Industrialization,
Nanjing, China
{zhuzhu0020,djingwang,shoushan.li}@gmail.com, gdzhou@suda.edu.cn

Abstract. In this paper, we define a novel task named interactive gender inference, which aims to utilize interactive text to identify the genders of two interactive users. To address this task, we propose a two stage approach by well incorporating the dependency among the interactive samples sharing identical users. Specifically, we first apply a standard four-category classification algorithm to get a preliminary result, and then propose a global optimization algorithm to achieve better performance. Evaluation demonstrates the effectiveness of our proposed approach to interactive gender inference.

Keywords: Gender inference · Social media

1 Introduction

The vigorous growth of social media in recent years, such as Twitter and Facebook, has produced an unprecedented amount of user-generated data. The open availability of such data provides an excellent opportunity to research on latent demographic features of online users. Gender inference is such a research issue which aims to identify the gender of a user. Due to its wide applications in social media analysis, gender inference has attracted more and more attention in recent years (Schler et al. 2006, Mukherjee and Liu 2010, Tang et al. 2011, Ciot et al. 2013).

However, previous studies mainly focus on gender inference on a single user. In fact, social media not only provides a platform for a single user to publish his own experiences or opinions, but also offers an opportunity for the users in a community to interact, e.g. allowing a user to comment on a certain message. This interactive process normally yields some kind of interactive text which is helpful to infer the genders of the involved users. Let's consider following interactive text as an example:

E1: User **A**: *Thanks for your cool necklace for my birthday. Miss you*
User **B**: *I miss you too, my dear wife. I'll be home soon*

While it may be difficult to infer the gender of user **A** from the text by user **A** only, it is easy to infer the gender of user **A** from the text by user **B** due to the existence of "*my dear wife*", and also the gender of user **B**.

A. Liu et al. (Eds.): DASFAA 2015 Workshops, LNCS 9052, pp. 252–258, 2015.
DOI: 10.1007/978-3-319-22324-7_24

In this study, we focus on jointly inferring the genders of users involved in an inter-action. For short, we name it interactive gender inference. Interactive gender inference has additional applications, e.g. helping understand interpersonal communication mech-anisms in sociology research (Dev et al. 2014) and driving AI research in human-machine interaction.

A straightforward approach to interactive gender inference is to cast it as a four-category classification problem, since four categories, i.e., *male to male*, *male to female*, *female to female* and *female to male* (denoted as *mm*, *mf*, *ff* and *fm*) can be naturally inferred from the interactive text. The problem with the four-category classi-fication approach is that it ignores the dependency among the users and the interactive text. For instance, in one case, we may predict one sample from **A** to **B** as *mf*, while in another case, we may predict another sample from **A** to **C** as *ff*. Obviously, these two predictions are contradictory.

In this paper, we address interactive gender inference with the focus on all the comments from a user to another user as the interactive text to infer the genders of the involved two users. Specifically, motivated by the constraint that the gender of a user should keep the same in his all involved interactions, we propose a two-stage classifi-cation approach to interactive gender inference. In the first stage, we perform standard four-category classification to infer the genders of the involved two users. In the second stage, we propose a global optimization algorithm to benefit from the above constraint.

2   Related Work

In the last decade, most of related studies in gender inference deal with either blog text (Schler et al. 2006) or email text (Mohammad et al. 2011). For instance, Schler et al. (2006) exploit the difference in writing style and content between *male* and *female* bloggers to determine an author's gender. Similar to other classification tasks, most of such studies focus on exploring effective features to improve the performance (Mukherjee et al. 2010, Peersman et al. 2011, Gianfortoni et al. 2011).

More recently, with the rapid growth of social media, more and more researchers turn to micro-blog text. Burger et al. (2011) describe the construction of a large multi-lingual dataset labeled with the genders of Twitter users. Miller et al. (2012) identify the genders of Twitter users using Perceptron and Naive Bayes with selected n-gram features. Ciot et al. (2013) conduct the first assessment of the latent attribute inference in languages beyond English, focusing on gender inference of Twitter users.

3   A Two-Stage Approach

Generally, the users and user interactions in micro-blog can be represented as a graph, i.e., $G = (V, E)$ where V is the set of $|V| = N$ users and $E \subset V \times V$ is a set of $|E| = M$ interactive edges among users. Here, an interactive edge $e_{i,j} \in E$ is directed and repre-sented by the comment text from $v_i \in V$ to $v_j \in V$. In this paper, our objective is to learn a model to infer the interactive gender of each edge $e_{i,j} \in E$, namely interactive gender.

In this paper, interactive gender is represented as a triple $(e_{i,j}, Rr_{ij}, Pp_{ij})$, where $e_{i,j} \in E$ is an edge; $Rr_{ij} \in YY$ is the label associated with the edge $e_{i,j}$ and $YY = \{mm, mf, fm, ff\}$ is the category set. Pp_{ij} is the probability vector obtained by an algorithm for inferring the interactive gender. Generally, an interactive gender could also be inferred from the respective user gender, defined as a triple (v_i, r_i, p_i), where $v_i \in V$ is a node; $r_i \in Y$ is a label associated with the node, and $Y = \{male, female\}$. For convenience, we assume $r_i = 1$ if the label is *male*; Otherwise, $r_i = 0$. p_i is the probability vector.

3.1   Stage 1: Four-Category Classification

Formally, the objective of four-category classification is illustrated as follows:

$$f(x) \rightarrow y \text{ Where } y \in YY \text{ and } YY = \{mm, mf, fm, ff\} \tag{1}$$

In this first stage, this predictor is used to determine the interactive genders of all the edges in the test data. That is to say, we obtain all the triples $(e_{i,j}, Rr_{ij}, Pp_{ij})$ where $e_{i,j}$ is a sample in the test data. Specifically, Pp_{ij} contains the probabilities of the sample belonging to each category, i.e.,

$$Pp_{ij}=<p(Rr_{ij} = mm), p(Rr_{ij} = mf), p(Rr_{ij} = ff), p(Rr_{ij} = fm) > \tag{2}$$

3.2   Stage 2: Global Label Optimization

The objective of global label optimization is to minimize the difference between the true gender label of each user and the inferred gender from the first stage. In general, each user node is involved in two types of edges: (a) the node appears in the left side of the edge; (b) the node appears in the right side of the edge. Figure 1 shows the node sets connected to the node v_i.

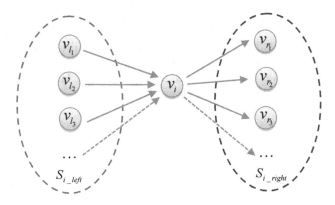

Fig. 1.   The node sets connected to the node v_i where S_{i_right} denotes the node set that contains the nodes from the right side and S_{i_left} denotes the node set that contains the nodes from the left side.

(a) When user v_i appears as the left node in the edge, the overall difference between the true label r_i and the inferred gender label is given as follows:

$$\sum_{j \in S_{i\_right}} \left(r_i - \tilde{r}_{ij} \right)^2 \tag{3}$$

Where $\tilde{r}_{ij}$ is inferred from the interactive gender of the edge $e_{i,j}$ according to following inference rule

$$\tilde{r}_{ij} = \begin{cases} 1 & \text{if } Rr_{ij} = mm \text{ or } Rr_{ij} = mf \\ 0 & \text{Otherwise} \end{cases} \tag{4}$$

(b) When user v_i appears as the right node in the edge, the overall difference between the true label r_i and the inferred gender label is given as follows:

$$\sum_{k \in S_{i\_left}} \left(r_i - \tilde{r}_{ki} \right)^2 \tag{5}$$

Where $\tilde{r}_{ki}$ is inferred by the interactive gender of the edge $e_{k,i}$ according to following inference rule,

$$\tilde{r}_{ki} = \begin{cases} 1 & \text{if } Rr_{ki} = mm \text{ or } Rr_{ki} = fm \\ 0 & \text{Otherwise} \end{cases} \tag{6}$$

When all users with both left and right connections are considered, our objective is to minimize the overall difference as follows,

$$\min \sum_{i=1}^{N} \left(\sum_{j \in S_{i\_right}} \left(r_i - \tilde{r}_{ij} \right)^2 + \sum_{k \in S_{i\_left}} \left(r_i - \tilde{r}_{ki} \right)^2 \right) \tag{7}$$

where N is the number of all users. To solve the above optimization problem, we first compute the partial derivative of r_i

$$\sum_{j \in S_{i\_right}} \left(2 \cdot r_i - 2 \cdot \tilde{r}_{ij} \right) + \sum_{k \in S_{i\_left}} \left(2 \cdot r_i - 2 \cdot \tilde{r}_{ki} \right) \tag{8}$$

Then, the best value of r_i to minimize formula (8) is the one making formula (8) equal 0. Thus, we get

$$r_i = \frac{1}{\left| S_{i\_right} \right|} \cdot \sum_{j \in S_{i\_right}} \tilde{r}_{ij} + \frac{1}{\left| S_{i\_left} \right|} \cdot \sum_{k \in S_{i\_left}} \tilde{r}_{ki} \tag{9}$$

From this formula, we can see that the optimization value of r_i is the average value of the gender labels inferred from both the left and right connected nodes.

The final decision on the user gender is made according to the value of r_i, i.e.,

$$\text{assign } v_i \rightarrow male \text{ if } r_i > 0.5 \text{ Otherwise } v_i \rightarrow female \qquad (10)$$

Subsequently, the final decision on the interactive gender is made according to the user gender.

4   Experimentation

Data Setting. The data is collected from Sina Micro-blog (http://weibo.com/), one of the most famous Micro-blog platforms in China. From the website, we crawl user homepage which contains user messages (e.g. *name, gender, verified type*), and corresponding comments. Overall, we get a data set of 53,675 users. From this data set, we select 20191 users as the training group and 9339 users as the test group (these two groups have no interactions between each other). Furthermore, we omit those users with less than 10 comments. Table 1 shows the statistics about the final data set.

Table 1. Statistics about the data set

#	Training Data	Test Data
mm	2883	1109
mf	4462	1599
ff	10954	3395
fm	4596	1591
Total	22895	7694

Features. Each interactive sample is treated as a bag-of-words and transformed into a binary vector encoding the presence or absence of one word feature. To get the word features in Chinese text, we use ICTCLAS (http://www.ictclas.org/ictclas_down-load.aspx) to perform word segmentation on the Chinese text. Apart from these basic features, we include two kinds of complex features, F-measure and POS sequence pattern features, which yield the state-of-the-art performance in gender inference (Mukherjee and Liu, 2010).

Classification Algorithm. We use the maximum entropy (ME) algorithm which is implemented with the public tool, Mallet Toolkits (http://mallet.cs.umass.edu/).

Evaluation Measurement. The performance is evaluated by both the F1-score in each category and the macro average of all F1-scores (Macro-F1).

Experimental Results. Figure 2 compares the performance, where **Baseline** (4-category) means the four-category classification approach, i.e., the one employed in the first stage of our approach. From the figure, we can see that complex features, such as F-measure and POS sequence pattern features, significantly improve the performance. We can also see that our two-stage approach significantly outperforms the **Baseline** approach consistently. This indicates the dependency of interactive texts in user genders and the effectiveness of our global optimization algorithm.

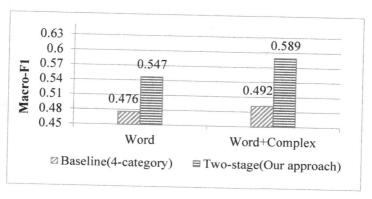

Fig. 2. Performance comparison on interactive gender inference

5   Conclusion

In this paper, we address interactive gender inference with a two-stage approach. In the first stage, we utilize a standard four-category classification method to perform preliminary prediction. In the second stage, we propose a global optimization algorithm to benefit from the dependency among interactive texts in user genders. Evaluation on a large data set from Micro-blog platform shows the effectiveness of our two-stage approach over a strong baseline.

Acknowledgments. This research work has been partially supported by three NSFC grants, No. 61273320, No.61375073, No.61331011, and Collaborative Innovation Center of Novel Software Technology and Industrialization.

References

Burger, J., Henderson, J., Kim, G., Zarrella, G.: Discriminating gender on twitter. In: Proceedings of EMNLP 2011, pp. 1301–1309 (2011)

Ciot M., Sonderegger, M., Ruths, D.: Gender inference of twitter users in non-english contexts. In: Proceedings of EMNLP 2013, pp. 1136–1145 (2013)

Dev, H., Ali, M.E., Hashem, T.: User interaction based community detection in online social networks. In: Bhowmick, S.S., Dyreson, C.E., Jensen, C.S., Lee, M.L., Muliantara, A., Thalheim, B. (eds.) DASFAA 2014, Part II. LNCS, vol. 8422, pp. 296–310. Springer, Heidelberg (2014)

Gianfortoni P., Adamson, D., Rosé, C.: Modeling of stylistic variation in social media with stretchy patterns. In: Proceedings of EMNLP 2011, pp. 49–59 (2011)

Mukherjee, A., Liu, B.: Improving gender classification of blog authors. In: Proceedings of EMNLP 2010, pp. 207–217 (2010)

Miller, Z., Dickinson, B., Hu, W.: Gender prediction on twitter using stream algorithms with N-gram character features. Int. J. Intell. Sci. **2**(4), 143–148 (2012)

Mohammad, S., Yang, T.: Tracking sentiment in mail: how genders differ on emotional axes. In: Proceedings of the 2nd Workshop on Computational Approaches to Subjectivity and Sentiment Analysis, pp. 70–79 (2011)

Peersman, C., Daelemans, W., Van Vaerenbergh, L.: Predicting age and gender in online social networks. In: Proceedings of SMUC 2011, pp. 37–44 (2011)

Schler, J., Koppel, M., Argamon, S., Pennebaker, J.: Effects of age and gender on blogging. In: Proceedings of AAAI 2006, pp. 199–205 (2006)

Tang, C., Ross, K., Saxena, N., Chen, R.: What's in a name: a study of names, gender inference, and gender behavior in facebook. In: Xu, J., Yu, G., Zhou, S., Unland, R. (eds.) DASFAA Workshops 2011. LNCS, vol. 6637, pp. 344–356. Springer, Heidelberg (2011)

Joint Sentiment and Emotion Classification with Integer Linear Programming

Rong Wang[1,2], Shoushan Li[1,2(✉)], Guodong Zhou[1,2], and Hanxiao Shi[3]

[1] Natural Language Processing Lab, School of Computer Science and Technology,
Soochow University, Suzhou, China
[2] Collaborative Innovation Center of Novel Software Technology and Industrialization,
Nanjing, China
{wangrong2022,shoushan.li}@gmail.com, gdzhou@suda.edu.cn
[3] School of Computer Science and Information Engineering, Zhejiang Gongshang University,
Hangzhou, China
hxshi@mial.zjgsu.edu.cn

Abstract. As two foundational tasks in sentiment analysis, sentiment classification and emotion classification have been considered separately and studied independently in the literature. In this paper, we put forward an Integer Linear Programming (ILP)-driven joint learning approach to leveraging the relationship between these two tasks. Empirical study verifies the appropriateness and effectiveness of our proposed approach to joint sentiment and emotion classification.

Keywords: Sentiment classification · Emotion classification · Integer linear programming (ILP)

1 Introduction

During the last decade, sentiment analysis has attracted considerable attention in multiple research communities, such as natural language processing (NLP), data mining and social media (Pang and Lee 2008, Liu 2012). As two foundational tasks in sentiment analysis, sentiment classification is concerned with predicting coarse-grained sentimental orientations (e.g., *positive* or *negative*) towards a topic, and emotion classification is concerned with predicting fine-grained personal emotions (e.g., *happy*, *sad*, *surprise*, *fear*) expressed by a human being (Quan and Ren 2009). Both have their value in a wide range of real-life applications, such as opinion mining and psychological analysis.

Although both sentiment classification and emotion classification have been well explored in the literature, they are normally addressed separately. This largely ignores the dependency between sentimental orientations and personal emotions. For example, if we can determine the personal emotion of the sentence in Example 1 to be *happy*, we can easily infer the sentimental orientation of the sentence to be *positive*, while, if we can determine the sentimental orientation of the sentence in Example 1 to be *positive*, we can easily eliminate those very *negative* emotions, e.g., *angry* and *disgust*, although the exact emotion is hard to infer.

© Springer International Publishing Switzerland 2015
A. Liu et al. (Eds.): DASFAA 2015 Workshops, LNCS 9052, pp. 259–265, 2015.
DOI: 10.1007/978-3-319-22324-7_25

Example 1: *I am so happy to get this book. I really like it.*

Given the close relationship between sentimental orientations and personal emotions, it is natural to leverage preciously annotated data from one task to help another task. In this paper, we explore joint sentiment and emotion classification in better exploiting the annotated data from either task. Specifically, we design some constraints between sentimental orientations and personal emotions and perform global inference on the outputs from both the sentiment and emotion classifiers with Integer Logical Programming (ILP).

The remainder of this paper is organized as follows. Section 2 overviews related work on both sentiment classification and emotion classification. Section 3 presents our approach to joint sentiment and emotion classification. Section 4 evaluates the proposed approach. Finally, Sect. 5 gives the conclusion.

2   Related Work

Sentiment classification has been extensively studied in the last decade since the pioneering work by Pang et al. (2002). Earlier studies on this research issue could refer to two comprehensive surveys by Pang et al. (2008) and Liu (2012). Recent studies on sentiment classification mainly focus on the solving the sparse data problem in machine learning-based approaches to sentiment classification, such as unsupervised learning (Ou et al. 2014), semi-supervised (Zhou et al. 2013), cross-domain (Li et al. 2013a), and cross-lingual sentiment classification (Li et al. 2013b). Our work follows the same spirit but leverages the resources from a related task, i.e., emotion classification.

Emotion classification is likewise a hot research topic in the data mining and natural language processing communities. One main group of such studies is about resource construction, such as emotion lexicon building (Xu et al. 2010) and sentence-level or document-level corpus construction (Quan and Ren 2009). Another main group of related studies is about the supervised learning approaches to emotion classification (Alm et al. 2005, Purver and Battersby 2012).

Works on joint learning based on sentiment and emotion classification are rare. The only one exception we find is the work by Gao et al. (2013) which uses an extra annotated data with both sentiment and emotion annotation to estimate the transformation probabilities to help the two tasks. However, this type of annotated data is not available in most real-life applications. In contrast, in this study, our approach to joint learning is based on ILP, which needs no extra annotated data.

3   Joint Sentiment and Emotion Classification

Our basic idea to joint sentiment and emotion classification is to leverage the close relationship between sentimental orientations and personal emotions. Specifically, ILP is utilized to achieve global optimization in capturing such relationship in the outputs of sentiment and emotion classifiers via some constraints. Figure 1 illustrates our joint learning framework.

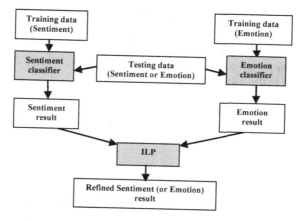

Fig. 1. The framework of joint sentiment and emotion classifications with ILP

Let x be the feature vector of a testing sample. Since the text representation in both sentiment and emotion classification is sometimes the same, e.g., bag-of-words representation, the testing sample x is capable of being classified by either the sentiment classifier or the emotion classifier but with different output results.

In sentiment classification, the testing sample is classified to a sentiment category, e.g., *positive* or *negative*. Suppose that the sentiment category of the i-th testing sample is y_i and $y_i \in \{0, 1, 2\}$ where 0 denotes the *neutral* category, 1 denotes the positive category and 2 denotes the negative category. In order to use binary values to represent the result, we use three different variables, i.e., $y_{0i} \in \{0, 1\}$, $y_{1i} \in \{0, 1\}$, $y_{2i} \in \{0, 1\}$, to denote whether the *neutral*, *positive*, or *negative* category is decided by the classifier. For instance, $y_{0i} = 0$ means the classification result is not *neutral*. Therefore, the testing result is denoted by a vector, i.e., $< y_{0i}, y_{1i}, y_{2i} >$. In addition to the category labels, the statistic classifier often provides the posterior probability information, i.e.,

$$P_y =< P_{y_{0i}=1}, P_{y_{0i}=0}, P_{y_{1i}=1}, P_{y_{1i}=0}, P_{y_{2i}=1}, P_{y_{2i}=0} > \tag{1}$$

The corresponding label information is represented by

$$Y =< y_{0i}, 1 - y_{0i}, y_{1i}, 1 - y_{1i}, y_{2i}, 1 - y_{2i} > \tag{2}$$

In emotion classification, the testing sample is classified to an emotion category, e.g., *happy* or *sad*. Suppose that the emotion category of the i-th testing sample is z_i and $z_i \in \{0, 1, 2, \ldots, 8\}$ where numbers 0–8 denote the categories of *neutral, joy, love, expect, surprise, anger, anxiety, hate, sorrow* in this study. The probability and label results are:

$$P_z =< P_{z_{0i}=1}, P_{z_{0i}=0}, P_{z_{1i}=1}, P_{z_{1i}=0}, \cdots, P_{z_{8i}=1}, P_{z_{8i}=0} > \tag{3}$$

and

$$Z =< z_{0i}, 1 - z_{0i}, z_{1i}, 1 - z_{1i}, \ldots, z_{8i}, 1 - z_{8i} > \tag{4}$$

Our objective of joint learning is to adjust the label result given the probability result, together with some constraints. The objective function is designed to make the label result similar to the probability result as much as possible. We use *Cosine* method to measure the similarity between the label and probability vector and thus our objective becomes to maximize the following formula:

$$\text{Max} \quad \frac{< P_y, P_z > \cdot < Y, Z >}{\left| < P_y, P_z > \right| \left| < Y, Z > \right|} \tag{5}$$

Given the probability results from the two classifiers, the term $\left| < P_y, P_z > \right|$ is a fixed value. Additionally, we find that whatever the final label result is, the term $\left| < Y, Z > \right|$ is also a fixed value under the assumption that there is only one sentiment label and one emotion label assigning to a testing sample. Therefore, our objective function becomes:

$$\text{Max} \quad < P_y, P_z > \cdot < Y, Z > \tag{6}$$

This is exactly an integer linear programming problem. In detail, the objective function is:

$$\text{Max} \quad \sum_{k=0}^{2} P_{y_{ki}=1} \cdot y_{ki} + \sum_{k=0}^{2} P_{y_{ki}=0} \cdot (1 - y_{ki}) + \sum_{l=0}^{8} P_{z_{li}=1} \cdot z_{li} + \sum_{l=0}^{8} P_{z_{li}=0} \cdot (1 - z_{li}) \tag{7}$$

Subject to:
(C1) Integer constraint:

$$y_{ki} \in \{0, 1\} \text{ and } z_{li} \in \{0, 1\} \tag{8}$$

(C2) Single label constraint:

$$\sum_{k=0}^{2} y_{ki} = 1, \text{ and } \sum_{l=0}^{8} z_{li} = 1 \tag{9}$$

(C3) Neutral emotion constraint: When the testing sample is classified as a *neutral* sample in sentiment classification, the emotion labels of the sample must be the emotion label of *neutral*, i.e.,

$$y_{0i} = z_{0i} \tag{10}$$

(C4) Positive emotion constraint: When the testing sample is classified as a *positive* sample in sentiment classification, the emotion labels of the sample must be in the emotion label set of {*joy, love, expect, surprise* }, i.e.,

$$y_{1i} = z_{1i} + z_{2i} + z_{3i} + z_{4i} \tag{11}$$

(C5) Negative emotion constraint: When the testing sample is classified as a *negative* sample in sentiment classification, the emotion labels of the sample must be in the emotion label set of {*anger, anxiety, hate, sorrow*}, i.e.,

$$y_{2i} = z_{5i} + z_{6i} + z_{7i} + z_{8i} \tag{12}$$

4 Experimentation

Experimental Settings

- **Data Set:** Our experiment is performed on the Ren-CECps corpus (Quan and Ren 2009), which contains 34,603 sentences. Each sentence is annotated with a sentiment label and an emotion vector (sometimes contains multiple emotion labels). In our experiment, we only consider the majority emotion.
- **Features:** Each sentence is treated as a bag-of-words and transformed into binary vectors encoding the presence or absence of word unigrams.
- **Classification Algorithm:** The maximum entropy (ME) classifier implemented with the public tool, Mallet Toolkits (http://mallet.cs.umass.edu/) is employed in all our experiments. The posterior probabilities belonging to the categories are also provided in this tool.
- **Implementation**: From the corpus, we randomly select 3500 sentences as the test data for both sentiment and emotion classification. Among the remaining data, we randomly select 2000, 3500, and 7000 sentences as the training data for sentiment classification and select another 2000, 3500, and 7000 sentences as the training data for emotion classification. The ILP is solved by the tool named lp_solve 5.5.2.0 (http://web.mit.edu/lpsolve_v5520/).

Experimental Results

In this section, we compare the following approaches to joint sentiment and emotion classification.

- **Single_Sentiment**: using only the sentiment classifier to obtain the sentiment result
- **Single_Emotion**: using only the emotion classifier to obtain the emotion result
- **Joint_Sentiment**: performing joint learning with both the sentiment and emotion classifiers to obtain the sentiment result
- **Joint_Emotion**: performing joint learning with both the sentiment and emotion classifiers to obtain the emotion result

Figure 2 shows the results of different approaches when different sizes of training data are employed. From this figure, we can see that our joint-learning approach robustly outperforms single task classifiers across different sizes of labeled data on either sentiment classification or emotion classification. On average, they increase the accuracy from 0.635 to 0.654 in sentiment classification and the accuracy from 0.322 to 0.335 in emotion classification.

Experimental Results

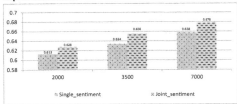
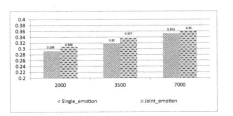

Fig. 2. Accuracy performances of different classification approaches over different sizes of labeled data

5   Conclusion

In this paper, we propose a novel approach to joint sentiment and emotion classification. Our approach mainly leverages ILP to globally infer sentiment and emotion categories from the outputs of the two classifiers. Experimental results demonstrate that, compared to individual classifiers, our joint learner consistently achieves better performance.

Acknowledgments. This research work has been partially supported by three NSFC grants, No. 61273320, No. 61375073, No. 61331011, one NSF grant of Zhejiang Province No. Z1110551 and Collaborative Innovation Center of Novel Software Technology and Industrialization.

References

Alm, C., Roth, D., Sproat, R.: Emotions from text: machine learning for text-based emotion prediction. In: Proceedings of EMNLP 2005, pp. 579–586 (2005)

Gao, W., Li, S., Lee, S., Zhou, G., Huang, C.: Joint learning on sentiment and emotion classification. In: Proceeding of CIKM 2013, pp. 1505–1508 (2013)

Li, S., Xue, Y., Wang, Z., Zhou, G.: Active learning for cross-domain sentiment classification. In: Proceeding of IJCAI 2013, pp. 2127–2133 (2013a)

Li, S., Xue, Y., Wang, Z., Lee, S., Huang, C.: Data quality controlling for cross-lingual sentiment classification. In: Proceeding of IALP 2013, pp. 125–128 (2013b)

Liu, B.: Sentiment Analysis and Opinion Mining (Introduction and Survey). Morgan & Claypool Publishers, San Rafael (2012)

Ou, G., Chen, W., Li, B., Wang, T., Yang, D., Wong, K.-F.: CLUSM: An unsupervised model for microblog sentiment analysis incorporating link information. In: Bhowmick, S.S., Dyreson, C.E., Jensen, C.S., Lee, M.L., Muliantara, A., Thalheim, B. (eds.) DASFAA 2014, Part I. LNCS, vol. 8421, pp. 481–494. Springer, Heidelberg (2014)

Pang, B., Lee, L.: Opinion mining and sentiment analysis. Found. Trends Inf. Retrieval **2**(12), 1–135 (2008)

Pang, B., Lee, L., Vaithyanathan, S.: Thumbs up? sentiment classification using machine learning Techniques. In: Proceedings of EMNLP 2002, pp. 79–86 (2002)

Purver, M., Battersby, S.: Experimenting with distant supervision for emotion classification. In: Proceedings of EACL 2012, pp. 482–491 (2012)

Quan, C., Ren, F.: Construction of a blog emotion corpus for chinese emotional expression analysis. In: Proceedings of EMNLP 2009, pp. 1446–1454 (2009)

Xu G., Meng, X., Wang, H.: Build chinese emotion lexicons using a graph-based algorithm and multiple resources. In: Proceeding of COLING 2010, pp. 1209–1217 (2010)

Zhou, S., Chen, Q., Wang, X.: Active deep learning method for semi-supervised sentiment classification. Neurocomputing **120**, 536–546 (2013)

Mining Wikileaks Data to Identify Sentiment Polarities in International Relationships

Arpit Jain and Arnab Bhattacharya[✉]

Department of Computer Science and Engineering,
Indian Institute of Technology, Kanpur, Kanpur, India
{arpitj,arnabb}@iitk.ac.in

Abstract. The infamous Wikileaks cables are a large-scale resource for analyzing international relationships. We use sentiment analysis on this dataset to extract opinion polarities in the international scenario. We use an unsupervised approach based on standard sentiment lexicon with modifiers to mine opinion polarities among the cables to and from embassies/consulates of USA. Sharp changes in opinion polarities are mapped to international events happening around the time of the cable at the location of the embassy/consulate, and a positive/negative correlation is drawn. The dataset consists of 232,410 cables from 1966 up to October 2009 concerning 272 embassies and consulates across the world. The top 28 of the spikes/dips in polarity changes coming from 20 embassies/consulates are then evaluated. Our results show that there is a strong correlation (76 %) between our findings and sentiments surrounding actual events. For example, our study was able to correctly identify suicide terrorist attacks outside the American embassy in Casablanca. It could also highlight a cable that referred to a terrorist who was later arrested in New Delhi possessing secret documents related to Indian Army.

1 Introduction

Wikileaks (https://www.wikileaks.org/) is an international, online, non-profit organization which publishes secret information, news leaks and classified media from anonymous sources. The group has released a number of significant documents called "cables" which have become front-page news items. A confidential text message exchanged between a diplomatic mission, like an embassy or a consulate, and the foreign ministry of its parent country is called a *cable*. For simplicity, a cable can be considered as a telegram or an email with fields such as sender, receiver, subject, body content, priority with additional attributes such as tags and confidentiality of the message. One of the important reasons for this study lies in finding correlation between official dialogue among embassies and newspaper reports during that period of time.

While opinion mining and sentiment analysis [1] have become increasingly popular over the last few years, surprisingly the Wikileaks dataset has not been worked upon by many. We attempt to alleviate this shortcoming by analyzing the Wikileaks cables in terms of sentiments and correlate them with actual events.

© Springer International Publishing Switzerland 2015
A. Liu et al. (Eds.): DASFAA 2015 Workshops, LNCS 9052, pp. 266–272, 2015.
DOI: 10.1007/978-3-319-22324-7_26

To the best of our knowledge, this is the first attempt to study the Wikileaks dataset from the perspective of computational linguistics. The aim is to see how much the sentiments expressed in the private cables match that of the prevalent sentiments expressed through public news channels.

The dataset consists of 232,410 cables from 1966 up to October 2009 concerning 272 embassies and consulates across the world. The data is released by Wikileaks and is accessible from https://wikileaks.org/cablegate.html.

2   Related Work

The Wikileaks corpora, despite being a large dataset, has not been studied in depth by the research community so far except a few [2]. The text of the cables have been studied in depth manually by news organizations/journalists and geo-politicians [3] without resorting to computational linguistics. The importance of Wikileaks data and its implications to various issues such as journalism [4], transparency [5,6] and hacking [7] have been highlighted.

Many techniques have been proposed for determining sentiment polarity from textual data including mapping the variations of sentiments over parameters such as topics/categories [1,8] or entities [9], drawing correlation with past events/occurrences with the help of timelines [10], and even making future predictions [11].

Opinion mining over real time data has become a field of growing interest. Analysis of seemingly personal opinions expressed on the microblogging platform Twitter has been effective in predicting election results [9] as well as creating a sentiment classifier using the corpus itself [12]. As news articles contain opinions about public entities, studies have been done to identify the polarities expressed in them [10,13].

3   Approach

To identify the sentiments in a cable, we work with only the text in the body and do not use the header and metadata fields.

First, we use NLTK's (http://www.nltk.org/) TreeBank word tokenizer to tokenize each cable text and remove extraneous spaces, tabs and indentation. We then use the POS Tagger (Parts-of-speech tagger) from NLTK again to tag the resultant text. The description of the tags are as per the Penn Treebank project (http://www.cis.upenn.edu/~treebank/).

The words are next mapped to sentiments using the SentiWordNet database (http://sentiwordnet.isti.cnr.it) and SentiStrength dataset (http://sentistrength. wlv.ac.uk). While the SentiWordNet defines a list of words and their corresponding sentiments (such as "happy" is positive, etc.), the SentiStrength uses different qualifiers including inverters, incrementers and decrementers to modify the strength of the sentiments. An inverter (such as "not") reverses the polarity of the sentiment (e.g., "not happy"). An incrementer (such as "very") increases the sense of the sentiment (e.g., "very happy") while a decrementer (such as "mildly") decreases the sense (e.g., "mildly happy").

We use these to mark our POS-tagged text to sentiments. The *average* sentiment score of all the cables within a time period is then computed for each embassy/consulate. A local maximum/minimum in the sentiment score timeline is designated as a *spike/dip*.

4   Results

The time periods for which the average sentiment scores were computed were not uniform. The reason is that there were very few cables before 2000 and only a moderate number of cables between 2000 and 2005. Consequently, the entire time from 1966 to 1980 was included as one time period. Similarly, the periods 1980–1989, 1990–1994, and 1995–1999 were considered to be single time periods. Each year from 2000 to 2005 was a single time period as well. From 2006 onwards, the number of cables became very large. Therefore, each month from January, 2006 onwards was analyzed as a single time period. The data was analyzed up to October, 2009. This produced a total of 56 time periods. For every embassy/consulate, a graph showing the average sentiments for each of the time periods was produced. If there was no cable for a particular embassy/consulate for a particular time period, the sentiment score was taken to be 0.

Of the 272 graphs obtained, we considered the top 28 polarity flips (dips and spikes in the graphs) for mapping our results with sentiments expressed in the cables. The results are summarized in Table 1. The number of *distinct* embassies/consulates appearing in these top 28 flips is 20. Note that the polarity flips were considered on an absolute scale, i.e., a negative sentiment of -0.34 is equivalent to a positive sentiment of 0.34.

Our finding that the majority of the sentiments are negative is in broad agreement with another independent study [2]. In general, diplomats are critical of other countries. Of the top 28 flips, only one consulate (Dusseldorf) shows an overall positive sentiment.

The cables from which these sentiments originated were then looked up manually. For each such cable, a *correlation* is drawn with respect to the newspaper reports around that time or event. The correlations were categorized as (i) very positive, (ii) positive, (iii) neutral (no correlation), (iv) negative, and (v) very negative.

We next discuss some of these results in more detail. To access the text of the cable, please prepend the text https://wikileaks.org/cable/ to the corresponding URL (Fig. 1).

Consulate at Casablanca, Morocco: There are three time periods involving the consulate at Casablanca that shows strongly negative sentiments.

The first one in January, 2006 (2006/01/06CASABLANCA45.html) talks about the Visas Viper program. The program is used to maintain a "watch list" over known and suspected international terrorists. It has been widely discussed in media [14].

Table 1. Average sentiments across individual embassies/consulates.

Embassy/Consulate	Sentiment	Period	Correlation
Casablanca	−1.341	Sep, 2006	++
Port Au Prince	−1.161	May, 2007	+
Buenos Aires	−0.961	Jul, 2009	+
Casablanca	−0.925	Jan, 2006	++
Port Louis	−0.813	Apr, 2006	0
Reykjavik	−0.625	Feb, 2007	++
Johannesburg	−0.619	Nov, 2008	−
Valletta	−0.604	May, 2009	0
Sao Paulo	−0.563	Oct, 2009	0
Buenos Aires	−0.562	Aug, 2009	++
Riga	−0.558	Apr, 2006	+
Tirana	−0.540	Jul, 2007	+
Naha	−0.537	Sep, 2007	−
Athens	−0.537	May, 2009	+
Port Au Prince	−0.536	Sep, 2006	++
Buenos Aires	−0.533	May, 2009	++
Caracas	−0.530	1980–1990	+
Bujumbura	−0.530	Dec, 2008	−
Tijuana	−0.510	Jan, 2009	+
Halifax	−0.500	Aug, 2009	0
Naples	−0.500	May, 2009	++
Buenos Aires	−0.495	1990–1995	+
Naha	−0.490	Jan, 2008	+
Lagos	−0.490	Mar, 2008	++
Naha	−0.489	Jun, 2008	++
Casablanca	−0.488	Jan, 2007	++
Belmopan	−0.479	Apr, 2008	++
Dusseldorf	+0.477	Jul, 2008	+

The second cable in September, 2006 (2006/09/06CASABLANCA1020.html) continues to discuss the Visas Viper program including the names of quite a few individuals. This generates the highest negative sentiment among all the cables.

In January, 2007 there were two cables (2007/01/07CASABLANCA10.html and 2007/01/07CASABLANCA16.html). While the first one expressed distrust about Morocco's family code, the second one was critical about child labor.

Embassy at Port Louis, Mauritius: The embassy at Port Louis shows up once in April, 2006 (2006/04/06PORTLOUIS210.html). The cable talks about

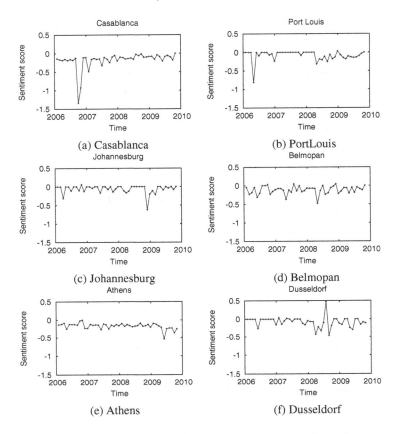

Fig. 1. Sentiment timelines for individual embassies/consulates.

human rights and UNSC sanctions. The correlation shown is neutral. In other words, the cable did not express any significant sentiment, either positive or negative.

Consulate at Johannesburg, South Africa: The cable (2008/11/08 JOHANNESBURG180.html) that shows a large negative sentiment having originated at the consulate at Johannesburg discusses about a changing visa procedure. However, the cable is mostly positive as it talks about easing of visa regulations. As a result, the correlation with the automatically computed sentiment score is negative.

Embassy at Belmopan, Belize: The cable in April, 2008 (2008/04/08 BELMOPAN192.html) from the embassy at Belmopan talks about a particular individual and his alleged links with well-known terror organizations across the world. The same terrorist was later arrested in New Delhi, India in 2012 for trying to smuggle secret military documents pertaining to the Indian army to a Pakistani agent [15].

Embassy at Athens, Greece: paginationThe embassy at Athens shows a large overall negative sentiment in May, 2009 owing to many factors. These include an attack by domestic terrorist groups on a branch of Eurobank in Athens on May 12, 2009 (2009/05/09ATHENS799.html), the feared growth of radicalization (2009/05/09ATHENS901.html), possible prosecution of pirates (2009/05/09ATHENS875.html), confirmation of the first H1N1 flu virus case (2009/05/09ATHENS835.html), and Greek views on the GAERC (2009/05/09-ATHENtml).

Consulate at Dusseldorf, Germany: The only cable to have a significantly positive sentiment was from the consulate at Dusseldorf in July, 2008 (2008/07/08DUSSELDORF34.html). It summarizes how the term "locust" originally used disapprovingly about equity capital firms has lost much of its negative meaning and has even come to be described positively as model investors.

4.1 Evaluation

We used a total of 28 flips to evaluate our method. The sentiments of these 28 flips were manually labeled by the authors themselves *before* they were passed through our automatic sentiment evaluation method.

To evaluate the method quantitatively, we assigned the following scoring scheme to the correlations obtained: (i) ++: 4, (ii) +: 3, (iii) 0: 2, (iv) −: 1, (v) −−: 0. Thus, if the sentiment of a cable correlates strongly positively with that of the news articles, then a score of 4 is counted, and so on.

Out of a maximum possibility (perfect correlation) of 112, our tool scored 85. Thus, it can be argued that our correlation results show an accuracy of 76 %, which is a boost of 52 % over the random score of 56.

4.2 Issues

Some of the issues we found with this approach, however, are as follows. There is no support for sarcasm detection. As the cables being examined were from official embassies/consulates, the presence of sarcasm in the body text is generally not expected. Nevertheless, there might still be a few cases of sarcastic tone. Moreover, the names of individuals/entities which hold meaning in the English language (e.g., "Mr. Happy") cannot be handled. Further, since we consider the average score of the sentiment for a period of time, a single cable can greatly affect the overall score if the number of cables exchanged during that period is extremely small.

4.3 Crowdsourcing

Since there is no ground truth, it is challenging to evaluate our results. The best way is to seek human opinions for each of the cases. To this end, we have started a crowdsourcing project that collects user opinions about a cable. The crowdsourcing platform is realized by developing a Google Chrome extension, available from http://goo.gl/GHKiZt. Whenever a user visits a Wikileaks site using the Chrome browser, an automatic pop-up will request her to input her opinion about the cable she is browsing through. For all other sites, the extension will remain dormant.

5   Conclusions

In this paper, we used sentiment analysis on Wikileaks Cablegate data to understand the changes in political relations between USA and different countries over time. Our results show that there is a strong correlation between our findings and the actual events reported in the cables.

References

1. Pang, B., Lee, L.: Opinion mining and sentiment analysis. Found. Trends Inf. Retr. **2**(1–2), 1–135 (2008)
2. Unsupervised clustering of people, places, and organizations in wikileaks cables with NLP cues. http://nlp.stanford.edu/courses/cs224n/2011/reports/beyangl-caoxuwen.pdf
3. Springer, S., Chi, H., Crampton, J., McConnell, F., Cupples, J., Glynn, K., Warf, B., Attewell, W.: Leaky geopolitics: the ruptures and transgressions of wikileaks. Geopolitics **17**(3), 681–711 (2012)
4. Eldridge, S.: Beyond wikileaks: Implications for the future of communications, journalism and society. Digital Journalism 3
5. Roberts, A.: Wikileaks: the illusion of transparency. Int. Rev. Admin. Sci. **78**(1), 116–133 (2012)
6. Heemsbergen, L.J.: Designing hues of transparency and democracy after wikileaks: vigilance to vigilantes and back again. New Media and Society (2014)
7. Lindgren, S., Lundström, R.: Pirate culture and hacktivist mobilization: the cultural and social protocols of wikileaks on twitter. New Media Soc. **13**(6), 999–1018 (2011)
8. Mullen, T., Collier, N.: Sentiment analysis using support vector machines with diverse information sources. In: EMNLP, vol. 4, pp. 412–418 (2004)
9. Tumasjan, A., Sprenger, T.O., Sandner, P.G., Welpe, I.M.: Predicting elections with twitter: what 140 characters reveal about political sentiment. In: ICWSM, vol. 10, pp. 178–185 (2010)
10. O'Connor, B., Balasubramanyan, R., Routledge, B.R., Smith, N.A.: From tweets to polls: Linking text sentiment to public opinion time series. In: ICWSM, vol. 11, pp. 122–129 (2010)
11. Asur, S., Huberman, B.: Predicting the future with social media. In: Web Intelligence and Intelligent Agent Technology (WI-IAT), vol. 1, pp. 492–499 (2010)
12. Pak, A., Paroubek, P.: Twitter as a corpus for sentiment analysis and opinion mining. In: LREC (2010)
13. Godbole, N., Srinivasaiah, M., Skiena, S.: Large-scale sentiment analysis for news and blogs. In: ICWSM, vol. 7 (2007)
14. The cable state department official: Visas viper cable just the tip of the iceberg. http://thecable.foreignpolicy.com/posts/2010/01/04-state_department_offici
15. Pak spy caught with army secrets. http://zeenews.india.com/news/nation/pak-spy-caught-with-army-secrets_78832

Extracting Indoor Spatial Objects from CAD Models: A Database Approach

Dazhou Xu[1], Peiquan Jin[1,2(✉)], Xiaoxiang Zhang[1], Jiang Du[1], and Lihua Yue[1,2]

[1] School of Computer Science and Technology,
University of Science and Technology of China, Hefei 230027, China
jpq@ustc.edu.cn
[2] Key Laboratory of Electromagnetic Space Information,
Chinese Academy of Sciences, Hefei 230027, China

Abstract. With the increasing development of indoor positioning technologies such as Wifi and RFID, indoor location based services (LBSs) has been a hot topic in recent years. Differing from GPS-based outdoor LBSs, we lack sufficient indoor maps which are the foundation of indoor LBSs. In this paper, we present a database approach to extract indoor spatial objects, e.g., rooms and doors, from CAD models, and then transform them into an indoor moving-object database. With this mechanism, we are able to efficiently generate indoor maps and support indoor-space queries. In addition, we implement a prototype system to demonstrate the feasibility of our proposal. It shows that our approach has a high precision on extracting indoor spatial objects and can support indoor spatial queries effectively.

Keywords: Indoor maps · CAD model · Indoor moving-object database · Extraction

1 Introduction

With the development of indoor positioning technologies and various portable devices, it is necessary to provide location based services in indoor spaces [1]. For example, shopping malls intend to find the hotspots in buildings and thereby can adjust the deployment of shops and provide better services for customers. This calls for the research on moving object data management in indoor spaces.

One fundamental issue in indoor moving-object management is the construction of indoor-space maps [2]. This is different from outdoor spaces, where we have sufficient digital maps such as Google Maps and city road-network maps. For indoor spaces, there are no existing solutions for automatically generating indoor maps. Although Google Maps start to provide indoor maps, it requires users to manually upload indoor maps, which is not feasible for most users. An intuition approach is to manually draw the floor plan of an indoor space (for example, PALMAP (http://www.palmap.cn/)

A. Liu et al. (Eds.): DASFAA 2015 Workshops, LNCS 9052, pp. 273–279, 2015.
DOI: 10.1007/978-3-319-22324-7_27

is a company providing indoor maps), but this is much time-consuming and costly and the drawing of indoor spaces cannot be directly used by indoor location-based services.

In this paper, we present a database approach to automatically extract indoor spatial objects and generate indoor maps from CAD (Computer-Aided Design) models. Our work is motivated by the observation that CAD is commonly used in indoor-space design, resulting in a great number of CAD files depicting structure of buildings. Therefore, our basic idea is to automatically extract indoor spatial objects from such CAD models, based on which we can construct indoor maps efficiently. Further, we store the extracted objects in indoor moving-object databases. Thus, we can support indoor queries on the extracted indoor spatial objects effectively. In summary, we make the following contributions in this paper.

(1) We propose a database approach to extracting indoor spatial objects from CAD models. With this approach, indoor spatial objects can be automatically extracted and transformed into an indoor moving-object database.
(2) We develop a Web-based prototype system to demonstrate the feasibility and effectiveness of our proposal. It provides visualization of CAD models as well as the extracted indoor maps. An interactive interface for evaluating indoor spatial queries is also implemented.

2 Problem Statement

CAD models are typically represented by the Drawing Exchange Format (DXF) [3]. A DXF file stores pairs of codes and associated values, where a code indicates the value type associated with the code. All the pairs in a DXF file are organized into seven sections, namely HEADER, CLASSES, TABLES, BLOCKS, ENTITIES, OBJECTS, and THUMBNAILIMAGE. Here, the ENTITIES section contains all graphical objects in the drawing and we will focus on this section.

The original data in DXF files are low-level graphical elements including *points*, *lines*, *arcs*, *polylines*, and *circles*. However, one indoor spatial object may involve many graphical elements. For example, a room usually involves several lines or polylines. Thus, the challenging problem is how to find appropriate graphical elements for indoor spatial objects. Due to the large number of graphical elements in a DXF file, it is not trivial to find the right elements that describe an indoor spatial object. For example, in Fig. 1, the walls of a room are separated by some arcs, lines, and white squares, and each wall is depicted by two parallel lines. Therefore, it is not feasible to simply use a boundary-tracing method [5] to extract rooms from CAD models.

As a result, the formal problem can be defined as follows.

Given a CAD model (DXF file) depicting the structure of an indoor space, return a set of rooms, doors, and the topological relationship between doors and rooms, where each room is represented as a polygon, each door is represented as a point, and each topological relationship is a pair of (door, room) indicating that a door is connected with a room.

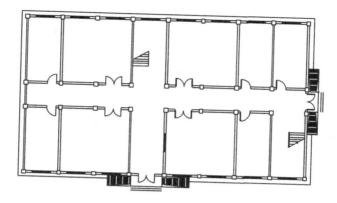

Fig. 1. An example of CAD models

3 Our Method

3.1 Basic Idea

We focus on the extraction of rooms and doors from CAD models. As doors are usually represented by arcs, we can simply extract all arcs as door candidates and further make a refinement to remove those candidates that are not connected with rooms (after rooms have been recognized).

Thus, in the following we concentrate on the extraction of rooms. There are two problems regarding this issue. First, a room involves too many lines in CAD models. For example, a wall may consist of two or even more parallel lines, which is commonly used in CAD to indicate the thickness of a wall. Second, rooms are not always rectangles and there may be more than doors associated with one room. Thus, if we employ a rule-based approach to determine rooms, there will be duplicated rooms and have to remove them.

The basic idea of our approach is to first employ a line-reduction preprocessing to simplify the line set that composes of a room. For this purpose, we perform a geometric clustering on the lines in a CAD model and simplify the parallel lines within one cluster into a single line. After that, we use a line-extending technique to divide the entire indoor space into a set of geometric shapes. Finally, we use an MBR-based method to remove duplicated rooms.

3.2 Algorithm

In order to detect indoor spatial objects from CAD models, we define the following concepts.

Definition 1 (*point-point adjacency*). Two points are adjacent if their Euclidian distance is below a predefined threshold α.

$$p \overset{adj}{\leftrightarrow} q \Leftrightarrow dist(p,q) < \alpha$$

Definition 2 (*point-line adjacency*). A point p is adjacent to a line $ln(s, e)$ if there is a *point-point adjacency* relation between ln's start point or end point and p.

$$p \overset{adj}{\leftrightarrow} ln \Leftrightarrow \left(p \overset{adj}{\leftrightarrow} s \ \ or \ \ p \overset{adj}{\leftrightarrow} e \right)$$

Definition 3 (*line-line adjacency*). Let $ln_1(s_1, e_1)$ and $ln_2(s_2, e_2)$ be two lines. ln_1 and ln_2 are adjacent if there is a *point-line adjacency* relation between an end point of one of the two lines and another line.

$$ln_1 \overset{adj}{\leftrightarrow} ln_2 \Leftrightarrow \left(s_1 \overset{adj}{\leftrightarrow} ln_2 \ or \ e_1 \overset{adj}{\leftrightarrow} ln_2 \right) \ or \ \left(s_2 \overset{adj}{\leftrightarrow} ln_1 \ or \ e_2 \overset{adj}{\leftrightarrow} ln_1 \right)$$

Definition 4 (*parallel lines*). Let $ln_1(s_1, e_1)$ and $ln_2(s_2, e_2)$ be two lines. ln_1 and ln_2 are parallel such that they satisfy the following condition.

$$ln_1 \overset{par}{\leftrightarrow} ln_2 \Leftrightarrow dot\_product(ln_1, ln_2) - dist(s_1, e_1) \cdot dist(s_2, e_2) < \beta$$

In this paper, we compute the distance d between two parallel lines as follows.

Definition 5 (*near lines*). Two parallel lines are near if the distance between them is below a predefined threshold ε.

Algorithm 1. *room-extraction*

Input: *DXF*: the CAD model.

Output: *P*: the set of extracted rooms.

```
1:  RL ← lines (DXF); // extract line set from DXF
2:  NL ← near-line-clustering(RL); // near-line-based clustering
3:  L ← simplify (NL); // merging lines in the same cluster into a single one
    /* grow lines to generate polygons */
4:  for each line l ∈ L do
5:  |   grow l until both end points of l intersects with another lines in L;
6:  end for
7:  P ← get all polygons from L; // except the bounding polygon of the floor
    /* remove duplicated polygons */
8:  M ← Constructing MBRs for each polygon in P;
9:  for each two MBRs m₁, m₂ ∈ M do
10: |     if (m₁ contains m₂) and (m₁'s room has more doors than m₂'s)then
11: |        remove m₁ from M;
12: |        remove the polygon bounded by m₁ from P;
13: |     end
14: end for
15: return P;
end room-extraction
```

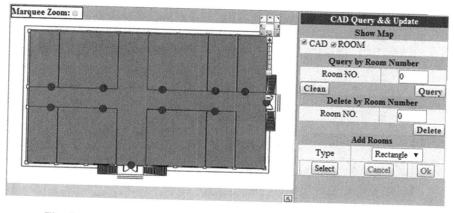

Fig. 2. Extracted rooms and doors from the CAD model shown in Fig. 1

Algorithm 1 shows the detailed procedure of extracting rooms. We first obtain line clusters through a near-line-based clustering step. Then, we simplify the lines in each cluster by merging near lines into a single line. Next, we construct polygons using a line-growing technique, and finally we remove duplicated polygons by constructing MBRs for all polygons. After extracting rooms, we continue to refine door candidates that are extracted from the arcs in the CAD model. Those candidates not connected with any rooms are removed from the door set. Consequently, we obtain a set of rooms, doors, as well as the connecting relations between rooms and doors. As an example, Fig. 2 shows the visualized rooms and doors extracted from the CAD model shown in Fig. 1.

3.3 Integration into Indoor Moving-Object Databases

Next, we integrate the extracted indoor spatial objects into an indoor object database. Current relational database systems do not provide inherited support for complex object representation and storage. However, object-relational database systems such as Oracle and PostgreSQL offer user-defined type extensions. Therefore, we extend indoor spatial types in object-relational database systems to allow storing the extracted indoor spatial objects.

We extend three indoor spatial types on Oracle, which are *Indoor Position, Indoor Space,* and *Indoor Geometry*. Note that our extension is based on Oracle Spatial that provides SDO_GEOMETRY for representing spatial data. The *Indoor Position* type is designed to represent the locations of a moving object in indoor spaces, which is not used in this paper. The *Indoor Space* type describes rooms, doors, as well as the topological relationship between doors and rooms. It is represented as a triple (*RoomSet, DoorSet, room-door*). Here, *RoomSet* is the set of room identifiers, *DoorSet* is the set of door identifiers, and *Room-Door* represents the topological relationship between *RoomSet* and *DoorSet*. The *Indoor Geometry* type stores the exact geometric information about rooms and doors. An *Indoor Geometry* value stores an object identifier and

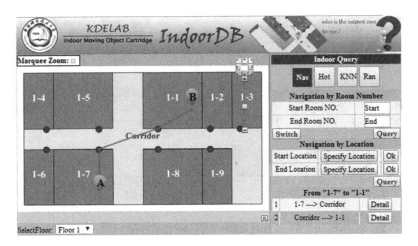

Fig. 3. Running indoor LBS queries on the extracted indoor data

its geometric information, with the form (*ObjectID, Geo*), where *ObjectID* is a room ID or a door ID and *Geo* is an SDO_GEOMETRY type supported by Oracle Spatial.

4   Preliminary Results

We build a Web-based prototype system to demonstrate our design. Our implementation is based on IndoorDB [4], which is an indoor moving-object database cartridge extended on Oracle. IndoorDB offers an SQL-based interface to insert indoor spatial objects into a database, thus we can read the extracted indoor spatial objects and construct on-the-fly SQL statements to import objects into the database. After that, we can run SQL statements on the imported indoor spatial objects to evaluate different kinds of LBS queries. For example, Fig. 3 shows the result of running a navigation query on the indoor spatial objects extracted from Fig. 1.

We also measure the effectiveness of our proposal on a real complex CAD model that describes the indoor structure of the building of the School of Computer Science and Technology in our university. The building consists of seven floors (0 ~ 6), a basement floor (−1), and 61 rooms. We get the CAD model from the archives of our university and manually check the precision of our algorithm for the CAD model of each floor. The results are shown in Table 1, indicating that our approach can reach an average precision of 94.1 %.

Table 1. Extracting results of a building

Floor	−1	0	1	2	3	4	5	6
Total rooms	38	21	21	19	19	21	31	26
Rooms correctly detected	36	20	20	18	18	20	28	24
Precision (%)	94.7	95.2	94.7	94.7	94.7	95.2	90.3	92.3

5 Conclusion

In this paper, we present a database approach to extract indoor spatial objects from CAD models and further integrate them into an indoor moving-object database. Our design provides a practical and economical way to generate indoor maps, and thus can offer fundamental supports to indoor LBSs. In the future, we will design a more general algorithm that can extract more kinds of objects from CAD models.

Acknowledgements. This work is supported by the National Science Foundation of China (61379037, 61472376, and 61272317) and the OATF project funded by University of Science and Technology of China.

References

1. Jin, P., Zhang, L., Zhao, J., et al.: Semantics and modeling of indoor moving objects. Int. J. Multimedia Ubiquit. Eng. **7**(2), 153–158 (2012)
2. Turner, E., Zakhor, A.: Floor plan generation and room labeling of indoor environments from laser range data. In: GRAPP, pp. 22–33 (2014)
3. Autodesk, DXF Reference (2012). http://images.autodesk.com/adsk/files/autocad_2012_pdf_dxf-reference_enu.pdf
4. Li, Q., Jin, P., Zhao, L., Wan, S., Yue, L.: IndoorDB: extending oracle to support indoor moving objects management. In: Meng, W., Feng, L., Bressan, S., Winiwarter, W., Song, W. (eds.) DASFAA 2013, Part II. LNCS, vol. 7826, pp. 476–480. Springer, Heidelberg (2013)
5. Schafer, M., Knapp, C., Chakraborty, S.: Automatic generation of topological indoor maps for real-time map-based localization and tracking. In: IPIN, pp. 1–8 (2011)

Incremental Class Discriminant Analysis on Interval-Valued Emitter Signal Parameters

Xin Xu[1](✉), Wei Wang[2], and Jiaheng Lu[3]

[1] Science and Technology on Information System Engineering Laboratory,
Nanjing 210007, China
flora.xin.xu@gmail.com
[2] State Key Laboratory for Novel Software and Technology,
Nanjing University, Nanjing 210093, China
ww@nju.edu.cn
[3] Key Laboratory of Data Engineering and Knowledge Engineering,
Renmin University, Beijing 100872, China
jiahenglu@gmail.com

Abstract. Emitter signal parameter analysis has been widely recognized as one crucial task for communication, electronic reconnaissance and radar intelligence analysis. However, the parameter measurements are characteristic of uncertainty in the form of intervals. In addition, the measurements are typically accumulated continuously. Existing machine learning methods for interval-valued data are unfit in such a case as they generally assume a uniform distribution and are restricted to static data analysis. To address the above problems, we bring forward an incremental class discriminant analysis method on interval-valued emitter signal parameters. Experimental results have validated its effectiveness.

Keywords: Emitter identification · Class discriminant analysis · Signal processing

1 Introduction

Class discriminant analysis on emitter parameters has always been recognized as indispensable for communication, electronic reconnaissance and radar intelligence analysis. For instance, in electronic reconnaissance, reconnaissance strategies could be made and associate reconnaissance activities could be carried out after emitter identification with the discriminating signal parameters. The emitter working modes could also be inferred from the these discriminative signal parameters.

However, the measurements of emitter signal parameters are typically characteristic of uncertainty and continuous growth. These two problems have imposed

This work was supported by National Natural Science Foundation of China (No. 61402426, 61373129) and partially supported by Collaborative Innovation Center of Novel Software Technology and Industrialization.

A. Liu et al. (Eds.): DASFAA 2015 Workshops, LNCS 9052, pp. 280–285, 2015.
DOI: 10.1007/978-3-319-22324-7_28

great challenges for class discriminant analysis of emitter signal parameters. Firstly, the measurements of signal parameters are uncertain, typically interval-valued and complying with a certain normal distribution (Fig. 1(a)). Such uncertainty may be resulted from the unstable working status of transmitter circuit, environmental noises or other unknown interference sources. Secondly, the interval-valued signal parameter measurements were accumulated continuously (Fig. 1(b)). In full spectrum perception, the amount of received signals from various kinds of emitters could be explosive.

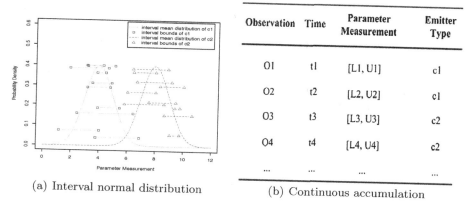

(a) Interval normal distribution

Observation	Time	Parameter Measurement	Emitter Type
O1	t1	[L1, U1]	c1
O2	t2	[L2, U2]	c1
O3	t3	[L3, U3]	c2
O4	t4	[L4, U4]	c2
...	...	...	...

(b) Continuous accumulation

Fig. 1. Interval-valued emitter signal parameter data

Few machine learning methods are fitful for incremental interval-valued emitter signal parameter analysis as they either deal with precise valued data only and process interval-valued data statically under an assumption of uniform distribution. In emitter parameter analysis, the uniform assumption does not hold. The distribution of emitter signal parameters is approximately normal instead. Inspired by the above problems, we bring forward an incremental class discriminant analysis method on interval-valued emitter signal parameters.

2   Related Work

Quite a number of machine learning methods have been put forward to address the uncertainty of interval-valued data. In addition, symbolic data analysis [4] has been proposed to extend the classical data models to take into account of the interval-valued information. The representatives discriminant interval-valued data analysis strategies include point value replacement [1], p-box [2] and Hausdorff distance methods [3]. However, existing interval-valued machine learning methods typically assume that the variables are uniformly distributed in the interval. Nor did they consider the incremental learning of interval-valued variables.

3   Method

In this section, we formally present our incremental class discriminant analysis method on interval-valued emitter signal parameters.

Suppose the interval-valued emitter data set E is composed of a set of continuously accumulating observations. The current number of observations, signal parameters and emitter types are denoted as M, N and K respectively. There are M_k number of observations in each emitter type c_k. Each observation o_m, $1 \leq m \leq M$, is consisted of N number of interval-valued parameter measurements, $\{I_{mn}\}_{1 \leq n \leq N}$, an associated emitter type c_m and a time stamp. Each interval-valued measurement $I_{mn} = [L_{mn}, U_{mn}]$ is consisted of a lower bound L_{mn} and an upper bound U_{mn}. The lower bound L_{mn} and upper bound U_{mn} are the minimum and maximum measurement respectively among w independent measurements of parameter p_n from observation o_m.

The w measurements are assumed to comply with the same interval normal distribution $\mathcal{N}(\mu_{mn}, \sigma_{mn}^2)$. We also assume that the mean values of parameter interval distributions for observations in the same class c_k, $\{\mu_{mn}\}_{m \in \Omega k}$ ($1 \leq k \leq K$), comply with the same class normal distribution $N(\mu_{kn}, \sigma_{kn}^2)$ as well and that $\sigma_{mn} \approx \sigma_{kn}$.

$$\begin{cases} x_{mn} \sim \mathcal{N}(\mu_{mn}, \sigma_{mn}^2) \\ \mu_{mn} \sim \mathcal{N}(\mu_{kn}, \sigma_{kn}^2) \\ \sigma_{mn} \approx \sigma_{kn} \end{cases} \tag{1}$$

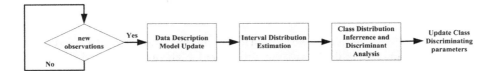

Fig. 2. Outline

Based on a specified data description model, our incremental class discriminant analysis method on interval-valued emitter signal parameters carries out interval distribution estimation, data description model update, class distribution inference and class discriminant analysis in an incremental way, as shown in Fig. 2.

3.1   Interval Distribution Estimation

Given a newly-arrived interval-valued measurement $I_{mn} = [L_{mn}, U_{mn}]$ of signal parameter p_n from observation o_m, the lower bound measurement L_{mn} corresponds to the smallest order statistic while the upper bound measurement U_{mn}

corresponds to the largest order statistic. The order statistics of standard normal random variables have been approximated [6]. Therefore, we have:

$$
\begin{cases}
E(\frac{L_{mn}-\mu_{mn}}{\sigma_{mn}}) = E(1,w) = -\Phi^{-1}(\frac{1-\alpha}{w-2\alpha+1}) \\
E(\frac{U_{mn}-\mu_{mn}}{\sigma_{mn}}) = E(w,w) = -\Phi^{-1}(\frac{w-\alpha}{w-2\alpha+1})
\end{cases}
\tag{2}
$$

As a result, the estimated $\widetilde{\mu_{mn}}$ and $\widetilde{\sigma_{mn}}$ could be estimated as below:

$$
\begin{cases}
\widetilde{\mu_{mn}} = \frac{L_{mn}+U_{mn}}{2} \\
\widetilde{\sigma_{mn}} = \frac{U_{mn}-L_{mn}}{4}(\frac{1}{\Phi^{-1}(\frac{1-\alpha}{w-2\alpha+1})} - \frac{1}{\Phi^{-1}(\frac{w-\alpha}{w-2\alpha+1})})
\end{cases}
\tag{3}
$$

For each newly-arrived observation o_m at a new time point t_Q, we incrementally estimate their mean value and variance in the interval normal distribution.

3.2   Data Description Model Update

To facilitate the incremental learning, we define the data description model as a mean value matrix Σ_μ, a variation matrix Σ_{σ^2} and a class distribution vector V. The elements of the data description model are all initialized as zeros.

Each element of the two-dimensional mean value matrix $\Sigma_\mu(k, n)$ is defined as the sum of estimated mean values for signal parameter p_n from all the observation o_m in type c_k, $\Sigma_\mu(k, n) = \sum_{o_m \in \Omega_k} \widetilde{\mu_{mn}}$, $1 \leq n \leq N$ and $1 \leq k \leq K$. Each element of variation matrix $\Sigma_{\sigma^2}(k, n)$ is defined as the sum of estimated variations for signal parameter p_n from all the observation o_m in emitter type c_k, $\Sigma_{\sigma^2}(k, n) = \sum_{o_m \in \Omega_k} \widetilde{\sigma_{mn}}^2$, $1 \leq n \leq N$ and $1 \leq k \leq K$. Each element $V(k)$ of class distribution vector V is defined as the number of observations in emitter type k, $V(k) = M_k$, $1 \leq k \leq K$.

For each newly-arrived observation o_m at a new time point t_Q, we incrementally update the data description model with its estimated interval distributions for each signal parameter.

3.3   Class Distribution Inference and Discriminant Analysis

As the M_k interval-valued measurements in each emitter type are independent with each other. Upon the M_k estimated mean values and variances for each signal parameter p_n in each emitter type c_k, we can further infer the class normal distribution $\mathcal{N}(\widetilde{\mu_{kn}}, \widetilde{\sigma_{kn}}^2)$ as follows:

$$
\widetilde{\mu_{kn}} = \frac{\sum_{m \in \Omega_k} \widetilde{\mu_{mn}}}{M_k} = \frac{\Sigma_\mu(k, n)}{V(k)}
\tag{4}
$$

$$
\widetilde{\sigma_{kn}} = \sqrt{\frac{\sum_{m \in \Omega_k} \widetilde{\sigma_{mn}}^2}{M_k}} = \sqrt{\frac{\Sigma_{\sigma^2}(k, n)}{V(k)}}
\tag{5}
$$

The discriminating power of each signal parameter p_n for each emitter type pair $c_u - c_v$ could be then evaluated by the welch t-test accordingly.

The t statistic is calculated by $t_{uvn} = \frac{\Sigma_\mu(u,n)/V(u) - \Sigma_\mu(v,n)/V(v)}{\sqrt{\frac{\Sigma_{\sigma^2}(u,n)}{V(u)^2} + \frac{\Sigma_{\sigma^2}(v,n)}{V(v)^2}}}$. The correspond-

ing degree of freedom is computed as $df_{uvn} = \frac{(\frac{\Sigma_{\sigma^2}(u,n)}{V(u)^2} + \frac{\Sigma_{\sigma^2}(v,n)}{V(v)^2})^2}{\frac{\Sigma_{\sigma^2}(u,n)^2}{V(u)^4(V(u)-1)} + \frac{\Sigma_{\sigma^2}(v,n)^2}{V(v)^4(V(v)-1)}}$. As long

as the associated p-value $pval_{uvn}$ of the welch t-statistic t_{uvn} is below 0.05, parameter p_n is assumed to be discriminating for type pair $c_u - c_v$.

4   Results

We evaluated the effectiveness of our weighted class discriminant evaluation method on interval-valued emitter signal parameters on a series of synthetic data sets. During experiments, we fixed the number of observations in each emitter type as the same. The default synthetic data set was composed of 50000 ($M = 50\,k$) observations from five ($K = 5$) different emitters, $10k$ observations each type. The original number of signal parameters was initialized as ten ($N = 10$).

Given an estimated value of mean or standard deviation of a class normal distribution for a certain signal parameter on a certain emitter type, we define the absolute error as the absolute difference between the estimated value and the underlying true value. During experiments, we varied the number of observations per emitter type and calculated the corresponding absolute errors for each parameter on each emitter type. Under each parameter setting, we simulated the experiments 1000 times and generated a boxplot for the absolute errors. It turned out that the absolute errors of signal parameters from each emitter type all tend to converge to zero, as illustrated in Fig. 3.

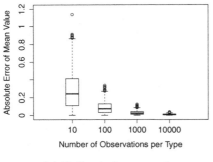

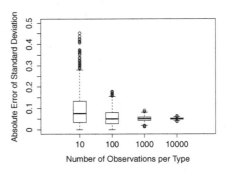

(a) Estimated mean value   (b) Estimated standard deviation

Fig. 3. Boxplot of absolute errors

5   Conclusion

In this work, we have brought forward a novel incremental class discriminant analysis method on interval-valued emitter signal parameters to address the problems of uncertainty and rapid growth of emitter data set. Our method is not only robust to the uncertainty of interval-valued parameter measurements, but also able to carry out class discriminant parameter analysis incrementally. Simulated experiments have indicated the effectiveness of our method. These merits enable our method to be applied promisingly in the field of emitter identification.

References

1. Neto, E.A.L., de Carvalho, F.A.T.: Centre and range method to fitting a linear regression model on symbolic interval data. Comput. Stat. Data Anal. **52**, 1500–1515 (2008)
2. Utkin, L., Destercke, S.: Computing expectations with continuous p-boxes: univariate case. Int. J. Approx. Reason. **50**, 778–798 (2009)
3. De Carvalho, F., De Souza, R., Chavent, M., Lechevallier, Y.: Adaptative hausdorff distances and dynamic clustering of symbolic interval data. Pattern Recog. Lett. **27**(3), 167–179 (2006). Elsevier
4. Noirhomme-Fraiture, M., Brito, P.: Far beyond the classical data models: symbolic data analysis. Stat. Anal. Data Min. **4**(2), 157–170 (2011)
5. Asharaf, S., Murty, M.N., Shevade, S.K.: Rough set based incremental clustering of interval data. Pattern Recog. Lett. **27**(6), 515–519 (2006)
6. Royston, J.P.: Expected normal order statistics (exact and approximate). J. Royal Stat. Soc. Ser. C (Appl. Stat.) **31**(2), 161–165 (1982)

Visualization Tool for Boundary Image Matching Based on Time-Series Data

Seongwoo Moon[1], Sanghun Lee[1], Bum-Soo Kim[2], and Yang-Sae Moon[1(✉)]

[1] Department of Computer Science, Kangwon National University, Chuncheon, Korea
{swmoon,sanghun,ysmoon}@kangwon.ac.kr
[2] Institute of Industrial Management, KAIST, Daejeon, Korea
bskim@mozart.kaist.ac.kr

Abstract. In this paper, we propose a visualization tool for boundary image matching based on time-series matching techniques. The proposed tool works as a client-server model that first converts boundary images to time-series data and then exploits efficient time-series matching techniques. The client shows the matching results with different charts or graphs to provide various viewpoints. The server efficiently performs the time-series matching using the multidimensional index, which supports a huge number of image time-series. By adapting visualization techniques on time-series matching, we can easily and intuitively understand the boundary matching results as well as time-series matching results. In particular, our polar chart, which represents 1-D time-series to 2-D boundary images, may give a strong intuition of understanding various trends of time-series data. We provide five different visualization methods, and we believe that those methods will be very helpful to understand the matching results of image time-series.

1 Introduction

Visualization is a technique to communicate meaningful data by using images, charts, diagrams, or animations [1]. The visualization has been widely used in many applications in science, education, information knowledge, engineering, medicine, etc. In particular, it is very effectively used in data mining applications to represent the results of SNA(social network analysis), text mining, or time-series matching. Therefore, through the visualization, we can improve efficiency of recognition, understanding, analysis, and process of the data [1].

In this paper, we design and implement a visualization tool for boundary image matching [2] which exploits time-series matching techniques. In general, the original image matching cannot support a huge number of data images. To support a large-scale image database, we develop the matching system in the time-series domain rather than the image domain. For this, we first convert boundary images into time-series data [2]. We then build a multidimensional index not only for supporting a huge number of data images but also for using efficient time-series matching techniques. That is, by using the time-series data, we not only efficiently support the large-scale boundary image database but also represent the matching results as various charts or graphs. Likewise, by adopting

© Springer International Publishing Switzerland 2015
A. Liu et al. (Eds.): DASFAA 2015 Workshops, LNCS 9052, pp. 286–292, 2015.
DOI: 10.1007/978-3-319-22324-7_29

time-series matching techniques, the proposed visualization tool provides time-series viewpoints as well as image viewpoints for the results of boundary image matching. Our visualization tool provides five different charts and graphs to provide a variety of viewpoints for the matching results, and we can intuitively understand the actual difference between query and data images (or image time-series). Readers are referred to a simple demo in [3] for our actual implementation.

2   Related Work

Time-series matching is the problem of finding time-series data similar to the given query time-series [4]. In this paper, we deal with the k-NN search [5] that finds k nearest time-series to the query and use the Euclidean distance [4] as the similarity measure.

In recent years, there have been a lot of visualization efforts on a time-series data. Adar et al. [6] visualize the process of time-series matching to help the naïve users who have a little knowledge on time-series matching. They also graphically represent the distance matrix of DTW(dynamic time warping) [7] distance. Weber et al. [8] visualize various multidimensional time-series data to the spiral chart with various colors, thickness, and concentrations for intuitive understanding. However, these two papers focus on only the visualization of pure time-series themselves, but they have no effort on visualization for comparing and analyzing the matching results. Based on this observation, in this paper we propose a visualization tool that visualizes not only pure time-series data but also the matching results. Also, our visualization tool focuses on boundary image matching rather than time-series matching and provides five different charts and graphs for easy and intuitive understanding.

3   Design and Implementation of the Visualization Tool

Our visualization tool for boundary image matching works on the following client-server environment. We implement the client on Microsoft Visual Studio 2010 in Windows 7 operating system, and we use C# language with MSChart. We implement the server on a Linux machine with CentOS 5.9 and use C language for implementing the matching engine.

We have used a real data set consisting of 12,000 time-series of length 360, which are converted from the image data by using CCD [2]. We extract six features from a time-series of length 360 using PAA (piecewise aggregate approximation) [4], and we use the R*-tree [2] as the multidimensional index.

The overall boundary image matching framework works in the client-server model as shown in Fig. 1. First, the client sends a user selected time-series, which is converted from a query image and the coefficient k of k-NN to the server. Second, the server sends the k results, which are retrieved from the database using the index, to the client. Finally, the client visualizes the results, which are received from the server, to different charts or graphs.

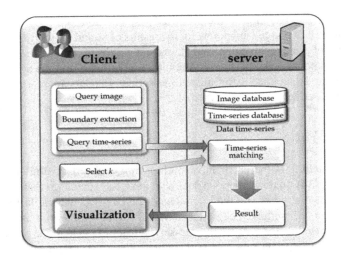

Fig. 1. Overall framework of the proposed visualization system.

Figure 2 is an initial screen of our boundary image matching visualization tool. In the figure, "Open" of Part Ⓐ is a button for selecting a query image(time-series), "k of k-NN" is a selection box for selecting the coefficient k, and "Search" is a button for searching results by sending a query with k to the server. Part Ⓑ shows the information of matching results, which are received from the server, where each column shows the similarity ranking between the query and result images, the image number, and its actual distance. Part Ⓒ is a visualization frame that shows the boundary image matching results to the five different charts and graphs.

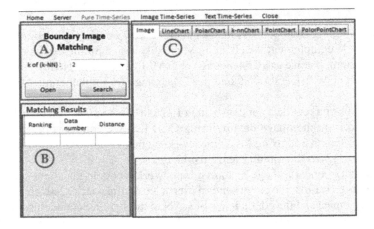

Fig. 2. Initial screen of the client visualization tool.

Figure 3 shows real execution screens of Fig. 2. The above left image of Part Ⓒ in Fig. 3 is a user selected image, and the above right image is the best similar image from the k results. Bottom images are thumbnails of the k result images in the order of similarity distance. If the user clicks a thumbnail image, the above right image is changed to the clicked image, and the line and polar charts are redrawn for the clicked image as shown in Figs. 4 and 5.

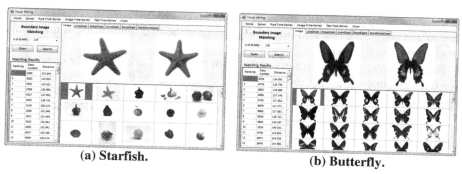

(a) Starfish.   **(b) Butterfly.**

Fig. 3. Screenshots of boundary image matching results.

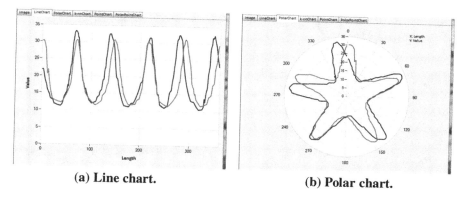

(a) Line chart.   **(b) Polar chart.**

Fig. 4. Chart representation of Starfish results.

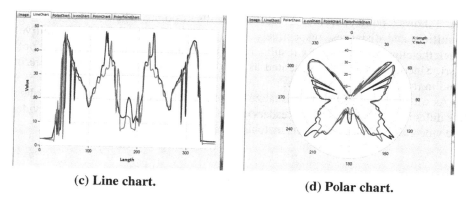

(c) Line chart.   **(d) Polar chart.**

Fig. 5. Chart representation of Butterfly results.

Figures 4(a) and 5(a) show line charts of query and resulting time-series, respectively. The line chart is one of the most widely used in the time-series visualization, and we thus provide the line chart as the basic chart. Using the line chart, we can intuitively measure the difference in between query and resulting time-series data.

Figures 4(b) and 5(b) show the visualization of query and resulting time-series by using the polar chat. Using the polar chart, we can easily infer the original boundary of an image. We can also intuitively compare the two time-series of query and data images.

Figure 6 shows the visualization of the similarity distances between the query and k data (results) time-series. By showing the k-NN distances to the graph, we can easily find the change of relative distances since we can note the change of inclines from the graph. Furthermore, we may use the graph to find an optimal k in the k-NN search.

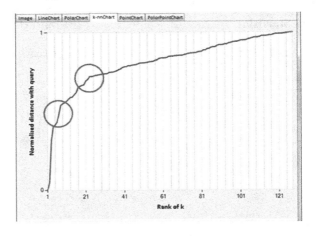

Fig. 6. Similarity distances from the query to k-NN results.

Figures 7 and 8 represent the retrieved results as the point in the charts. First, Fig. 7 shows a point chart that draws the k results as points, where x- and y-axes are identifiers and distances, respectively. Using this point chart, we can intuitively check the clustering trend of k results. Next, Fig. 8 redraw Fig. 7 by using the polar chart. Since the chart is represented as a circular form, we can see the clustering trend in the different perspective.

As we explained earlier, we visualize boundary and time-series matching results as five different charts and graphs. Readers are referred to [3] to see a simple demo video. The video uses the images and its matching results explained in this paper.

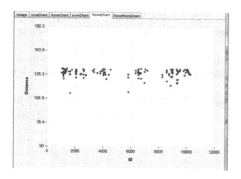

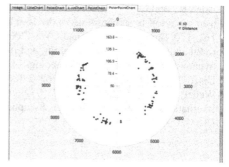

Fig. 7. Point chat for clustering trend.   **Fig. 8.** Polar chart for clustering trend.

4   Conclusions

In this paper, we design and implement a visualization tool that provides boundary image matching in large image time-series databases. Contributions are summarized as follows. First, we present a client-server framework that makes it feasible to maintain a huge number of image time-series in the server and at the same time to provide the user-friendly interface in the client. Second, we provide the line and polar charts for the matching results so as to easily understand the similarity of the matching results. In particular, in the polar chart, we can intuitively compare two time-series as their boundary images. Third, we provide two point charts so as to easily find the clustering trend of the matching results. We believe that our visualization tool will be very helpful to analyze and understand the matching results of image time-series. As the future work, we will consider the visualization of scaling-invariant and rotation-invariant image matching in the time-series domain.

Acknowledgement.   This research was supported by Basic Science Research Program through the National Research Foundation of Korea (NRF) funded by the Ministry of Science, ICT and future Planning(NRF-2014R1A2A2A01002548).

References

1. Wikipedia. http://en.wikipedia.org/wiki/Visualization
2. Kim, B.-S., Moon, Y.-S., Choi, M.-J., Kim, J.: Interactive noise-controlled boundary image matching using the time-series moving average transform. Multimedia Tools Appl. **72**(3), 2543–2571 (2014)
3. Moon, Y.-S., Lee, S., Kim, B.-S., Moon, Y.-S.: Boundary image matching visualization demonstration, January 2015. http://dke.kangwon.ac.kr/~demo
4. Moon, Y.-S., Lee, B.S.: Safe MBR-transformation in similar sequence matching. Inf. Sci. **270**, 28–40 (2014)
5. Dong, W., Moses, C., Li, K.: Efficient k-nearest neighbor graph construction for generic similarity measures. In: Proceedings of the 20th International Conference on World Wide Web, pp. 577–586. Hyderabad, India, March/April 2011

6. Adar, E., Weld, D.S., Bershad, B.N., Gribble, S.D.: Why we search: visualizing and predicting user behavior. In: Proceedings of the 16th International Conference on World Wide Web, pp. 161–170. Banff, Alberta, Canada, May 2007
7. Rakthanmanon, T. et al.: Searching and mining trillions of time series subsequences under dynamic time warping. In: Proceedings of the 18th International Conference on Knowledge Discovery and Data Mining, pp. 262–270. Beijing, China, Aug. 2012
8. Weber, M., Alexa, M., Muller, W.: Visualizing time-series on spirals. In: Proceedings of the International Conference on IEEE Symposium on Information Visualization, pp. 7–13, San Diego, CA, Oct. 2001

Performance Analysis of Hadoop-Based SQL and NoSQL for Processing Log Data

Siwoon Son[1], Myeong-Seon Gil[1], Yang-Sae Moon[1](✉),
and Hee-Sun Won[2]

[1] Department of Computer Science, Kangwon National University,
Chuncheon, South Korea
{ssw5176,gils,ysmoon}@kangwon.ac.kr
[2] Electronics and Telecommunications Research Institute, Daejeon, South Korea
hswon@etri.re.kr

Abstract. Recently, many companies and research organizations are seeking scalable solutions by using Hadoop ecosystems. The log data management with large-scale and real-time properties is one of the appropriate application on top of Hadoop. In this paper, we focus on SQL and NoSQL choices for building Hadoop-based log data management system. For this purpose, we first select major products supporting SQL and NoSQL, and we then present an appropriate scheme for each product by considering its own characteristics. All the schema are for real-time monitoring and analyzing the log data. For each product, we implement insertion and selection operations of log data in Hadoop, and we analyze the performance of these operation. Analysis results show that MariaDB and MongoDB are fast in the insertion, and PostgreSQL and HBase are fast in the selection. We believe that our evaluation results will be very helpful for users to choose Hadoop SQL and NoSQL products for handling large-scale and real-time log data.

1 Introduction

Hadoop is a distributed environment system, which is widely used in a variety of applications since it is efficient at high volume data processing. In particular, nowadays we can easily get a huge volume of system logs, sensor data, stream data, and social data due to their fast production rates, and there are also many needs of new applications for utilizing those big data. In this paper, we focus on handling of log data in the Hadoop system. This is because the log data with various types and structures are continuously collected from the most of running systems, and Hadoop is very suitable for analyzing those big data.

Hadoop is generally used in conjunction with various software programs rather than used alone [1, 2], and we often configure the Hadoop system with RDBMS (relational database management system) or NoSQL to increase the efficiency of data processing and retrieval. The log management system considered in this paper uses both RDBMS (referred to as SQL) and NoSQL, where SQL is used for real-time monitoring of hot logs while NoSQL is used for analyzing or mining of accumulated log data. For this purpose, we first choose the open-source SQL and NoSQL products that are known to

A. Liu et al. (Eds.): DASFAA 2015 Workshops, LNCS 9052, pp. 293–299, 2015.
DOI: 10.1007/978-3-319-22324-7_30

be very efficient, and we then conduct extensive performance tests using those products. The products to be used for performance test are as follows:

- SQL: MariaDB, PostgreSQL, FireBird
- NoSQL: MongoDB, HBase, Riak

Through the performance analysis of the above products, in this paper we try to find meaningful results that can be used for constructing an efficient Hadoop system for maintaining and analyzing large-scale and real-time log data. Our performance evaluation results are summarized as follows: (1) in SQL, MariaDB is the best in insertions, and PostgreSQL is the best in selections; (2) in NoSQL, MongoDB is the best in insertions, and HBase is the best for selections.

2   Related Work

Log data contain the events and the times that they occur, and it is important to intuitively represent log data as the valuable information only. There are various methods of collecting logs, where the most common way is to install a log collection program to each node in the Hadoop system. Ganglia monitoring system [3] is a recent monitoring system that collects the log data resources from multiple computer systems for the purpose of log analysis and data mining. In this paper, we collect real-time log data using the Ganglia monitoring system. Figure 1 shows an example of actual log data collected by Ganglia.

```
dtlOne,sd-dev1,bytes_out,5164.80,20140813160000
dtlOne,sd-dev1,cpu_idle,99.8,20140813160000
dtlOne,sd-dev2,os_name,Linux,20140813160000
dtlOne,DTLNMS,tcp_listen,24,20140813160000
```

Fig. 1. An example of log data collected by Ganglia.

Hadoop [4] itself is not suitable for real-time monitoring of log data since it has no ability of processing real-time data generated quickly and continuously. Also, it is not suitable for storing a large volume of data with various structures and retrieving/analyzing those complex data for different purposes. To resolve this problem, a supplementary data management platform is required to store the log data into HDFS, retrieve/analyze those data, and perform the real-time monitoring operation.

Real-time monitoring generally uses RDBMS(SQL) since it requires fast retrieval time, and there is no need to store large-scale data. SQL refers to the software system for operating and managing relational tables, each of which consists of one or more properties, called attributes. Relational databases store various types of data as tables having consistent schema structures, and thus, they can easily represent relationships among data and support complex computations. In general, SQL products are divided into commercial and open source ones. Commercial examples are Oracle DBMS,

MS SQL, and IBM DB2; open source examples are SkySQL MariaDB, EnterpriseDB PostgreSQL, and Borland FireBird.

Recently, NoSQL is actively studied to store and analyze a huge volume of semi-structured or unstructured data. NoSQL is free from the constraints of schema, and it is possible to process large data rapidly. It can also store much more volume of data compared with the traditional RDBMS. According to the data representation, NoSQL can be classified into key/value-based, document-based, column-based, and graph-based models. The key/value-based model stores the data as the (key, value) form, and it thus supports only a single key representation. Redis, Riak, and Voldemort are major products of the key/value model. The document-based model uses an ID of a document as a key to process the query of search condition, which is similar to RDBMS. The representative document-based model is MongoDB, which has the highest market share among NoSQL products. The column-based model represents the data as name, value, and time stamp, and it processes only 'INSERT' operation without 'UPDATE'. The most commonly used column-based model is Cassandra and HBase. The graph-based model represents the data as nodes and relationships (edges), but it is not appropriate for dealing with big log data since all the data must be store in a singleserver.

3   Performance Evaluation of SQL Products

Most SQL products use the same or similar schema structures, and in this paper we design only one table scheme for log data and use it for all SQL products to be evaluated. Figure 2 shows the log table structure which consists of five columns collected by Ganglia, and we use this table for evaluating all three SQL products.

Log_Table				
cluster_name varchar(20)	host_name varchar(20)	log_key varchar(50)	log_value varchar(50)	time_stamp timestamp

Fig. 2. The table scheme of SQL products.

3.1   Insertion Operation in SQL

We use SQL to provide real-time monitoring of the log data. Real-time monitoring needs to insert small-sized log data periodically and frequently rather than large-sized log data at once. In this paper, we design and implement this insertion operation using MapReduce. We use Hadoop environment, it consists of a single NameNode and eight DataNodes. The MapReduce algorithm works as follows: (1) it reads lines one by one from the log file, where each line contains the log data of one minute for each data node; (2) it reorganizes the line to fit the scheme of Fig. 2; and (3) it stores the log record to SQL through a JDBC driver. We maintain the log data at each 1 min (33 kB), 30 min (1.2 MB), and 1 h (2.2 MB), respectively.

Figure 3(a) shows the insertion operation time in SQL according to the size of log data. As shown in the graph, MariaDB is slightly faster than PostgreSQL, and FireBird is worse than other products. In our knowledge, it is because MariaDB uses a light-weight structure in its database engine.

3.2   Selection Operation in SQL

For monitoring the real-time log data, we needs lookup operations that retrieves log records from relational databases with different types of the log data. Selection operation used in the experiment is to select the CPU usage of "sd-dev1" host in the order of time. Log data for the CPU usage is composed of "cpu_user", "cpu_system", "cpu_wio", and "cpu_idle". The data stored in SQL has 887,040 log records generated in a day (51 MB), and we can retrieve the 5,760 CPU usage information using the selection operation.

Figure 3(b) shows the time required for processing the given query of selection operations. As shown in the graph, the order of selection performance is PostgreSQL, MariaDB, and FireBird. Unlike the insertion, PostgreSQL shows the best performance. Considering Figs. 3(a) and (b), MariaDB and PostgreSQL are superior to FireBird, and we choose MariaDB for our log management project.

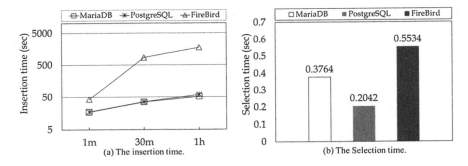

Fig. 3. Comparison of execution times in SQL.

4   Performance Evaluation of NoSQL Products

4.1   Schema Design

We need to design different DB schema for different NoSQL products because each NoSQL has its own characteristics. We perform the NoSQL evaluation in eight servers to carry out insertion/selection operations in a distributed environment.

MongoDB: MongoDB [6] is a major product of NoSQL with the document-based model. The data structure of MongoDB is represented by multiple documents with separate IDs, and we can store the data into a key/value form without the schema constraints of the documents. In this paper, we represent the log data as the keys of cluster_name, host_name, log_key, log_value, and time_stamp, and we store the measured values to the corresponding keys. We use string data types for default values except time_stamp. The time_stamp is expressed as a date type, since it is easy to support time-related operations.

HBase: HBase [7] is a NoSQL product with the column-based model and uses Hadoop storage structures. Time_stamp is used as a row key on HBase because each log is an event that occurs in a specific time. Thus, all the log data that occur at the same time can be represented in a single row. To distinguish the host that generates different logs, we configure the column family with cluster_name and host_name and host_name, and the log data that occur on the same host is stored in each column family. Please, note that the log_key of log data becomes a column qualifier, and the log_value becomes its cell value.

Riak: As a NoSQL product of the key/value-based model, Riak [8] divides the data into buckets, and the data is expressed as the form of (key, value) pairs within each bucket. In this paper, we use the bucket name as a combination of cluster_name and host_name to store the log data, and time_stamp and log_key are used as a key within the bucket. The key, however, just plays a role in distinguishing the log data, and Riak uses the secondary index with range queries to access the real data. We use the log data generated at a specific time as the value of the model. The value expression is based on JSON, and the time_stamp is designated as the secondary index.

4.2 Insertion Operation in NoSQL

NoSQL for the log data management system inserts a large volume of log data at a batch mode. In this paper, we implement the insertion operation as follows: it first reads the log data from HDFS, and it stores the data into a NoSQL DB through JDBC drivers without MapReduce. We use Hadoop environment, it consists of a single NameNode and eight DataNodes, equally to SQL insertion. The log data of one hour (2.2 MB), one day (51 MB), and ten days (508 MB) are used in the insertion operation. Compared with Sect. 3.1, we note that these data are much bigger than those of the insert operation in SQL.

Figure 4(a) shows a comparison graph of the insertion times for NoSQL products according to the data size. As shown in the graph, MongoDB shows the fastest insertion time compared to other two NoSQL products, and in particular, it is 3.5 times faster than HBase.

4.3 Selection Operation in NoSQL

To analyze big log data stored NoSQL, we often needs to select only partial data by many different requests of different users. To compare the required time for extracting

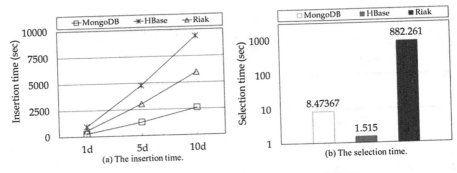

Fig. 4. Comparison of execution times in NoSQL DB.

the data from each NoSQL with its specific characteristics, we design the selection using only common operations applied to all three NoSQL products. The selection operation used in NoSQL is the change of CPU usage on three hosts, "dtlOne", "sd-dev1", and "sd-dev2" from 2014-08-12 16:01:00 to 2014-08-22 16:00:00. There are total 8,870,400 log data items stored in NoSQL, and 172,800 CPU items are extracted as a result of the selection operation.

Figure 4(b) shows a comparison graph on the selection time in three NoSQL products. As shown in the figure, HBase significantly outperforms Riak and MongoDB. Considering Figs. 4(a) and (b), we choose HBase for our log management project since the selection is generally much more frequently used in the log management system.

5   Conclusions

In this paper, we first present appropriate schema of RDBMS (SQL) and NoSQL for real-time monitoring and analyzing of log data. We then design and implement a process of reading the huge log data from Hadoop for SQL and NoSQL products. We also analyze the performance of the insertion and selection operations for each of SQL and NoSQL products. Among SQL products, MariaDB is the best one for the insertion while PostgreSQL is the best one for the selection. Among NoSQL products, MongoDB and HBase are the best ones for insertion and selection operations, respectively. We believe that our study can be practically applicable to construct a real-time log management and analysis system by providing the experimental evaluation results, which SQL or NoSQL products are superior to other ones in various Hadoop environments.

Acknowledgment.   This research was funded by the MSIP(Ministry of Science, ICT & Future Planning), Korea in the ICT R&D Program 2014.

References

1. Boulmakoul, A., Karim, L., Laarabi, M.H., Sacile, R., Garbolino, E.: MongoDB-Hadoop distributed and scalable framework for spatio-temporal hazardous materials data warehousing. In: Proceedings of the 7th Int'l Congress on Environmental Modelling and Software (iEMSs), San Diego, CA, vol. 3, pp. 2255–2267, June 2014
2. Vora, M.N.: Hadoop-HBase for large-scale data. In: Proceedings of Int'l Conference on Computer Science and Network Technology, Harbin, China, vol. 4, pp. 601–605, Dec. 2011
3. Shvachko, K., Kuang, H., Radia, S., Chansler, R.: The hadoop distributed file system. In: Proceedings of the 26th IEEE Symposium on Mass Storage Systems and Technologies (MSST), Lake Tahoe, Nevada, pp. 1–10, May 2010
4. Massie, M., Li, B., Nicholes, B., Vuksan, V.: Monitoring with Ganglia. O'Reilly Media Inc, Sebastopol (2012)
5. Rabl, T., Sadoghi, M., Jacobsen, H.-A., Gómez-Villamor, S., Muntés-Mulero, V., Mankowskii, S.: Solving big data challenges for enterprise application performance management. Proc. VLDB Endowment 5(12), 1724–1735 (2012)
6. Wang, X., Chen, H., Wang, Z.: Research on improvement of dynamic load balancing in MongoDB. In: Proceedings of the 11th IEEE Int'l Conference on Dependable, Autonomic and Secure Computing (DASC), Chengdu, Sichuan, China, pp. 124–130, Dec. 2013
7. HBase. http://hbase.apache.org/
8. Riak. http://basho.com/riak/

SVIS: Large Scale Video Data Ingestion into Big Data Platform

Xiaoyan Guo[✉], Yu Cao, and Jun Tao

EMC Labs China, Beijing, China
{xiaoyan.guo,yu.cao,simon.tao}@emc.com

Abstract. Utilizing big data processing platform to analyze and extract insights from unstructured video streams becomes emerging trend in video surveillance area. As the first step, how to efficiently ingest video sources into big data platform is most demanding but challenging problem. However, existing data loading or ingesting tools either lack of video ingestion capability or cannot handle such huge volume of video data. In this paper, we present *SVIS*, a highly scalable and extendable video data ingestion system which can fast ingest different kinds of video source into centralized big data stores. *SVIS* embeds rich video content processing functionalities, e.g. video transcoding and object detection. As a result, the ingested data will have desired formats (i.e. structured data, well-encoded video sequence files) and hence can be analyzed directly. With a highly scalable architecture and an intelligent schedule engine, *SVIS* can be dynamically scaled out to handle large scale online camera streams and intensive ingestion jobs. *SVIS* is also highly extendable. It defines various interfaces to enable embedding user-defined modules to support new types of video source and data sink. Experimental results show that *SVIS* system has high efficiency and good scalability.

1 Introduction

With rapid development of big data technologies, more and more vertical industries, such as Internet, Finance, Healthcare, are placed into big data processing platform to extract hidden values from their data by utilizing state-of-the-art big data analytics [1]. Video surveillance analytics is another emerging trend to extract insights from unstructured video stream data [2]. Surveillance videos are in nature of great magnitude, and hence require the powerful processing capability of big data platform.

However, loading or ingesting video surveillance data, the first step before data can be analyzed by big data platform, is particularly challenging [3]. Firstly, unstructured video data is of diverse types of sources (e.g. online camera streams, historical video files) and encoding formats (e.g. H.264, MJPEG, MPEG4). How to automatically and efficiently transcode and transform video data to be tackleable by big data platform remains as a challenging problem. Secondly, video data is of huge volume. The resolution of IP cameras is increasing rapidly. There are extremely huge number of online cameras in city-wide surveillance systems.

© Springer International Publishing Switzerland 2015
A. Liu et al. (Eds.): DASFAA 2015 Workshops, LNCS 9052, pp. 300–306, 2015.
DOI: 10.1007/978-3-319-22324-7_31

As such, ingesting systems must be scalable and efficient enough to handle huge volume of video data and intensive ingestion workloads. Thirdly, certain video content analytics work need to be conducted within ingestion systems in order to obtain timely analytics results, e.g. real-time determining potential incidents. As a result, it is desirable or even necessary that the ingestion system has real-time video processing functionalities.

There are many existing systems in big data ecosystem to realize data loading or data ingestion work. Flume [4], Sqoop [5], Kafka [6], Scribe [7] and Chukwa [8] are all distributed, reliable and available systems to efficiently ingest large amounts of data, especially log data, from different data sources into centralized big data stores, such as HDFS, HBase, relational SQL database, etc.

However, all these systems are not perfectly suitable to address video ingestion jobs. They lack video processing capability, and hence it is difficult to ingest video data especially video streaming data directly from IP cameras. They do not have good flexibility and extendability so that it is difficult to extend them to support new types of data source and data sink. These systems usually utilize batch loading, emphasizing more on ingestion throughput rather than latency. While in video surveillance area, real-time ingestion is a desired feature.

In this paper, we thereby present *SVIS*, a scalable and extendable video data ingestion system which can fast ingest diverse video sources into big data stores. *SVIS* integrates rich video processing functionalities. It is able to transcode and transform video data of different source types, and then directly pipeline them to video analytics applications. *SVIS* can easily scale out to support large scale video surveillance systems. It can be conveniently extended to support new video sources and data sinks.

Our principal contributions in this paper are summarized as follows.

1. We design and build a highly scalable video ingestion framework, which is able to handle huge volume of video data and extremely intensive ingestion jobs.
2. We integrate rich video content processing functionalities into our ingestion system, e.g. transcoding video into frame sequence files and detecting objects within video frames. As a result, data ingested into big data store is exactly of the desired format that can be analyzed in the straight-forward way. Real-time analytics at edge nodes can also be easily realized.
3. We define various extendable interfaces to support user-defined modules. As a result, the users can develop their own modules and embed them into the system to support specific video sources and sinks.
4. We define a video ingestion DSL (Domain-Specific Language) to allow users to define their own ingestion jobs by declarative interfaces. We build an intelligent schedule engine to guarantee efficiency and fault-tolerance of the ingestion jobs.
5. We conduct comprehensive experiments to demonstrate the efficiency and scalability of our video ingestion system in two scenarios of ingesting large historical video files and intensive online camera streams.

2   Video Data Ingestion System

The *SVIS* video data ingestion system moves huge volume of video data from different sources into a centralized repository. Figure 1 shows the deployment topology of *SVIS* system in a real video surveillance system. Data sources of *SVIS* system are not only historical video files but also online video streams, e.g. online IP cameras. *SVIS* system ingests video data not only within campus local network but also from wide area network. It is able to ingest video data into diverse data sinks, such as HDFS, structured database, and NoSQL data store. It is also capable to directly ingest and pipeline video data to online video analytics applications.

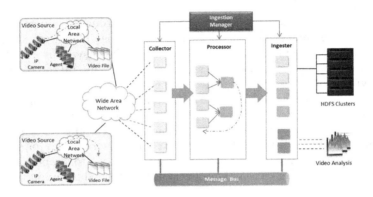

Fig. 1. System architecture

Ingestion Job Abstraction. To enhance system scalability and provide flexible and extendable interfaces, we abstract the ingestion job into three sub-tasks, which encapsulates the same kind of work unit into reusable modules. The three abstractive ingestion modules are *collector*, *processor* and *ingester*.

The collectors implement drivers to collect data from different types of data sources. They poll data from external video sources and then pipeline them to other modules. The first module in an ingestion job must be a collector.

The processors include rich built-in video processing functionalities, such as video transcoding, object detection and feature extraction. They receive video data from the collectors and perform video processing and video analytics at the edge node. The analytics results of processors are delivered to the ingesters or other processors. The processor is optional in an ingestion job. There are also multiple processors in a job to conduct a series of video processing tasks.

The ingesters implement drivers to write data into different data stores. They receive video data or analytics results from collectors or processors, and then ingest them into desired data sinks. The ingester is always the last module in an ingestion job.

In our video ingestion system, a video ingestion job is represented as a DAG, where the vertices are the instances of abstractive ingestion modules (i.e. collectors, processors, and ingesters), the edges are the established message queues to transfer data between two instances. An ingestion job is first scheduled into a DAG by the schedule engine and then executed accordingly. As such, the video data flows from video sources to collectors, processors and ingesters and finally arrives in data sinks.

System Overall Architecture. The middle of Fig. 1 depicts the high-level system architecture of *SVIS* video ingesting system, which consists of the following three major functional components: *ingestion manager, ingestion workers* and *message bus*.

Ingestion manager is responsible for scheduling job execution, launching ingestion jobs, monitoring runtime status of ingestion jobs and managing the whole ingestion cluster. By effectively scheduling and distributing an ingestion job to multiple ingestion workers and monitoring their status, it guarantees high scalability, load balance and fault-tolerance of the ingestion system.

Ingestion workers are responsible for executing the ingestion jobs actually. An ingestion worker is a node to execute one of the three ingestion modules mentioned in above section, i.e. collector, processor and ingester. The ingestion workers are scheduled by the ingestion manager. An ingestion job is represented as a DAG. For each vertex, the ingestion manager launches a corresponding ingestion worker to do the task. The ingestion workers can be dynamically added or removed according to the ingestion workload. It guarantees good flexibility and scalability of the ingestion system.

The ingestion workers share and transfer data via a configured messaging queue, the message bus shown in Fig. 1. The message bus provides multiple messaging patterns according to different ingestion topologies. It guarantees great reliability and high availability of the ingestion data.

The complete workflow of an ingestion job is described as follows. The ingestion manager accepts the ingestion request and ingestion job definition from the user. It then schedules and breaks down the ingestion job into a DAG ingestion topology, i.e. independent ingestion tasks, which can be directly executed by the ingestion workers. According to the ingestion topology, the ingestion manager launches a series of ingestion workers and distributes the ingestion tasks to them. The ingestion workers understand the task definitions and execute them in coordination of the ingestion manager. The ingestion workers transfer data via message bus and ingest video data into desired data sinks. During the ingestion, ingestion manager periodically collects tuntime status of each ingestion worker and monitor the ingestion progress. Once a task is failed, it restarts the corresponding ingestion worker and re-execute the task.

Video Ingestion Extendable API. There are varied video sources and video sinks in different video surveillance systems. It is impossible to implement a one-size-fits-all ingestion system to support all the data sources and sinks. As such, we abstract the video ingestion job into three fundamental modules, i.e. collector, processor and ingester. For each module, we define abstractive interfaces for the

```
collector:                              ingester:
    - name: collector1                      - name: ingester1
      sourceType: videoFile                   sinkType: gfxd
      moduleType: VideoFileFrameCollector     moduleType: FaceGFXDIngester
      parallelism: 1                          parallelism: 1
      outputQueue: FaceDetectionQ1            inputQueue: FaceDetectionQ2
      property:                               property:
          host:                                   jdbcURL:
              - "10.62.98.123"                        - "jdbc:gemfirexd://phd10:1527/"
          filePath:                               tableName:
              - "/root/videoclips/7fgate.mp4"         - "FaceDetectionCam4"
                                                  persistProperty:
                                                      - "HDFS"
processor:                              queuepipe:
    - name: processor1                      - name: FaceDetectionQ1
      moduleType: FaceDetectionProcessor      type: rabbitMQ
      parallelism: 1
      inputQueue: FaceDetectionQ1           - name: FaceDetectionQ2
      outputQueue: FaceDetectionQ2           type: rabbitMQ
```

Fig. 2. DSL example of the ingestion job

users to implement user-defined modules to extend the system to support new specific types of data source and data sink.

For collector module, the users need to implement the interface to obtain data from data source. For ingester module, the users need to implement the interface to write data into corresponding data sink. For processor module, the users need to implement the interfaces to process and analyze the video data in order to gain intermediate results. After defining a module, the user can embed it into the ingestion system. The module will then be available and can be included in an ingestion job.

Video Ingestion DSL. We define a video ingestion DSL (Domain-Specific Language) to allow users to define their own ingestion jobs by declarative interfaces. With the declarative language, the users can define an ingestion job in a quite straight-forward way. The users can also be able to dynamically submit or stop an ingestion job easily.

With the declarative language, the users need to define the properties of collectors, ingesters and desired processors, desired parallelism degree, and queue pipe topology among these modules. All the definitions can be configured in a YAML or JSON configuration file. Figure 2 shows an example of YAML configuration file of ingesting a video file into HDFS and conducting face detection processing in the meantime.

3   Experimental Study

Experiment Setup. We setup a 10-node video ingestion cluster on 10 physical servers with 16 GB RAM and Xeon E5-2640@2.00 GHz 4 core CPU. They were connected by 1 Gbps network. We deployed 1 ingestion manager node, 8 ingestion worker nodes, and 1 RabbitMQ server node serving as the message bus.

There were two types of video sources: video files and online camera streams. We generated video files of different encodings and sizes. We also setup 20 IP

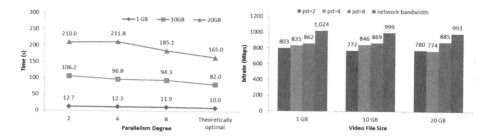

Fig. 3. Experimental results of ingesting video files into HDFS

cameras to provide online RTSP stream data. These cameras were configured with FPS of 20 and frame resolution of 800*600.

There were two types of data sinks in our experiments: HDFS and Gemfire XD [9]. The write speed of HDFS was about 50 MB/s. Gemfire XD is a main-memory based, distributed low latency data store for structured data and key-value data. HDFS and Gemfire XD were both deployed in a 10-node cluster.

Efficiency of Ingesting Large Video Files. We conducted experiments to ingest large volume of video files into HDFS and studied ingestion speed and efficiency. We generated three video files with size of 1 GB, 10 GB and 20 GB and parallel ingested them with different parallelism degrees. Greater parallelism degree indicates that more ingestion workers are employed by *SVIS* ingestion system. Figure 3 shows the experimental results: left figure shows the total ingestion time of each file, right figure shows the achieved ingestion bitrate of each file.

Due to write speed of HDFS and network bandwidth, the largest throughput is bounded at 1000Mbps. From the experimental results, we can find that with a proper parallelism degree, the ingestion system can reach particularly high ingestion throughput that is about 85 % of the optimal throughput.

Latency and Scalability of Ingesting Intensive Online Camera Streams. We conducted experiments to ingest online camera streams into Gemfire XD. We increased ingestion workload (by increasing the ingested camera number from 1 to 20) and evaluated the latency and throughput of our ingestion system. Figure 4 shows the experimental results: left figure shows the average ingestion latency and total FPS achieved for each workload, right figure shows the achieved detail FPS of each camera when ingesting the 20-camera workload.

The latency represents the time interval of a video frame between generated by the camera and ingested into Gemfire XD. While camera number and total workload are increasing, the average latency is slightly increasing. However, the largest latency is bounded at 40ms, which is less than the generation interval of frames (i.e. 1000 ms / 20fps = 50 ms). As such, our ingestion system can achieve real-time ingestion capability that is able to catch up with the generation speed of camera frames.

In addition, the total FPS is linear increasing along with the increasing of ingestion workload (i.e. camera number). It proves that there is no frame loss during the ingestion. We can hence conclude that the overall performance of our *SVIS* video ingestion system is linearly scalable.

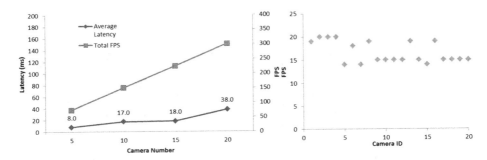

Fig. 4. Experimental results of ingesting online camera streams into gemfire XD

References

1. Han, Hu, Wen, Yonggang, Chua, Tat-Seng, Li, Xuelong: Toward scalable systems for big data analytics: a technology tutorial. IEEE Access **2**, 652–687 (2014)
2. Devasena, C.L., Revath, R., Hemalatha, M.: Video surveillance systems - A survey. IJCSI Int. J. Comput. Sci. Issues **8**(4), 1 (2011)
3. Intel: Extract, Transform, and Load Big Data with Apache Hadoop. White Paper (2013)
4. Apache Flume. http://flume.apache.org/
5. Apache Sqoop. http://sqoop.apache.org/
6. Apache Kafka: A high-throughput distributed messaging system. http://kafka.apache.org/
7. Scribe. http://sourceforge.net/projects/scribeserver/
8. Apache Chukwa. https://chukwa.apache.org/
9. Pivotal Gemfire XD. http://www.pivotal.io/big-data/pivotal-gemfire-xd

A³SAR: Context-Aware Spatial Augmented Reality for Anywhere, Anyone, and Analysis

Benjin Mei[1], Dehai Liu[1], Xike Xie[2], Jinchuan Chen[3]([✉]), and Xiaoyong Du[1]

[1] School of Information, Renmin University of China, Beijing, China
{meibenjin,liudehai,duyong}@ruc.edu.cn
[2] Department of Computer Science, Aalborg University, Aalborg, Denmark
xkxie@cs.aau.dk
[3] Key Laboratory of Data Engineering and Knowledge Engineering,
Renmin University of China, MOE, Beijing, China
jcchen@ruc.edu.cn

Abstract. Internet and camera equipped mobile devices with versatile capabilities and inexpensive costs make it possible for a Spatial Augmented Reality (SAR) platform. In general, the SAR is to enhance the sensing of the real world by combining overlay data on top of the view. It supports applications in medical management, urban projects, and online gaming etc. Nevertheless, existing systems are mostly focus on visualizing formatted information on mobile devices. They are short in exploiting the profound nature of the reality. In this paper, we propose a novel context-aware platform, called A³SAR, which augments the realities under three contexts: Anywhere Augmentation, Anyone Augmentation, and Analysis Augmentation. A³SAR aims at seamlessly integrating the virtual and real worlds by incorporating emerging technologies from different dimensions. For the Anywhere Augmentation, the overlay is constructed based on the semantics extracted from websites with geospatial information (e.g., upcoming shuffles at a bus station, or the history of an antique in museum). For the Anybody Augmentation, the overlay is built according to users preferences and profiles (e.g., for a piano, visualizing music for a player, but visualizing maintenance instructions for a tuner). More than just loading pre-existing information, the Analysis Augmentation also provides analytical data dynamically (e.g., visualizing the most endangered spots in a fire accident). However, challenges rise in several aspects: (1) efficiency for handling concurrent service requests, especially analytical tasks; (2) overlay accuracy regarding noisy information; (3) semantic extraction from heterogeneous sources. We propose a series of technical solutions: we design an intelligent engine for efficient analytical overlays; we improve the calibration accuracy by addressing spatial imprecision; we tackle the heterogeneous modeling problem by considering a semantic web based solution. The real and the virtual, two worlds in parallel, have intersected at SAR, and are converged within A³SAR.

A. Liu et al. (Eds.): DASFAA 2015 Workshops, LNCS 9052, pp. 307–312, 2015.
DOI: 10.1007/978-3-319-22324-7_32

1 Introduction

Spatial Augmented Reality (SAR in short) is a research area aiming at reinforcing the real world with overlay information on top of views. The proliferation of intelligent mobile terminals makes SAR possible in the pervasive computing era. Particularly, a mobile phone, with the support of GPS, camera, gyroscope, and compass, can generate overlaying information over the screen. For example, when travelling to a new city, people are immersed by the large volume of new information. The augmented information can serve as a travel assistant. Thus, between the real and virtual worlds, SAR constructs a mediate world in order to enrich users perception of the reality.

Existing systems, such as Layar (Netherlands, 2008) and Yelp (US, 2009), focus on how to visualize formatted information based on users' locations and orientations. The methods are short in exploiting the profound nature of the reality, and neglect the concrete context for users' applications [7]. In this paper, we propose a context-aware augmented reality platform, which consists three progressive parts: Anywhere Augmentation, Anybody Augmentation, and Analysis Augmentation.

- Anywhere Augmentation refers to the location-based context-aware SAR. The augmented overlay varies with respect to the change of locations. For example, if a user points the phone to a building, the screen displays facilities inside, as well as others' ratings and comments. For another example, if a user is choosing clothes, by scanning the watermark, the handheld screen shows advices from designer, possible accessories recommendation, recent fashion trends, etc.
- Anybody Augmentation refers to the user-based context-aware SAR. The augmented overlay is privately customized according to users preferences and profiles. For example, if a tuner points the phone at a piano, he probably expects the overlay for the parameters and instructions. But if a player does that, he would prefer a music sheet.
- Analysis Augmentation refers to the analytical context-aware SAR. Instead of loading pre-existing information from repository, the overlay offers advanced knowledge about the surrounding environment. For example, if one wants to purchase flats, which are attributed by floors, prices, areas, decorating conditions, the service can rank the flats based on the dominance relationship and visualize the result on screen [1,6].

We would like to address the technical challenges involved in constructing such a platform. First, the location and users semantics are to be abstracted from heterogeneous information sources, varying from the formatted to semi-structured. Second, the spatial information obtained is often noisy and imprecise [4]. Third, concurrent online analytical requests require high performance of the server in terms of execution efficiency, elasticity, and throughput.

A^3SAR, which is designed to address above challenges, consists of a mobile end and a service cloud, as shown in Fig. 1. The cloud is constituted by three layers: (1) the semantic information extractor, which elaborates location and user models from website repositories and social networks, respectively; (2) the

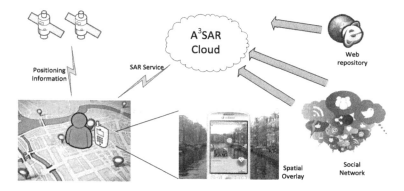

Fig. 1. A³SAR platform

extracted information is then kept and indexed in a cloud storage system; (3) the overlay generator which calculates overlay information to be displayed on users handheld screen upon receiving a mobile request. The mobile end visualizes the overlay information by a user-friendly interface.

The progressive context-aware SAR has not been well studied. The proposed work combines the knowledge from cloud computing, spatial databases, and computer vision. Moreover, with the aid of the A3SAR, we believe that a significant progress can be achieved for augmented reality.

In the following, we would like describe the architecture of A³SAR in Sect. 2, highlight the technical details of the spatial index in Sect. 3, and conclude the paper in Sect. 4.

2   System Overview

Figure 2 describes the architecture of A³SAR, which contains the following four modules.

- **Overlay Generator.** The overlay generator is versatile in offering different kinds of services. It is the core part of the entire system, which involves a set of operations for various services.
- **Spatial Augmented Index.** According to our previous studies [2,3,8,9], the spatial queries, especially the spatial augmented queries and analytical tasks (e.g., numerical augmented queries, and textual augmented queries), are expensive to evaluate. To provide efficient execution of cloud services, we are going to design a spatial augmented index. We would also like to evaluate how the index can be initialized, and quickly updated upon insertion and deletion of objects.
- **Semantic Information Extractor.** This module is to extract the semantic patterns from the webpage repositories and social networks. The semantic information includes spatial information and users profiles and preferences. The extracted patterns are updated incrementally in accordance with the updates of the crawled website repository.

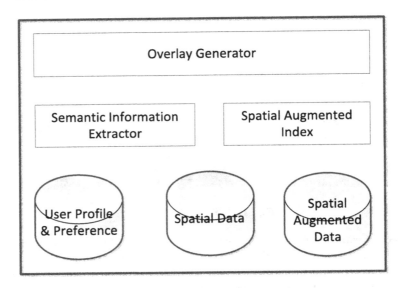

Fig. 2. A³SAR architecture

The Overlay Generator fulfills fundamental queries and analyzing tasks, including range query, k nearest neighbor query, and spatial ranking analysis, etc., for different types of augmented spatial data. In Sect. 3, we will investigate how the proposed Spatial Augmented Index could facilitate the computation of those queries. With the help of Spatial Augmented Index, the Overlay Generator can support the following new queries and analysis in basic SAR applications. For example, in SAR navigation, the overlay is generated for the objects retrieved in the query range.

1. Augmented range query [2]: what are the objects that located within the query range specified by the given query point.
2. Augmented k nearest neighbor query [3,8]: what are the k objects that are the closest to the query point.
3. Augmented spatial ranking [9]: if a ranking function is defined as the number of spatial objects dominated by the specified object, the spatial ranking retrieves top-k objects with the highest ranking scores.

Efficient algorithms are developed to compute the above queries, based on the Spatial Augmented Index.

The goal of the Semantic Information Extractor is to harvest large volume of entities and their relationships from the Internet. Information extraction contains several sub-tasks and is not an easy job. Fortunately, there are some open source information extraction software such as Mallet and DBpedia Spotlight. Our work benefit much from these existing systems. Moreover, the results of information extractor cannot be perfectly correct. We alleviate this imprecision with the help of large knowledge bases like Linked Open Data, YAGO and BTC.

We try to recognize the extracted entities from the knowledge bases and utilize the information in the knowledge base to revise and enrich the extracted information.

All these modules, the Spatial Augmented Index, the Overlay Generator, and the Semantic Information Extractor, are integrated with a cloud storage system. On top of the cloud system, we design a set of SAR services for geographical applications. For example, we provide sampled services, such as SAR navigation service, SAR statistical service, SAR ranking service, and so on. We will also develop a mobile app for visualizing the overlay information.

3   Spatial Index

The performance of A³SAR depends heavily on the efficiency of spatial augmented index. We now discuss its technical details in this part.

The overlay generator is the core part of the cloud service. So we first study the overlay primitives can be efficiently materialized. In [9], we have implemented an R-tree structure, enabling the ranking analysis. In experiments done in [9], the query efficiency is improved by over 75 % and the construction cost is within 2 s for 200k objects. While this is a significant gain, there is still room for improvement. In particular, we adopt peer-to-peer and parallel computing techniques and organize the spatial data into a layered index structure which is physically distributed among computing nodes. Basically, each query will be firstly redirected to a cluster through a top-layer index, then forwarded to specific computing node through a cluster-level index and then processed within a local index. The top two indices are distributed and implemented in a peer-to-peer mode.

The new index is elastic with respect to storage, and can support batch and incremental update. In the current index design, each tree node is augmented with an affiliated entity (e.g., histogram in [9] and posting list in [5]). Actually, the augmented entity can have a large size (e.g., a big posting list table). It costs extra costs for maintaining such structure. In this work, we design an elastic strategy so that the cloud storage can be easily enlarged or shrunk with respect to data volume adjustment.

For the tree structure proposed in [5,9], an update of an object incurs chain effect of the path to the tree root. To avoid the maintenance cost, we design a batch update strategy which keeps track of the updating data but only updates it if necessary. Additional data structure is also developed to support such strategy. We also design incremental update algorithm, where the update operations, such as insertions and deletions are handled by the index.

Furthermore, we consider two types of augmentation: numerical augmentation, and textual augmentation in the new index. In [9], the index nodes are augmented by numerical histograms, whereas in [5], the index nodes are augmented by posting list table (or reverse lists). It is beneficial to combine heterogonous augmentations in the index constructions, thus offers a unified structure for ease of the deployment in the cloud server. These two orthotropic augmented dimensions are combined in this new index.

4   Conclusion and Future Works

In this paper, we propose a novel context-aware platform, A^3SAR, which augments the realities under three contexts: Anywhere Augmentation, Anyone Augmentation, and Analysis Augmentation. We briefly introduce its main features, architecture, and the technical details of its spatial index. In the future, we will perform extensive experiments on real and synthetic data to verify the efficiency and effectiveness of our algorithms. We will also develop a mobile app to test the overall performance of the entire system.

This work is supported by the Fundamental Research Funds for the Central Universities, the Research Funds of Renmin University of China (No. 14XNLQ06). The authors would also like to thank Sa Shi-Xuan Research Center of Big Data Management and Analytics for its supports.

References

1. Borzsony, S., Kossmann, D., Stocker, K.: The skyline operator. In: 2001 Proceedings of 17th International Conference on Data Engineering, pp. 421–430 (2001)
2. Chen, J., Cheng, R.: Efficient evaluation of imprecise location-dependent queries. In: Proceedings of the 23rd International Conference on Data Engineering, ICDE 2007, pp. 586–595, The Marmara Hotel, Istanbul, Turkey, 15–20 April 2007 (2007)
3. Cheng, R., Chen, L., Chen, J., Xie, X.: Evaluating probability threshold k-nearest-neighbor queries over uncertain data. In: EDBT 2009, Proceedings of 12th International Conference on Extending Database Technology, pp. 672–683, Saint Petersburg, Russia, 24–26 March 2009 (2009)
4. Cheng, R., Kalashnikov, D.V., Prabhakar, S.: Querying imprecise data in moving object environments. IEEE Trans. Knowl. Data Eng. **16**(9), 1112–1127 (2004)
5. Cong, G., Jensen, C.S., Wu, D.: Efficient retrieval of the top-k most relevant spatial web objects. Proc. VLDB Endow. **2**(1), 337–348 (2009)
6. Huang, Z., Lu, H., Ooi, B.C., Tung, A.K.H.: Continuous skyline queries for moving objects. IEEE Trans. Knowl. Data Eng. **18**(12), 1645–1658 (2006)
7. Wagner, D., Schmalstieg, D.: History and future of tracking for mobile phone augmented reality. In: 2009 International Symposium on Ubiquitous Virtual Reality, vol. 0, pp. 7–10. IEEE, July 2009
8. Xie, X., Cheng, R., Yiu, M., Sun, L., Chen, J.: Uv-diagram: a voronoi diagram for uncertain spatial databases. VLDB J. **22**(3), 319–344 (2013)
9. Xie, X., Lu, H., Chen, J., Shang, S.: Top-k neighborhood dominating query. In: Meng, W., Feng, L., Bressan, S., Winiwarter, W., Song, W. (eds.) DASFAA 2013, Part I. LNCS, vol. 7825, pp. 131–145. Springer, Heidelberg (2013)

Semi-supervised Clustering Method for Multi-density Data

Walid Atwa and Kan Li[✉]

Beijing Engineering Research Center of High Volume Language Information Processing
and Cloud Computing Applications, School of Computer,
Beijing Institute of Technology, Beijing, China
Walid_mufic@yahoo.com, likan@bit.edu.cn

Abstract. Finding clusters is a challenging problem especially when the clusters are being of widely varied shapes, sizes, and densities. Density-based clustering methods are the most important due to their high ability to detect arbitrary shaped clusters. However, they are depending on two specified parameters (*Eps* and *Minpts*) that define a single density. Moreover, most of these methods are unsupervised, which cannot improve the clustering quality by utilizing a small number of prior knowledge. In this paper we show how background knowledge can be used to bias a density-based clustering method for multi-density data. Experimental results confirm that the proposed method gives better results than other semi-supervised and unsupervised clustering algorithms.

1 Introduction

DBSCAN is a density based clustering algorithm and its effectiveness for spatial datasets has been demonstrated in the existing literature [1]. However, there are two distinct drawbacks for DBSCAN and its extension methods: (1) the performances of clustering depend on two specified parameters. One is the maximum radius of a neighborhood (*Eps*) and the other is the minimum number of the data points contained in this neighborhood (*Minpts*). In fact these two specified parameters define a single density. Nevertheless, without enough prior knowledge, these two parameters are difficult to be determined; (2) with these two parameters for a single density, DBSCAN does not perform well to datasets with varying densities.

For example, in Fig. 1(a), DBSCAN fails to find the four clusters, because this dataset has four different densities and the clusters are not totally separated by sparse regions. In Fig. 1(b), DBSCAN discovers only the three small clusters and considers the other two large clusters as noises, or merges the three small clusters in one cluster to be able to find the other two large clusters. These problems occur due to using global values of the parameters (*Eps*, *Minpts*).

Semi-supervised clustering algorithms have been received a significant amount of attention in data mining and machine learning fields. Unlike traditional clustering algorithms, semi-supervised clustering (also known as constrained clustering) is a category of techniques that tries to incorporate prior information like pairwise constraints into the clustering algorithms. Pairwise constraints provide the supervision information like

© Springer International Publishing Switzerland 2015
A. Liu et al. (Eds.): DASFAA 2015 Workshops, LNCS 9052, pp. 313–319, 2015.
DOI: 10.1007/978-3-319-22324-7_33

(a)   (b)

Fig. 1. Clusters with varying densities

must-link (*ML*) and cannot-link (*CL*), where must-link constraint specifies that the pair of instances should be assigned to the same cluster, and cannot-link constraint specifies that the pair of instances should be placed into different clusters.

In this paper, we propose a semi-supervised clustering (called SemiDen) algorithm that discovers clusters of different densities and arbitrary shapes. The idea of the proposed algorithm is to partition the dataset into different density levels and compute the density parameters for each density level set. Then, use the pairwise constraints for expanding the clustering process based on the computed density parameters. Evaluating SemiDen algorithm on real datasets confirms that the proposed algorithm gives better results than other semi-supervised and unsupervised density based approaches. In summary, our contribution in this paper is clustering multi-density datasets and arbitrary shapes using pairwise constraints.

2   Clustering Multi-density Data

In this section, we propose a semi-supervised density-based clustering (SemiDen) algorithm that can find clusters of varying densities, shapes and sizes, even in the presence of noise and outliers. The proposed algorithm is divided into two main parts: (1) partitioning the dataset into different density levels; (2) using pairwise constraints for expanding the clustering process for each density level. We summarize our semi-supervised clustering (SemiDen) algorithm in Algorithm 1.

First, we describe the details of partitioning the dataset into different density levels. First our algorithm finds the k-nearest neighbors for each point in the given dataset. Based on the k-nearest neighbors, a local density function is used to find the density at each point. Where the local density function at point x is defined as the sum of the distances among the point x and its k-nearest neighbors, as shown in Eq. (1).

$$DEN(x) = \sum_{i=1}^{k} D(x, y_i) \tag{1}$$

where $D(x, y_i)$ is the Euclidean distance between point x and its k-nearest neighbors y_i.

$$D(x, y) = \sqrt{\sum_{j=1}^{n} (x_j - y_j)^2} \tag{2}$$

After computing the local density function for each data point, we sort them in ascending order and compute the density variation between each two adjacent points p_i and p_{i+1} denoted by $DENVAR(p_i, p_{i+1})$. Then, we get $DENVAR$ list (denoted by $DVList$) in which each element in $DVList$ is a density variation between two points in the dataset.

$$DENVAR\left(p_i, p_{i+1}\right) = \frac{\left|DEN\left(p_{i+1}\right) - DEN\left(p_i\right)\right|}{DEN\left(p_i\right)} \tag{3}$$

For datasets with widely varied densities, there will be some distinct variation depending on the densities of the data points. But for points in the same density level, the range of variation is small. Thus, we can acquire all density level sets by detecting these distinct variations of density.

Definition 1: (Density Level Set). Density level set (DLS) consists of points whose densities are approximately the same. In other words, the density variations of the data points within the same DLS should be relatively small. Points p_i and p_j belong to the same DLS if they satisfy the following condition:

$$p_i, p_j \in DLS_k \ if \ DENVAR\left(p_i, p_j\right) \le \tau$$

where τ is a density variation threshold which divides a multi-density dataset into several density level sets.

We implement partitioning method on $DVList$. Given a density variation threshold τ (Definition 1), remove $DENVAR$ values which are bigger than τ out of $DVList$, then the points of remaining separated segments are considered as different density level sets. Here, we compute τ according to the statistical characteristics of the $DVList$ as follows:

$$\tau = E\left(DVList\right) + \sigma\left(DVList\right) \tag{4}$$

where E is mathematical expectation and σ is standard deviation of $DVList$. According to the $DVList$ values, there are only a small number of points with large $DENVAR$ values which are used to divide the dataset into different sets according to the threshold τ.

After partitioning the dataset into different density level sets, we need to find representative value of the parameters (Eps and $Minpts$). We initialize the parameter $Minpts$ as k-nearest neighbor and try to identify the value of parameter Eps for each density level. For a certain density level set (DLS), its corresponding Eps will be magnified by simply choosing the maximum DEN value. As we know, there are some points may correspond to border objects or noise and these points may have some influence on the Eps value. To deal with this problem, we compute Eps_i for DLS_i as follows:

$$Eps_i = maxDEN\left(DLS_i\right) \cdot \sqrt{\frac{medianDEN(DLS_i)}{meanDEN(DLS_i)}} \tag{5}$$

where *maxDEN*, *meanDEN* and *medianDEN* are the maximum, mean and the median density of DLS_i respectively.

Finally computing the *Eps* parameter for each density level and initializing the parameter *MinPts* as *k*-nearest neighbor, we use the pairwise constraints for expanding the clustering process for each density level as follows:

- **In Step 11(a):** we check if *Point* belongs to clusters or noise set. Where the key idea of density-based clustering is that for each point of a cluster the neighborhood of a given radius (*Eps*) has to contain at least a minimum number of points (*MinPts*). Therefore we compute the *Point's Eps*-neighborhood. If the number of points in *Eps*-neighborhood less than *MinPts*, adding *Point* to noise set.

- **In Step 11(b):** satisfy Must-link constraints. If *Point* belongs to a must-link constraints, all the points contained in this *Point* are assigned to the current cluster, so as to satisfy the must-link constraints.

- **In Step 11(c):** satisfy Cannot-link constraints. Before adding point *p* into the current cluster, we should ensure that the adding operation does not violate cannot-link constraints. If there is a point *q* in the current cluster and a pair $\{p, q\} \in CL$, adding *p* into the cluster will violate the cannot-link constraints, therefore the points *p* should not be assigned to the current cluster.

Algorithm 1. SemiDen

Input: A set of data points *X*; set of must-link constraints *ML*; the set of cannot-link constraints *CL*
Output: Several clusters and a set of noises;
Begin
  1. Compute local density function for each *Point* in *X*;
  2. Sort the density values in ascending order;
  3. Compute the density variation values between each two adjacent points;
  4. Partition the dataset into different density level set (*DLS*) according to Definition 1;
  5. **For** each density level set (*DLS$_i$*)
  6. Initialize *MinPts* as *k*-nearest neighbor;
  7. Initialize all points in *DLS$_i$* as *UNCLUSTERED*;
  8. Estimate the parameter *Eps$_i$*;
  9. *ClusterId* = 0;
  10. **For** each *Point* in *DLS$_i$*
  11. **If** *Point.Id* = *UNCLUSTERED*
      Compute *Point's Eps*-neighborhood *neighborhood*;
     a. **If** the number of points in *neighborhood* < *MinPts*
        Point.Id = *NOISE*;
      Else
        Point.Id = *ClusterId*;
     b. **If** *Point* has must-link constraints
       For each point *o* in *ML(Point, o)*
         o.Id = *ClusterId*;
     c. **For** each point *p* in neighborhood **do**
       If *p.Id* = *UNCLUSTERED* and does not violated cannot-link constraints
         p.Id = *ClusterId*;
  12. *ClusterId* = *ClusterId* +1;
  13. Return all clusters and a set of noises
End

---

3   Experimental Results

In this section we present two experimental results of SemiDen algorithm on a variety of datasets, including synthetic datasets and several real world datasets. We implement our algorithm in Java and work on a 2.4 GHz Intel Core 2 PC running windows XP with 2 GB main memory.

Besides the proposed algorithm, we also implemented some competing counterparts as well as the baseline methods listed below for comparison.

1. **APSCAN:** An unsupervised clustering algorithm that uses affinity propagation for clustering datasets with varying densities [2].
2. **HISSCUL:** A hierarchical semi-supervised density based clustering algorithm. HISSCLU use the parameters ρ and ξ to establish borders between clusters when there are no clear cluster boundaries. In order to maximally preserve the original cluster structure HISSCLU is recommended to set up with $\rho = 1.0$, $\xi = 0.5$ [3].
3. **C-DBSCAN**: A density based semi-supervised clustering algorithm that is based on DBSCAN for clustering datasets with arbitrary structures. C-DBSCAN depends on two specified parameters (*Eps* and *MinPts*). We set the parameters *Eps* = 0.5 and *MinPts* = 4 (default values in DBSCAN) [4].
4. **SSDBSCAN:** Semi-supervised density based clustering algorithm that automatically finds density parameters for each natural cluster in a dataset [5].

The experiments were performed on datasets from UCI repository (yeast, segment, digits-389 and magic). These datasets provide a good representation of different characteristics: numbers of samples are ranges from 1484 to 19,020, dimensionalities from 8 to 19, and number of clusters from 2 to 10.

Figure 2 shows the NMI results over the different number of pairwise constraints on the real datasets. It can be observed from Fig. 2 that our algorithm "SemiDen" generally performs better than the four other methods when the number of constraints increased (e.g. yeast, segment, and magic).

We also notice that the constraint based clustering algorithms generally outperform the traditional clustering algorithms. It can be seen from Fig. 2 that the performance of APSCAN in all datasets is constant value as it is unsupervised clustering algorithms. This tends to prove the utility of constraint based clustering algorithms over unsupervised approaches when expert knowledge is available.

To evaluate the efficiency of clustering algorithms, we compare the average CPU time consumption of each semi-supervised clustering algorithm, with different number of pairwise constraints as shown in from Fig. 3.

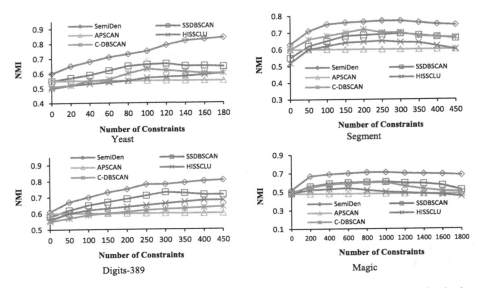

Fig. 2. Comparison of normalized mutual information over the different number of pairwise constraints

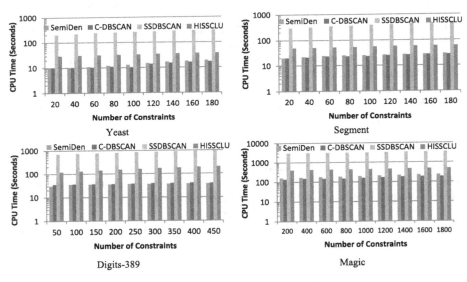

Fig. 3. Comparison of execution time for the semi-supervised clustering algorithms.

Acknowledgment. The Research was supported in part by Natural Science Foundation of China (No.60903071), National Basic Research Program of China (973 Program, No.2013CB329605), Specialized Research Fund for the Doctoral Program of Higher Education of China, and Training Program of the Major Project of BIT.

References

1. Ester, M., Kriegel, H.P., Sander, J., Xu, X.: A density based algorithm for discovering clusters in large spatial databases with noise. In: Proceedings of 2nd International Conference on Knowledge Discovery and Data Mining, pp. 226–231 (1996)
2. Chen, X., Liu, W., Qiu, K., Lai, J.: APSCAN: a parameter free algorithm for clustering. Pattern Recognit. Lett. **32**, 973–986 (2011)
3. Bohm, C., Plant, C.: Hissclu: a hierarchical density-based method for semi-supervised clustering. In: Proceedings of 11th International Conference on Extending Database Technology (2008)
4. Ruiz, C., Spiliopoulou, M., Menasalvas, E.: Density-based semi-supervised clustering. Data Min. Knowl. Discov. **21**, 345–370 (2010)
5. Lelis, L., Sander, J.: Semi-supervised density-based clustering. In: Proceedings of 8th IEEE International Conference on Data Mining, pp. 842–847 (2009)

Towards a Pattern-Based Query Language for Document Databases

Xuhui Li[1,2,3]($\boxtimes$), Zhengqi Liu[1], Mengchi Liu[1], Xiaoying Wu[1], and Shanfeng Zhu[3]

[1] State Key Lab of Software Engineering, Computer School, Wuhan University, Wuhan, China
lixuhui@whu.edu.cn, mengchi@sklse.org,
{charleslzq,xiaoying.wu}@gmail.com
[2] School of Information Management, Wuhan University, Wuhan, China
[3] Shanghai Key Lab of Intelligent Information Processing, School of Computer, Fudan University, Shanghai, China
zhusf@fudan.edu.cn

Abstract. Document databases are becoming popular, but how to present complex document query to obtain useful information from the document remains an important topic to study. In this paper, we describe the design issues of a pattern-based document database query language named JPQ, which uses various expressive patterns to extract and construct document fragments following a JSON-like document data model. It adopts tree-like extraction patterns with a coherent pattern composition mechanism to extract data elements from hierarchically structured documents and maintain the logical relationships among the elements. Based on these relationships, JPQ deploys a deductive mechanism to declaratively specify the data transformation requests and considers also data filtering on hierarchical data structure.

1 Introduction

Document databases are a kind of so-called NoSQL databases which store information in nested collections of records or rows. In recent years some document databases like MongoDB [2], CouchDB [3], OrientDB [4] and RavenDB [5] have been developed and are becoming popular. In practice, the dominant format for document database is JSON [1].

With the wide spread use of document databases, how to query JSON-like documents has become an important issue. Existing document databases usually provide lightweight query interfaces to satisfy basic query needs. Functional languages like Jaql [6], languages employing XPath-like expressions like JSONiq [7]

This research is supported by the NSF of China under contract No.61272110, No.61272275 and No.71420107026, the Open Fund. of Shanghai Key Lab. of Intelligent Info. Processing under contract No.IIPL-2011-002, the State Key Lab. of Software Engineering (SKLSE) under contract No.SKLSE20120907, and the China Postdoctoral Science Foundation under contract No. 2014M562070.

© Springer International Publishing Switzerland 2015
A. Liu et al. (Eds.): DASFAA 2015 Workshops, LNCS 9052, pp. 320–325, 2015.
DOI: 10.1007/978-3-319-22324-7_34

and SQL-like languages like UnQL [8] are suggested to query JSON-like documents recently, but they are not expressive and declarative enough in presenting complex queries.

We suggest a pattern-based query language named JPQ (standing for JSON-like Pattern Query). To begin with, the hierarchy in JSON-like document can be represented by a simplified model named JHM, using the following grammar.

```
<value>    ::= <atom> | <array> | <object>
<atom>     ::= <string> | <number> | <boolean> | empty | ...
<array>    ::= [ <value> (; <value>)* ]
<object>   ::= {<keyvalue> (, <keyvalue>)* }
<keyvalue> ::= <string>:<value>
```

A JHM document fragment is represented as a value which can be an atomic one like a string, an array of values, or an object comprising one or more key-value pairs. Compared to conventional object databases, JHM adopts no compound data type and, further, no predefined data schema. This essentially affects document queries in several ways. Firstly, it disables randomly accessing an arbitrary part of an object, thus an enumeration mechanism is required. Secondly, it does not use labels with a predefined semantics, thus querying key strings becomes an important way to find expected information . Finally, it supports the heterogeneity of array elements , thus users often need to manually align heterogeneous array elements to produce homogeneous results.

To handle these problems, JPQ adopts expressive patterns with coherent mechanisms to extract multiple document data elements simultaneously and construct document fragments in a flexible structure, and thus enable complex document queries be presented in a declarative way. Our work makes the following contributions: (a) to our knowledge our work is the first attempt to use tree-like patterns in querying JSON style document databases. (b) JPQ focuses on the expressiveness and is easy to present complex queries. (c) JPQ deploys a deductive rewriting mechanism to specify data transformation requests, which allows complex data construction to be presented declaratively.

As an example, We consider such a query on the following document fragment *univ* illustrating a scenario of a university human resource information system : find the schools in which more than one president works as a faculty member and list each president and corresponding school in an array.

```
{ "president":{"ID":"0001", ...},
  "executive-vice-president":{"ID":"0002",...},
  "vice-presidents":[{"ID":"0003",...};...]
  "schools":[{"name":"Computer School", "dean":{"ID":"0011", ...},
            "faculty":[{"ID":"0001", ...};...]},     ... ], ... }(univ)
```

2   The JPQ Language

A common JPQ program is composed of the **from**, the **construct** and the **where** clauses to present data extraction, construction and filtering respectively, as the following code fragment which handles the previous request shows:

```
from doc("univ")
    </"?president?":(($p1{"ID":$id1})|[($p2{"ID":$id2})]),
        {"schools":[{"name":$n, "faculty":[{"ID":$id3}]}]}>
construct {"results":[^[{"president":$p1,"school":$n}]|
                    ^[^[{"president":$p2,"school":$n}]]]}
where (id1=id3 par id2=id3) with count($id3) > 1
```

The *from* clause in a JPQ program uses one or more extraction statements to specify extraction requests. In *construct* clause, user can specify the format of query result. Additionally, if there is a *where* clause, the extracted data elements would be filtered by the conditions specified in it before actually be used to construct the final results.

2.1   Data Extraction

We can use p_k and p_v to denote the key-value pattern and the value pattern respectively, use v to denote the variables, use r_s and r_v to denote the string predicate and the value predicate respectively, and list the abstract syntax of the extraction patterns as below:

$$p_k ::= v{:}p_v \mid r_s{:}p_v \mid (vr_s){:}p_v \mid {}^*{:}p_v \mid p_k \mid p_k$$
$$p_v ::= v \mid r_v \mid {}^* \mid \{p_k, ..., p_k\} \mid [p_v] \mid < p_v, p_v > \mid p_v \mid p_v \mid /p_k \mid //p_v$$

As structural query languages usually do, JPQ requires users to be familiar with the structure of target documents and to tell the program where to find the expected information. To complete this task, JPQ provides basic patterns that are structurally composed of the variables, the predicates and the wildcard "*" with the structural operators like ":", "{}" or "[ ]". The variables and the predicates are to test whether a key or a value satisfies the restrictions indicated by the pattern, and if it does, to bind the key or value to the variable. Then, these variables are structurally organized according to the structure indicated by extraction statements. In our example, the *"name"* : *$n* part will bind the school names to the variable *$n*

To search target documents, JPQ introduces the enumeration patterns, using path operators "/" and "//" as the same way as XML query languages do. And to test a value with multiple predicates, JPQ introduces the conjunctive pattern $< p_1, ..., p_n >$ where a value is to be respectively matched with the patterns $p_1, ..., p_n$ and the results are combined as a tuple.

Moreover, we can use the optional pattern of the form $p_1 \mid p_2 \ldots \mid p_n$ to handle heterogeneity, so we can process data elements with different structure in the same way. In the example, we have a president, an execute-vice-president and an array of vice-presidents. They all have attribute *ID*, which is all we care about in this query. So we can use */$r* *"?president?"*:($<$*$p1, "ID":$id1* $\} >$ $\mid [<$*$p2, "ID":$id2* $\} >$ *]*) to extract their IDs and then bind them to the variable *$r*, ignoring their structural differences.

$$(p_1,\ldots, p_i, p_{i+1},\ldots, p_n) \hookrightarrow (p_1,\ldots, p_{i+1}, p_i,\ldots, p_n) \text{ (tuple-commutation)}$$
$$(p_1,\ldots, p_j, p_{j+1},\ldots, p_n) \hookrightarrow (p_1,\ldots, p_j, (p_{j+1},\ldots, p_n)) \text{ (tuple-association)}$$
$$(p_1|\cdots|p_i|p_{i+1}|\cdots|\cdots|p_n) \hookrightarrow (p_1|\cdots|p_{i+1}|p_i|\cdots| p_n) \text{ (option-commutation)}$$
$$(p_1|\cdots| p_j|p_{j+1}|\cdots|p_n) \hookrightarrow (p_1|\cdots|p_j|(p_{j+1}|\cdots|p_n)) \text{ (option-association)}$$
$$p \hookrightarrow (p, p) \text{ (tuple-duplication)}$$
$$(p, p') \hookrightarrow p \,/\, p' \text{ if } var(p) \cap var(p') = \phi \text{ (tuple-hiding)}$$
$$(p \,|\, p') \hookrightarrow p \text{ (option-hiding)}$$
$$[p]_{[q]} \hookrightarrow \,\hat{}\,[p]_{[q]} \text{ (array-flattening)}$$
$$(p, p' \,|\, p'') \hookrightarrow (p, p') \,|\, (p, p'') \text{ (option-tuple-distribution)}$$
$$(p, [p']_{[q]}) \hookrightarrow [(p, p')]_{[q]} \text{ if } var(p) \cap var(p') = \phi \text{ (array-tuple-distribution)}$$
$$[(p, p')]_{[q]} \hookrightarrow [([(p, p')]_{[q]}, p\%)]_{[p\%]} \text{ (array-tpl-folding)}$$

Fig. 1. Restructuring rules of matching term

2.2 Data Construction

Data construction in JPQ is presented with construction patterns which are essentially the function invocations on the **matching terms**. A matching term is an expression to specify the (transformed) structure of the matching results of an extraction pattern. The abstract syntax of matching term is

$$t ::= v \mid (t,t) \mid t|t \mid [t]_t \mid \,\hat{}\, [t]_t \mid t\% \qquad (1)$$

For the array term $[t]_{t'}$, t' is named the **index term** of t and will work as an unique index in this array.

Using matching terms, a construction pattern can be specified as a normal function invocation like $fun(t)$, which transforms the original extracted data to be of a structure specified in the construction pattern, following a set of predefined restructuring rules. These restructuring rules, as Fig. 1 shows, constitute our deductive rewriting system, which also determines whether a *construct* clause or a *where* clause is legal or not. Moreover, construction pattern can embed certain constant values (such as key strings or values) and necessary notations (such as ":" and "{}") into the transformed matching term. As a result, JPQ can present the data transformation declaratively, as the construction statement in our example shows. For more details on rewriting system, please refer to our paper [9] and report [10].

2.3 Data Filtering

JPQ uses predicate conditions and compound conditions in the *where* clause to filter out the unwanted data elements. A predicate condition can be a simple predicate function invocation, a quantified condition or a composite condition combining subordinate conditions with the boolean connectives "**and**", "**or**" and "**not**". They all work in the same way as filtering conditions in other query languages do, so we won't discuss them in detail.

To handle complex query requests, JPQ introduces compound conditions. A compound condition c_1 par c_2 is a disjunctive condition where c_1 and c_2 are two conditions filtering values of different parts of an option and the filtering results are merged as the final result of the whole condition. The connective "par" is used to replace the connective "or" to avoid the inconsistency problem occurred in filtering option patterns. Another compound condition, c_1 with c_2, is used to filter hierarchical data elements. In the runtime, the predicate condition c_2 works on the result data elements filtered by the predicate c_1. Our example illustrates how to use these connectives.

For the details of the data filtering and other issues, please refer to our report [10] and [11].

2.4   Implementation

We have already implemented the core of JPQ with the support of Hadoop.

To handle data extraction, we build a tree representing specified JSON document hierarchy. Then, we will have to consider each node in this tree as the root of a potentially matched subtree which satisfies the conditions specified in the extraction pattern. For each atomic predicate in an extraction pattern, a mapper will be used to find the nodes which satisfy this predicate. Then the results will be sent to the reducers, each of which will use these information to determine whether the subtree rooted at a specific node of the JSON tree can be included in the result of data extraction or not.

The process of data construction will apply the same operations over multiple records and thus can be easily parallelized. So can the process of data filtering. The typical operations here, which might be transforming the records following a restructuring rule, or determining whether a node in the JSON tree satisfies a condition or not, will be applied to all elements through the map functions. Then the reduce functions will collect the results and output them.

3   Conclusion

In this paper, we introduced a new query language named JPQ. JPQ is a pattern-based functional language which adopts various expressive patterns to extract structural data elements from JSON-like documents and construct document fragments based on a deductive mechanism on data transformation. In comparison with the other query languages for JSON-like documents, JPQ exhibits many expressive and interesting features by the coherent pattern-based mechanisms in data extraction, data transformation and data filtering, as Fig. 2 shows.

Our study on the JPQ language is still in its preliminary stage. Currently we have finished the design of the language, and the core of JPQ has been implemented. However, the implementation of the complete version of the language on real data sets is still on going, and the performance on real and massive dataset will be evaluated in the near future. Also, we are extending the language with the update function so as to make it a full-fledged manipulation language. However, how to efficiently process complex queries with compound conditions is still a tough problem to be solved.

Languages \ Features	JPQ	Jaql	JSONiq	UnQL	Query API (MongoDB)
Full-fledged query language?	Yes	Yes	Yes	Yes	No
Data extraction approach	Pattern Matching	Function pipeline	Navigational function	Navigational operator	Function pipeline
Relationships among data elements	Conjunctive, Disjunctive, Aggregation	Not supported	Conjunctive	Conjunctive	Not supported
Schema-less query	Yes	No	Yes	Yes	Yes
Key information query	Pattern matching	Not supportd	Special function	Not supported	Not suported
Enumerating value content	Pattern matching	Not supported	Special function	Not supported	Special function
Handling heterogeneous data	Pattern matching	Special statement	Special statement	Not supported	Host program
Hierarchical data construction	Pattern matching	Nested program	Nested program	Not supported	Host program
Deductive data Transformation	Term rewriting system	Not supported	Not supported	Not supported	Not supported
Composite data filtering	Hierarchically pattern filtering	Not supported	Not supported	Not supported	Not supported

Fig. 2. Comparison of JSON-like document data query languages and interfaces

References

1. Crockford, D.: The application/json Media Type for JavaScript Object Notation (JSON), RFC 4627 (2006). http://www.ietf.org/rfc
2. Chodorow, K., Michael, D.: MongoDB: The Definitive Guide. O'Reilly Media, Sebastopol (2010)
3. Anderson, J.C., Jan, L., Slater, N.: CouchDB: The Definitive Guide: Time to Relax. O'Reilly Media, Sebastopol (2010)
4. Orientdb. http://www.orientdb.org/
5. RavenDB. http://ravendb.net/
6. Beyer, K., Ercegovac, V., Gemulla, R., Balmin, A., et al.: Jaql: a scripting language for large scale semistructured data analysis. Proc. VLDB Endow. **4**(12), 1272–1283 (2011)
7. Jonathan, R., Brantner, M., Florescu, D., et al.: XQuery for JSON, JSON for XQuery. XML Prague 63 2012 (2012)
8. Prestegarden, D.O.: UnQL: a query language for NoSQL databases. Ph.D. Dissertation., Norwegian University of Science and Technology (2012)
9. Li, X., Liu, M., Zhang, Y.: Towards a "more declarative" XML query language. In: Bringas, P.G., Hameurlain, A., Quirchmayr, G. (eds.) DEXA 2010, Part II. LNCS, vol. 6262, pp. 375–390. Springer, Heidelberg (2010)
10. Li, X., Liu, M., Zhu, S., Ghafoor, A.: XTQ: A Declarative Functional XML Query Language. CoRR abs/1406.1224 (2014)
11. Li, X., Liu, M., et al.: Design Issues of JPQ: a Pattern-based Query Language for Document Databases. Technical Report (2014). http://www.sklse.org:8080/jpq

Author Index

Printed in the United States
By Bookmasters